Photosynthetic Oxygen Evolution

Photosynthetic Oxygen Evolution

Edited by

H. Metzner

University of Tübingen
Tübingen, Germany.

1978

ACADEMIC PRESS
London   New York   San Francisco
A Subsidiary of Harcourt Brace Jovanovich, Publishers

ACADEMIC PRESS INC. (LONDON) LTD.
24/28 Oval Road,
London NW1

United States Edition published by
ACADEMIC PRESS INC.
111 Fifth Avenue
New York, New York 10003

Copyright © 1978 by
ACADEMIC PRESS INC. (LONDON) LTD.

All Rights Reserved

No part of this book may be reproduced in any form by photostat, microfilm, or any other means, without written permission from the publishers

Library of Congress Catalog Card Number: 77–93489
ISBN: 0–12–491750–X

Printed in Great Britain by
Whitstable Litho Ltd., Whitstable, Kent.

LIST OF CONTRIBUTORS

AKOYUNOGLOU, G.: Biology Department, Nuclear Research Center "Democritos", Athens (Greece).

ARGYROUDI-AKOYUNOGLOU, J.H.: Biology Department, Nuclear Research Center "Democritos", Athens (Greece).

ARNASON, T.: Department of Biology, Carleton University, Ottawa (Canada).

BRIN, G.P.: A.N. Bakh Institute of Biochemistry, USSR Academy of Sciences, Moscow (USSR)

CALZAFERRI, G.: Institut für anorganische, analytische und physikalische Chemie der Universität, CH-3000 Bern 9 (Schweiz)

CSATORDAY, K.: Institute of Plant Physiology, Biological Research Center, Hungarian Academy of Sciences, H 6701 Szeged (Hungary)

DROPPA, M.: Institute of Plant Physiology, Biological Research Center, Hungarian Academy of Sciences, H 6701 Szeged (Hungary)

DUJARDIN, E.: Laboratoire de Photobiologie, Département de Botanique, Université de Liège, 4000 Sart Tilman-Liège (Belgium)

DUNCAN, I.: The Royal Institution, London (Great Britain).

ETIENNE, A.-L.: Laboratoire de Photosynthèse du C.N.R.S., 91190 Gif-sur-Yvette (France).

FALUDI-DÁNIEL, A.: Institute of Plant Physiology, Biological Research Center, Hungarian Academy of Sciences, H 6701 Szeged (Hungary).

FUHRHOP, J.-H.: Gesellschaft für Biotechnologische Forschung mbH und Institut für Organische Chemie A der Technischen Universität, 3300 Braunschweig (BRD).

GARLAND, S.: Department of Biology, Carleton University, Ottawa (Canada).

GOVINDJEE: Department of Physiology and Biophysics, and Botany, University of Illinois, Urbana, Ill. 61801 (USA).

HARRIMAN, A.: The Royal Institution, London (Great Britain).

HOMANN, P.H.: Department of Biological Science and Institute of Molecular Biophysics, Florida State University, Tallahassee, Florida 32306 (USA).

LIST OF CONTRIBUTORS

HORVÁTH, G.: Institute of Plant Physiology, Biological Research Center, Hungarian Academy of Sciences, H 6701 Szeged (Hungary).

INOUE, Y.: Laboratory of Plant Physiology, The Institute of Physical and Chemical Research, Wako-Shi, Saitama 351 (Japan).

JAHNKE, H.: Robert Bosch GmbH, Forschungszentrum, 7016 Gerlingen (BRD)

JÁNOSSY, A.: Institute of Biophysics, Biological Research Center, Hungarian Academy of Sciences, H 6701 Szeged (Hungary).

JOUY, M.: Laboratoire de Photobiologie, Département de Botanique, Université de Liège, 4000 Sart Tilman-Liège (Belgium).

JUNGE, W.: Max-Volmer-Institut für Physikalische Chemie und Molekularbiologie, Technische Universität, 1000 Berlin 12 (BRD).

KAMEKE, E.v.: Institut für Chemische Pflanzenphysiologie, Universität Tübingen, 7400 Tübingen (BRD).

KATOH, S.: Department of Pure and Applied Sciences, College of General Education, University of Tokyo, Tokyo 153 (Japan).

KHANNA, R.: Department of Physiology and Biophysics, and Botany, University of Illinois, Urbana, Ill. 61801 (USA).

KLEEBERG, H.: Fachbereich Physikalische Chemie, Philipps-Universität, 3550 Marburg (BRD).

KLEVANIK, A.V.: Institute of Photosynthesis, USSR Academy of Sciences, Pushchino (USSR).

KLIMOV, V.V.: Institute of Photosynthesis, USSR Academy of Sciences, Pushchino (USSR).

KOBAYASHI, Y.: Laboratory of Plant Physiology, The Institute of Physical and Chemical Research, Wako-Shi, Saitama 351 (Japan).

KOENIG, F.: Max-Planck-Institut für Züchtungsforschung (Erwin-Baur-Institut), Abteilung Menke, 5000 Köln-Vogelsang (BRD).

KRASNOVSKY, A.A.: A.N. Bakh Institute of Biochemistry, USSR Academy of Sciences, Moscow (USSR).

KREUTZ, W.: Institut für Biophysik und Strahlenbiologie, Universität Freiburg, 7800 Freiburg i.Br. (BRD).

LAVOREL, J.: Laboratoire de Photosynthèse du C.N.R.S., 91190 Gif-sur-Yvette (France).

LICHTENTHALER, H.K.: Botanisches Institut (Pflanzenphysiologie) Universität Karlsruhe, 7500 Karlsruhe (BRD)

LUCK, W.A.P.: Fachbereich Physikalische Chemie, Philipps-Universität, 3550 Marburg (BRD).

MARKS, S.B.: Department of Chemistry, University of Illinois, Urbana, Ill. 61801 (USA).

MATHIS, P.: Service de Biophysique, Départment de Biologie, Centre d'Études Nucléaires de Saclay, 91190 Gif-sur-Yvette (France).

LIST OF CONTRIBUTORS

MENKE, W.: Max-Planck-Institut für Züchtungsforschung (Erwin-Baur-Institut) Abteilung Menke, 5000 Köln-Vogelsang (BRD).

METZNER, H.: Institut für Chemische Pflanzenphysiologie, Universität Tübingen, 7400 Tübingen (BRD).

MICHEL, J.M.: Laboratoire de Photobiologie, Département de Botanique, Université de Liège, 4000 Sart Tilman-Liège (Belgium).

PFISTER, K.: Botanisches Institut (Pflanzenphysiologie), Universität Karlsruhe, 7500 Karlsruhe (BRD)

PORTER, G.: The Royal Institution, London (Great Britain).

PULLES, M.P.J.: Department of Biophysics, Huygens Laboratory of the State University, Leiden (The Netherlands)

RADUNZ, A.: Max-Planck-Institut für Züchtungsforchung (Erwin-Baur-Institut), Abteilung Menke, 5000 Köln-Vogelsang (BRD)

RENGER, G.: Max-Volmer-Institut für Physikalische Chemie und Molekularbiologie, Technische Universität, 1000 Berlin 12 (BRD)

SARAI, A.: Department of Biology, Carleton University, Ottawa (Canada).

SATOH, K.: Department of Pure and Applied Sciences, College of General Education, University of Tokyo, Tokyo 153 (Japan).

SCHMID, G.H.: Max-Planck-Institut für Züchtungsforschung (Erwin-Baur-Institut), Abteilung Menke, 5000 Köln-Vogelsang (BRD).

SCHÖNBORN, M.: Robert Bosch GmbH, Forschungszentrum, 7016 Gerlingen (BRD).

ŠESTÁK, Z.: Institute of Experimental Botany, Czechoslovak Academy of Sciences, Praha (CSSR).

SHIBATA, K.: Laboratory of Plant Physiology, The Institute of Physical and Chemical Research, Wako-Shi, Saitama 351 (Japan).

SHUVALOV, V.A.: Institute of Photosynthesis, USSR Academy of Sciences, Pushchino (USSR).

SINCLAIR, J.: Department of Biology, Carleton University, Ottawa (Canada).

SIRONVAL, C.: Laboratoire de Photobiologie, Département de Botanique, Université de Liège, 4000 Sart Tilman-Liège (Belgium).

STEMLER, A.: Carnegie Institution of Washington, Department of Plant Biology, Stanford, Calif. 94305 (USA).

TIEN, H. Ti: Department of Biophysics, Michigan State University, East Lansing, Mich. 48824 (USA).

VAN GORKOM, H.J.: Department of Biophysics, Huygens Laboratory of the State University, Leiden (The Netherlands).

VIERKE, G.: Institut für Physikalische Biochemie und Kolloidchemie, Universität Frankfurt, 6000 Frankfurt 71 (BRD).

WEGMANN, J.: Institut für Chemische Pflanzenphysiologie, Universität Tübingen, 7400 Tübingen (BRD).

LIST OF CONTRIBUTORS

WYDRZYNSKI, T.: Department of Physiology and Biophysics, and Botany, University of Illinois, Urbana, Ill. 61801 (USA).

YAMAOKA, T.: Department of Pure and Applied Sciences, College of General Education, University of Tokyo, Tokyo 153 (Japan).

ZIMMERMANN, G.: Robert Bosch GmbJ, Forschungszentrum, 7016 Gerlingen (BRD).

PREFACE

In spite of intensive research some partial reactions of plant photosynthesis are still unknown. Whereas the path of carbon has been clearly elucidated meanwhile, the path of oxygen has still to be explained. There cannot be any doubt that the photosynthetic oxygen must be finally derived from water molecules. But the experimental data do not prove that there is a light-induced water decomposition ("photolysis").

There is general agreement that the photon energies absorbed by the sensitizer molecules can migrate within extended "antennae" before they reach a "reaction center". It is at these distinctive spots, where the first *chemical* reaction occurs. Apparently this primary step consists in an electron exchange between a donor - probably an excited sensitizer dimer - molecule and a still unidentified acceptor. Whereas the reduced acceptor handles its surplus negative charge *via* a chain of redox couples to the donor of a second light reaction, the oxidized primary donor has to regain its missing electron. To oxidize O^{--} anions to the oxidation state of molecular oxygen, four electrons have to be removed. Under continuous illumination the "waiting time" for a chlorophyll molecule between two photon absorptions is by orders of magnitudes longer than the survival time of the excitation energy. So there must be a transformation of the excitation energy into some form of a long-living intermediate.

Experiments with periodic light flashes demonstrate that the antennae need *four* flashes to release *one* oxygen molecule. This favours the assumption of a charge accumulation mechanism. On the other hand, it has long been realized that the O_2-releasing reaction requires the presence of organically bound manganese. Since this transition metal can exist in various oxidation states, there are many speculations on the natural Mn complex and on the possible mechanism by which it can store positive charges.

The next open question is the nature of the donor which reacts with the oxidized center. Here we have to regard experimental data on both oxygen and hydrogen isotope discrimination, which speak against the oxidative splitting of either free or bound water. Until now there is, however, no convincing alternative to the hypothesis

PREFACE

of H_2O decomposition.

Facing these problems, a Symposium on "Photosynthetic Oxygen Evolution" was invited to Tübingen. It was the intention of this meeting to present the meanwhile obtained data, to discuss their often controversial interpretations and to look for practicable new techniques to perform crucial experiments. To achieve this goal, photosynthesis specialists met with photo- and electrochemists and experts from other fields of physico-chemical research. This revealed the opportunity to recognize the present-day view on the structure of water and aqueous solutions, on the behaviour of irradiated redox systems and the properties of various model systems, including synthetic manganese complexes.

Most of the time was devoted to comparative studies on the light-induced charge exchange on the donor site of the photosynthetic electron transfer chain. These discussions included the proposal of new inhibitors and new aspects on the role of carbon dioxide. They also dealt with the structure of the photosynthetic apparatus and its ontogenetic evolution.

The success of the meeting encouraged the organizer to prepare all the contributions for printing. Publisher and editor expect that by this means they display the present state of knowledge and at the same time exhibit the most controversial problems. They do hope that the test will motivate biochemists, biophysicists and biologists to pursue a way which urgently asks for a close cooperation between different scientific disciplines.

The editor wants to thank all his colleagues, who agreed to have their contributions collected in a special volume. He extends his sincere gratitude to the sponsors of the Symposium - the Deutsche Forschungsgemeinschaft and the Ministry of Cultural Affairs of Baden-Württemberg - as well as to the coworkers of Academic Press Inc., London, both for retyping the manuscripts and for their effort to publish this volume at the earliest possible date.

Tübingen, September 1977

CONTENTS

List of Contributors v
Preface ix
List of Abbreviations xv

I. **Photoreactions in aqueous systems - physicochemical aspects**

LUCK, W.A.P. and H. KLEEBERG: Structure of water and aqueous systems. 1

CALZAFERRI, G.: Reduction degree in photoredox systems. 31

METZNER, H.: Oxygen evolution as energetic problem. 59

II. **Structural aspects**

KREUTZ, W.: On the structural arrangement of light reaction centres I-II in the photosynthetic membrane. 77

SCHMID, G.H., A. RADUNZ, F. KOENIG and W. MENKE: Characterization of the oxygen-evolving side of photosystem II by antisera to polypeptides and carotenoids. 91

YAMAOKA, T., K. SATOH and S. KATOH: Preparation of thylakoid membranes active in oxygen evolution at high temperature from a thermophilic blue-green alga. 105

HORVÁTH, G., M. DROPPA, A. JÁNOSSY, K. CSATORDAY and Á. FALUDI-DÁNIEL: Comparison of the O_2 evolving system in granal and agranal chloroplasts of maize. 117

III. **Electron and proton transport**

MATHIS, P.: Studies on the donor side of photosystem II by absorption spectroscopy. 125

VAN GORKOM, H.J., M.P.J. PULLES and A.-L. ETIENNE: Fluorescence and absorbance changes in TRIS-washed chloroplasts. 135

CONTENTS

KLIMOV, V.V., A.V. KLEVANIK, V.A. SHUVALOV and A.A. KRASNOVSKY: Reduction of pheophytin in the primary light reaction of photosystem II.   147

KOBAYASHI, Y., Y. INOUE and K. SHIBATA: Light-dependent binding of p-nitrothiophenol to photosystem II: oscillatory reactivity under illumination with flashes.   157

LICHTENTHALER, H.K. and K. PFISTER: New aspects on the function of naphthoquinones in photosynthesis.   171

HOMANN, P.H.: Oxygen-dependent photooxidations in photosystem II of isolated chloroplasts.   195

JUNGE, W. and W. AUSLÄNDER: Proton release during photosynthetic water oxidation: kinetics under flashing light.   213

IV. Oxygen evolution

RENGER, G.: Theoretical studies about the functional and structural organization of the photosynthetic oxygen evolution.   229

LAVOREL, J.: On the origin of damping of the oxygen yield in sequences of flashes.   249

GOVINDJEE and R. KHANNA: Bicarbonate: its role in photosystem II.   269

STEMLER, A.: Photosystem II activity depends on membrane-bound bicarbonate.   283

SINCLAIR, J., A. SARAI, T. ARNASON and S. GARLAND: Recent studies on photosystem II with the modulated oxygen electrode.   295

V. Role of manganese. Model systems

GOVINDJEE, T. WYDRZYNSKI and S.B. MARKS: Manganese and chloride: their roles in photosynthesis.   321

VIERKE, G.: Kinetics of deactivation of the charged state formed upon illumination on the oxidizing site of photosystem II in the presence of DCMU in *Chlorella* after extraction of membrane-bound manganese by NH_2OH.   345

KAMEKE, E. von and K. WEGMANN: Properties and function of two manganese-containing proteins from *Dunaliella* chloroplasts.   371

FUHRHOP, J.-H.: Molecular oxygen, light and metalloporphyrins.   381

HARRIMAN, A., G. PORTER and I. DUNCAN: Photoredox reactions of manganese.   393

CONTENTS

KRASNOVSKY, A.A. and G.P. BRIN: Photosensitization by titanium dioxide and zinc oxide: oxygen and hydrogen evolution ... 405

TIEN, H. Ti: Bilayer lipid membranes in aqueous media: incorporation of photosynthetic material from broken chloroplasts. ... 411

JAHNKE, H., M. SCHÖNBORN and G. ZIMMERMANN: Cathodic reduction of oxygen on chelates. ... 439

VI. Ontogenetic evolution of photosystem II

AKOYUNOGLOU, G. and J.H. ARGYROUDI-AKOYUNOGLOU: Photosystem II unit: assembly, growth and its interactions during development of higher plant thylakoids. ... 453

ŠESTÁK, Z.: Ontogenetic effects in oxygen evolution. ... 489

SIRONVAL, C., M. JOUY and J.M. MICHEL: Early photoactivity in illuminated, etiolated bean leaves. ... 495

DUJARDIN, E.: Emerson-like effect in protochlorophyll(ide) photoreduction in bean leaves. ... 511

Subject Index ... 521

LIST OF ABBREVIATIONS

AAS	atomic absorption spectroscopy
ACC	anthraquinonecyanine
ANT 2s	2-(3,4,5-trichloro)-anilino-3,5-dinitrothiophene
BLM	bilayer lipid membrane
BQ	p-benzoquinone
Chl	chlorophyll
Ch_T	chlorophyll triplet
CL	continuous light
CP	chlorophyll-protein-complex
DBMIB	2,5-dibromo-3-methyl-6-isopropyl-p-benzoquinone
DCMU	3-(3,4-dichlorophenyl)-1,1-dimethylurea
DCPIP	2,6-dichlorophenol indophenol
DMP	dimethylphthalate
DNB	m-dinitrobenzene
2,4-DNP	2,4-dinitrophenol
DPC	2,5-diphenylcarbazide
EDTA	ethylenediamine tetraacetic acid
EPR	electron spin resonance
FCCP	carbonylcyanide-p-trifluoromethoxyphenylhydrazone
HEPES	N-2-hydroxyethyl-piperazine-N-2-ethanesulfonic acid
FeCy	potassium ferricyanide
HQ	hydroquinone
LDC	light-dark cycle
MA	methylamine
MES	morpholinoethan sulfonic acid
MP^+	N-methylphenazinium cation
MV	methylviologen
NMR	nuclear magnetic resonance
NphSH	p-nitrothiophenol
OEC	oxygen evolving complex
OES	oxygen evolving system
Pc	phthalocyanine
PCB^-	C-phenyl-1,2-dicarba-undecaborate
PhD	phenylene diamine
PMS	phenazinemethosulfate
PS	photosystem
PTFE	polytetrafluor-ethylene
Q	quencher (primary acceptor of photosystem II)
Q	quinone

LIST OF ABBREVIATIONS

R	secondary acceptor of photosystem II
5-SAL	N-oxyl-4,4-dimethyloxazolidine derivative of 5-keto-stearic acid
SCE	standard calomel electrode
SDS	sodium dodecyl phosphate
SOD	superoxide dismutase
T-A	TRIS-acetone washed chloroplast membranes
TAA	dihydro-dibenzo-tetraazaannulene
Th	thionine
Th_T	thionine triplet
TPB^-	tetraphenylboron anion
TPP	tetraphenylporphyrine
Tricine	N-[tris(hydroxymethyl)-methyl]-glycine
TRIS	tris(hydroxymethyl)aminomethane
α	miss parameter
β	double hit parameter

STRUCTURE OF WATER AND AQUEOUS SYSTEMS

W.A.P. LUCK and H. KLEEBERG

*Fachbereich Physikalische Chemie,
Philipps-Universität Marburg (F.R.G.)*

Non-polar liquids

In search of a model for the liquid state, the hole model is simple and explains many properties [28]. One can think of a solid-like packed state. It expands with temperature on behalf of thermic vibrations and in addition contains holes, the concentration of which equals the number of molecules in the vapour phase, with which it is in equilibrium. With this quite powerful model one can understand Raoult's law: if we add extraneous molecules to our model-liquid, they occupy the holes; if the solved molecules do not have any vapour pressure, the pressure over the solution falls proportional to the amount of added particles. This model essentially holds for non-polar substances and is useful to calculate quantitatively the properties of non-polar liquids like heat of vapourization, specific heat or surface energy and surface tension [28]. To do this one needs an estimation of the intermolecular pair-potential. This can be taken from $3/2\ R \cdot T_c$ (T_c = critical temperature). This calculation can be done with liquid CH_4; its result is in reasonable agreement with the experimental heat content (H). If we compare the curves for methane and water (Fig. 1), which have comparable molecular weights, we observe large differences. Ice has a heat of sublimation of ~ -48.6 kJ·mol^{-1} ($= -11.6$ kcal·mol^{-1}). The heat of melting is

~-5.9 kJ·mol^{-1}(=-1.4 kcal·mol^{-1}). With further rise of the temperature the heat content gradually increases to comparatively high values. If we vapourize water at the melting point we need approximately 42 kJ·mol^{-1} (~10 kcal·mol^{-1}). We see, that the intermolecular interaction, which this plot describes in principal, is much higher with water than with methane. The reason for this totally different behaviour of water and all liquids with OH groups lies in the strong and angle-dependent intermolecular forces between protons and lone pair electrons of oxygen atoms. The angle can be easily understood, if we look at the model of a H_2O molecule. We have two

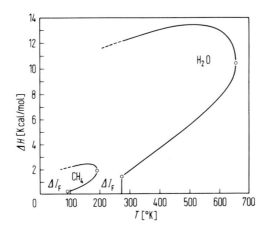

Fig. 1. Comparison of the heat content of methane and water.

positive centres at the protons and two negatively charged lone pair electrons; the connecting lines of these four charged spaces with the central oxygen closely form the angle of the tetrahedron. One can easily imagine, that the interaction with a second H_2O molecule very much depends on their relative orientation: if a proton approaches a lone pair electron attraction results, if a proton approaches another proton, the result will be repulsion, already at relative large distances. This phenomenon is not found with non-polar liquids. On behalf of this 3-

dimensional angle-dependent interaction we better understand the anomalous properties of water. Comparing water and methane we see, that water should boil at -80°C if it would not have any H bond interaction. Thus hydrogen bonds are fundamental for the water structure and the reason why we find living organisms on our planet.

As already mentioned, the interaction of H_2O molecules is angle-dependent and in the case of the approach of a proton and a lone pair electron - i.e. at H bond formation - it is maximal if the angle between the O-H bond-axis and the lone pair electron orbital is zero. If we fix H_2O molecules at this angle - the direction of lone orbitals may be symbolized by holes and that of the proton axis by a pin - we get the ice structure; it has a loose packing with four next neighbours contrary to non-polar substances, which may be regarded as a closest packing of spheres with about 12 first next neighbours. In ice we have six-membered rings of H bonded molecules, which are very stable. They give rise to the tridymite structure, which in certain directions has hexagonal symmetry. This finds a direct equivalent in the hexagonal symmetry of ice crystals grown under equilibrium conditions [17].

Structure of liquid water

Different from non-polar liquids orientation defects have to be taken into account in the liquid state of water. That means, that some of the hydrogen-bonds are opened - as given by the BOLTZMANN partition of energy - and the strong interaction (which totals approximately 17 kJ per H-bond) is reduced. So we can describe the liquid structure of water, if we are able to determine the amount of these orientational defects. This is possible by the use of infrared spectroscopy, especially in the overtone region [21, 26]. Fig. 2 gives the spectrum of HOD. For spectroscopic reasons partially deuterated water, which has the same liquid structure as normal water, is used.

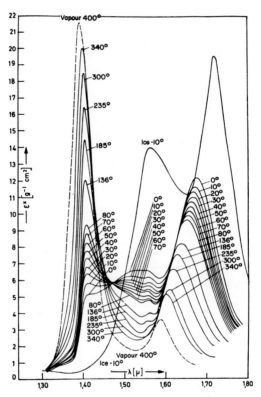

Fig. 2. Infrared spectra of HOD at various temperatures between -10°C (continuous line) and the critical temperature. Broken line: spectrum of water vapour (400°C).

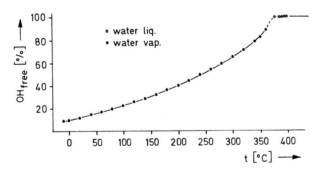

Fig. 3. Percentage of non H-bonded OH groups in liquid water as determined by the overtone method.

The spectra are recorded at different temperatures under saturation conditions. In ice the OH vibrations in the state of maximal H-bond interaction have an angle of 0 degree. Above the critical temperatures -i.e.>400°C - we find another frequency of absorbance, which corresponds to "free" - i.e. non H-bonded - OH groups. Between these temperatures intermediate situations are found. It was possible to develop a quantitative method to determine the concentration of orientation defects from the I.R. band intensity of these free OH groups [12, 13, 26]. Fig. 3 demonstrates the simple result. According to the spectroscopic data approximately 10% of free OH groups are produced during melting. If we compare the heat of melting (5.9 kJ·mol^{-1}) with the heat of sublimation (48.6 kJ·mol^{-1}) we find a good agreement of both results. If we further increase the temperature under equilibrium conditions, the amount of orientational defects gradually departs from the 10% value. At 100°C we find 22%, at 220°C 50% of free OH groups, i.e. only half of all H-bonds are broken.

Properties of liquid water

With the simple plot of Fig. 3 we easily succeed in describing a lot of properties of water quantitatively [15,21]. We will try this for a few examples: it has to be mentioned first, that the slope of this curve, i.e. its differential quotient, gives the amount of H bonds opened per degree. Multiplying the slope with the H-bond energy, we get the share of the H-bond on the specific heat (Table 1, see [15]). This is approximately half of the specific heat of water, which is ~75 kJ·mol^{-1}. After calculating the share of the H-bonds on the molar heat the comparison with other liquids shows that this part - and in consequence the total molar heat - is quite high in spite of the few degrees of freedom. By recalculating the specific heat per mole into gram-units, the value for water becomes surprisingly high; this is due to the large number of molecules per gram. It means, that water with its high heat capacity is an ideal liquid for operating a thermostat - especially a "movable" one like a mammal.

TABLE 1

Comparison of molar and specific heats and heat of vapourization of various substances and water

Substance	C_p			L_v		
	$J \cdot mol^{-1}$	$J \cdot g^{-1}$	t(°C)	$kJ \cdot mol^{-1}$	$kJ \cdot g^{-1}$	t(°C)
H_2O	75.3	4.18	20	44.0	2.45	20
CH_3OH	77.0	2.39	20	39.3	1.20	20
C_2H_5OH	107.6	2.34	20	42.3	1.00	20
$CH_3 \cdot CO \cdot CH_3$	124.3	2.13	20	31.0	0.55	20
$C_2H_5 \cdot O \cdot C_2H_5$	172.0	2.32	20	26.8	0.36	34
$CHCl_3$	112.6	0.92	20	31.4	0.26	20
CH_4	35.5(g)	2.22(g)	20	-	-	-
C_2H_6	71.6	2.39	-100	9.2	0.31	0
C_3H_8	96.7	2.18	- 53	15.1	0.35	20

By using the amount of free OH groups as given in Fig. 3 it is possible to calculate the molar heat of water at all temperatures up to the critical temperature [15,21].

The anomalous density maximum of water at 4°C can easily be explained: apart from the normal increase in vibrational volume with temperature, we find an increase of defects, which occupy less volume. Both effects counteract to give the mentioned maximum [15]. The heat of vapourization can be estimated, too. Evaporating one mole of water, we have to summarize all possible energetic interactions. We have a large amount of OH groups, which are H-bonded similar to ice. We have to know their number and their interaction

energy per molecule. To the multiplied value we have to
add the interaction energy of orientational defects, which
is given by the dispersion forces (~15 kJ·mol^{-1}) [18].
From the spectroscopically determined amount of defects we
may now calculate the heat of vapourization [15,29]. In
Fig. 4 the measured data are plotted as function of the
"reduced temperature", that is the water temperature T
divided by the critical temperature T_c. The bars give the
values calculated from the spectroscopically determined
amount of defects without using any adjusted constants.
Thus this spectroscopic method seems to be very useful in
order to calculate the properties of water quantitatively,
taking the simple model of open and closed H-bonds as a
basis. From Table 1 the heats of vapourization L_v of some
substances at 20°C may be compared. For water in spite of
its few intramolecular degrees of freedom it is abnormally
high, since every molecule may be H-bonded twice with
~17 kJ·mol^{-1} per H bond (=34 kJ·mol^{-1}) and in addition
possesses a dispersion energy of ~15 kJ·mol^{-1}. If we again
refer the values to gram units, the anomaly increases,

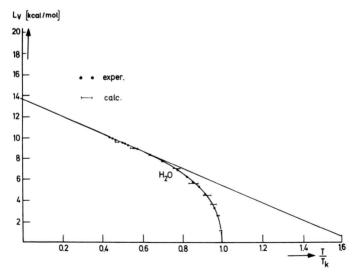

Fig. 4. Heat of vapourization of water. The horizontal bars indicate
the error after calculation by the spectroscopic method (from [15]).

since one gram of H_2O contains comparatively many molecules with their high interaction energy. So we, too, have an ideal cooling mechanism for the mentioned thermostat, if we take water as the substance with the maximal heat of vapourization and the minimum loss of weight.

To explain some other properties of water, we have to take into account the H bond energy in relation to the H bond angle β. Fig. 5 gives the spectroscopic result, here the ordinate is proportional to the interaction energy [22,23]. It has a minimum at an angle of β = 0° and slowly grows if the angle is increased. A consequence of this relation is e.g., that in crystal hydrates the H-bond angle lies between 0° and 15°, i.e. near the energy minimum [22,23]. With this plot an important property of the solid and liquid state may be shown: the co-operative mechanism. If we just consider a six-membered ring formed

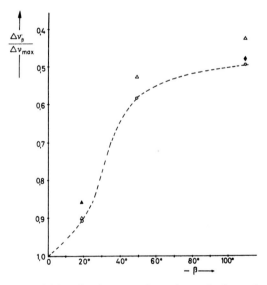

Fig. 5. Frequency shift ($\Delta\nu_\beta$) as a function of the H-bond angle β for different H_2O and CH_3OH associates as found by matrix spectroscopy (20 K) and in CCl_4 solutions (293 K).
o H_2O (matrix technique): Δ CH_3OH (matrix technique);
▲ R-OH (in CCl_4); ▼ C_2H_5OH (in CCl_4).
Ordinate normalized by dividing with the frequency shift ($\Delta\nu_{max}$) of linear H-bonds.

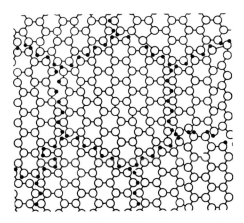

Fig. 6. Idealized cluster model of liquid water (at 0°C). Open circles: oxygen atoms; small full circles: hydrogen atoms of not H-bonded OH groups; dashes: H-bonds (see [12]).

by H_2O molecules with all H-bond angles of 0° - like in ice - and now remove one molecule, the remaining 5 molecules try to form H-bonds with each other again; but now the angle is approximately 10°, i.e. the H-bond energy is slightly lower. If we further increase the energy of this system, the probability to produce another defect is larger in an already deficient ring than in the still ordered parts of the lattice. This phenomenon is called the co-operative mechanism, i.e. the energy of the defects depends on the degree of order. The consequence is, that these defects are not statistically distributed. Due to this co-operativity we conclude that defective zones are formed as schematically drawn in Fig. 6 [12,13].

As the last property the surface tension σ of water may be discussed. It is given by the energy to enlarge the surface for one cm^2, which is the same as the often given force unit $dyn \cdot cm^{-1}$. In comparison with other liquids it is abnormally high for water; this is, of course, of importance in living organisms [29]. But we have to be aware that we compare cows and horses in this scale. If we consider molecular forces, we have to relate to one mole [28]. We may apply the equation

$$\delta_M = N_L^{2/3} \cdot V^{2/3} \cdot \sigma \qquad (1)$$

(V = molar volume)

After transformation into the usual energy unit of kJ·mol^{-1} the molar surface tension of water is still high in relation to e.g. alcohols, but the extremely high value per cm^2 is seen to be caused again by the large number of molecules per unit square. It is also possible to calculate the temperature dependence of the surface tension of water from our spectroscopic results (see ref. [17,29]).

In our simple model with the orientation defects concentrated in zones (Fig. 6), the circles are oxygen atoms, the lines are H-bonds and the closed circles are orientational defects. In an idealized model of H-bonded aggregates with ice structure - Raman spectra and X-ray data indicate that C_{2v}-symmetry dominates [30,33,34] - we can estimate the size of these aggregates with the spectroscopic values of orientation defects. These big aggregates are called "clusters"; Fig. 7 gives their mean size (Z) plotted against the reciprocal temperature. At the melting point we start with approx. 500 molecules per aggregate; at 50°C we still find more than one hundred, and at the boiling point (100°C) still 40 molecules make up one cluster. We can try to understand some more properties of water now. If we think of the solubility of hydrophilic substances or gases we expect, that they are primarily concentrated at the borders of these clusters. This is supported by the decrease of free OH groups as determined spectroscopically in aqueous solutions of NH_3 [27]. In agreement with this model Schröder assumed a two-step mechanism for the solubility of several gases [34]. First the solved molecules gather between the clusters; at higher vapour pressures over the solution - i.e. at higher concentration of gas in water - a second mechanism gains

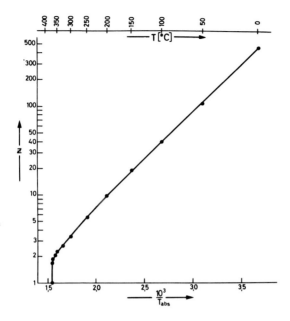

Fig. 7. Average cluster size (Z) in the simplified cluster model as calculated from the spectroscopically determined amount of free OH groups (from [18]).

in importance [32].

From our model (Fig. 6) the formation of gels seems to be easily understandable, too, if we consider the small amount of gel-forming substance being concentrated at the borders of these clusters. We must realize, however, that this picture is only an action shot of our model. In order to be at the energy minimum, the borders of the clusters try to form H-bonds. So there are several indications, that the relaxation time is of the order of 10^{-12} s, which one thinks to be the average lifetime of an H-bond [3]. Within these short time intervals the zones of orientational defects move through the liquid. Bringing chain-like molecules between these zones, one can imagine that this movement is reduced. Thus gel formation is induced. An example is the dye pseudo-isocyanine, which forms gels with 1 dye molecule per ~2000 molecules of water [24]. There are also well-known silica-gels with one SiO_2 per 330 H_2O-

molecules [20]. The viscosity, too, may be explained. Why then is water so little viscous in comparison to ice, in spite of its high amount of ice-like structures? The answer is, that the borders of the clusters may be easily displaced under shearing stress. The comparison with the measured activation energy for diffusion of ~20 kJ·mol^{-1} confirms [31], that for reorganization only one H-bond needs to be opened per cluster surface molecule; so we need only half the amount of the vapourization energy as activation energy. In particular we can spectroscopically prove that after addition of hydrochloric acid or sodium hydroxide the water bands increase in half-width [19]. This can be explained by the decrease in life-times of the aggregates. With 10^{-14}s they are of the same order as the frequency with which we measure.

Until now we only discussed an approximate but efficient "two species" model with open and closed H-bonds, but the spectroscopic results show, that we have to refine our water picture. The detailed bond analysis of HOD bands [21,26] shows that we have to assume at least three different OH bonds - corresponding to three different types of OH groups - the "free" OH (absorbing in the region 7000 - 7100 cm^{-1} or 1.400 μm - 1.429 μm), the linear ice-like H-bonds (absorption maximum 6400 cm^{-1} or 1.563 μm) and a third (intermediate) type (maximum at 6850 cm^{-1} or 1.46 μm). Investigations of H_2O matrix-spectra in solid N_2 indicate [22,23] that the intermediate band corresponds to H-bonds with different energetically unfavoured angles with a maximum around ~100 degrees. This angle-unfavoured state may be interpreted as resulting from intermediate positions during the formation of new linear H-bonds from orientation defects. The dependence of H-bond frequencies on the H-bond angle indicate, that the orientation of neighbours determines the interaction energy. On the other hand the presence of different organic H-bond acceptors like ether or C = O groups etc. could change it, too.

The energy minimum at H-bond angles near zero is a very important factor which determines many specific effects in biochemistry [14,16,19]. Fig. 8 gives the percentages of these three water species under saturation conditions as a function of T - as indicated by the HOD band analysis [26]. The possibility to apply a two-species model corresponds to the flat T maximum of the angle-unfavoured OH groups.

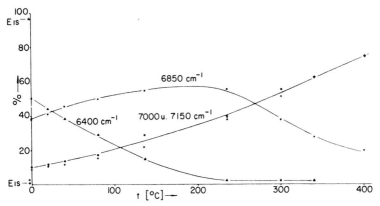

Fig. 8. Share of the different components (given in percent) of the HOD spectra between 0° and 400°C. 6400 cm^{-1} corresponds to ice-like H-bonded H_2O molecules; 6850 cm^{-1} to unfavourable H-bonds, 7000 and 7150 cm^{-1} to free OH groups (from [26]).

H-bonds with unfavourable angles are also present in reasonable amounts at temperatures between the melting point and the critical temperature. If we look at the temperature region around 100°C it is easily seen, that we have still 20% of ice-like H-bonds and approximately 50% which are unfavourable; the amount of free OH groups, however, is only slightly higher than 20%. Thus it cannot be expected to find any H_2O molecules which are not H-bonded as was often proposed [5,31]. We can be sure that the Nemethy-Scheraga model is wrong in respect to the not H-bonded molecules. At room temperature there are no demonstrable amounts of not H-bonded H_2O molecules. On the other hand we can say, that our spectra are always the result of all the differently H-bonded molecules present at the respective temperatures (compare Fig. 9).

Structure temperature

The temperature dependence of the H-bond structure in liquid water - as it can be determined by spectroscopic studies - induced the approximation-method using the water temperature as a measure for the H-bond balance. To characterize e.g. the small differences between the structure of liquid H_2O and D_2O one can remember that some T-dependencies of D_2O properties are similar to H_2O properties. They only show a shift of the T-scale to 3.5 - 4° higher T values which corresponds to the shift of the melting temperature of D_2O. For example gels melt at a ~3.8° higher temperature in D_2O than in H_2O [25]. Ribosomal RNA has a transition temperature which is by 4° higher in D_2O than in H_2O [9]. So we may say that D_2O has a "structure temperature" 4° below the experimental temperature. This means: the H-bond structure of D_2O at a certain temperature is similar to that of H_2O at a 4° lower temperature. The usefulness of this heuristic nomenclature can be demonstrated with biological systems like *Escherichia coli* [5]. The number of surviving cells at 52 or 55°C is always higher in D_2O than in normal water; the heat resistance difference corresponds to the difference in structure temperatures [5].

Aqueous electrolyte solutions

Our model also allows to consider the influence of electrolytes on the structure of water: Fig. 9 gives the 0.960 μm band of water at different temperatures. After addition of 1 mol of sodium sulphate - and density correction on the observed water concentration per cm^3 - the resulting band at an experimental temperature of 20°C is only slightly different from that of pure water. Under the same conditions sodium perchlorate, on the other hand,

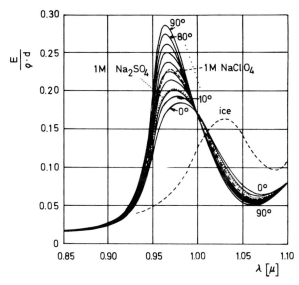

Fig. 9. Temperature dependence of the 960 nm overtone band of pure water in comparison with 1 M solutions of Na$_2$SO$_4$ and NaClO$_4$.

changes the spectrum considerably. The band resembles that of water at 40°C. In a first approximation one can quantitatively describe the influence of electrolytes by the determination of the structure temperature [10] as suggested by Bernal and Fowler qualitatively [1]. The structure temperature is that temperature of pure water, which has the same H-bond balance as the solution under investigation and therefore produces the same spectroscopic result. At room temperature sodium perchlorate would have a structure temperature of 37°, i.e. the structure of NaClO$_4$ solutions is comparable to that of water to 37°C. We would have more free OH groups, i.e. this water is more hydrophilic and can give salt-in effects on organic solutes and *vice versa* with salts of lower structure temperatures. With the structure temperature (T_{str}) we can characterize different electrolytes now. We find salt solutions with negative values for $T_{str} - T_{exp}$, the so-called "structure makers" and with positive values, the "structure breakers".

From our cluster model we may calculate the cluster size for the different structure temperatures and compare the induced change in cluster size. We find an increase in cluster size of electrolytes, which are "structure makers" and a decrease with the "structure breakers". The sequence of different salts, ordered by their T_{str} values, is more or less identical with the well-known Hofmeister or lyotropic ion series in colloid chemistry. This series can after 100 years of empiric knowledge be understood as a series which influences the water structure [10]. Additionally it has to be mentioned, that the transition from "structure makers" to "structure breakers" is temperature-dependent, but the sequence of the series always remains more or less the same. It is of interest, that the effect of sodium chloride is nearly zero and varies only little with temperature. Now we would like to test our considerations experimentally: Solubility changes of gases or organic solutes in water seem to be more sensitive to added ions than the infrared water bands are. We determined the partition coefficient of *p*-cresol between cyclohexane and water after addition of different concentrations of sodium sulphate [24]. Now we "salt out" cresol, and we can do as if the ions bind hydrate water strongly, so that this water of hydration is less able to solve organic molecules. Under the assumption that the remaining water is constant in respect to the partition coefficient we can calculate apparent hydration numbers. For a one molar Na_2SO_4 solution at 25°C we find a hydration number of approximately 20, which decreases with increasing temperatures [24]. This result agrees very well with what we find from our spectroscopic analysis. If our model is correct, we should get another plot for sodium chloride for example, which induces another change in the structure temperature. Indeed the partition coefficients of *p*-cresol between cyclohexane and NaCl solution only very little depend on the temperature, just as we would have predicted

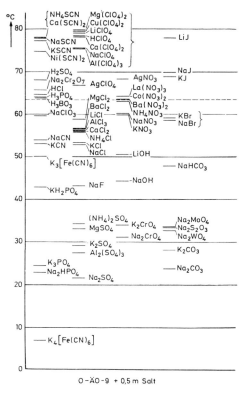

Fig. 10. Turbidity temperatures of aqueous solutions of nine-fold ethoxylated p-isooctyl-phenol (at 64°C). Data taken after addition of different salts to pure water up to a final concentration of 0.5 M. Dashed horizontal line: pure water.

from the spectroscopic results. The spectroscopic method is less sensitive to salt additions because at room temperature the water aggregates are so big that a change of the expansion of H-bonded water molecules changes the content of the "free OH" at the surface of clusters only at a small percentage. Another test for the model was tried by the measurement of the turbidity temperature of a solution of an octylphenyl residue to which a chain of 9 ethyloxides was bound (Fig. 10). Without any addition of salts at 64°C the solution became cloudy and separated into two phases. After addition of 0.5 moles of salt the turbidity point varies largely (Fig. 10). As in the other experiments, the order of salts was exactly that of the Hofmeister series. It should be emphasized, that the effect of the anions always exceeds that of the cations. So the Hofmeister ion series can be reduced to salt-induced changes in the water structure. We are quite confident that the rough measure, which the structure temperature is, gives a valuable description of the nature of electrolyte solutions. This includes effects on cosolutes, too. A biological example for the transition temperature of ribonuclease [4] after addition of different salts has been given. Again the effect establishes the same ion series. The spectroscopic structure temperatures of these additions at these transition temperatures have the tendency to lie nearer the transition of ribonuclease without any addition. The detailed analysis of salt effects indicates, that we have to distinguish between three different water bonds in aqueous electrolyte solutions: the primary (more or less permanent) hydration sphere, the secondary hydration sphere and normal (bulk) water. The properties of ternary systems: water, ions and organic solutes depend on the H-bond strength which the solute needs to become soluble, too. In the presence of strong acids or strong bases like OH^- the life-time of the H-bond can be another important parameter [11].

Interactions between water and organic molecules

Thinking on methods to investigate the H-bond state in aqueous solutions of organic molecules we have to remember that the H-bond state depends on the angle, which it forms with the accepting lone electron pairs. It is thus in the

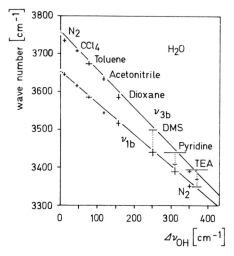

Fig. 11. Frequency of the symmetric (v_1) and asymmetric (v_3) vibration of water molecules with different acceptors (2 acceptors/H_2O) as a function of the uncoupled frequency ($v_{OH} = 1/2\ (v_3 + v_1)$).

region of smaller or larger charge density of the lone pair electrons of different acceptors as reflected by frequency shifts. If we consider the fundamental band of water after addition of different organic acceptors (Fig. 11), we see a frequency shift which depends on the pK_a-value of the base (Fig. 11). This investigation was carried out in CCl_4 solutions, where the H-bond angle may be freely chosen by the H_2O molecules. We can realize H-bonds in solutions of different strength by varying the acceptors, keeping the angle 0°. In both cases we find a frequency shift to higher wavelengths, i.e. lower wavenumbers with stronger H-bonds.

The interaction energy of water molecules can be changed by different organic acceptors, by different oriented water neighbours or by different ions. Detailed studies show that the strength of the electric field of interaction in all three cases covers a similar range. From the viewpoint of our spectroscopic experiences we can ask: could we combine the two big research fields on electrolyte solutions

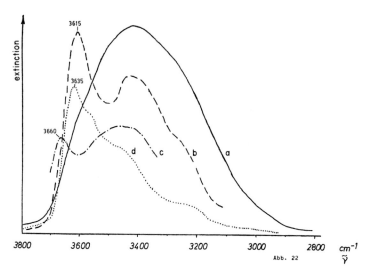

Fig. 12. Fundamental band of pure water in comparison with that of water inside different desalination membranes (all values for 20°C). a) pure water; b) H_2O in cellulose acetate (98% relative humidity); c) H_2O in porous glass (11% relative humidity); d) H_2O in polyimide (98% relative humidity).

and H-bonds to a common chapter of science? For the fundamental question in biology, how to open and close the helix of DNA whenever the organism needs it, we can discuss the following mechanism: local "temperature increase" of DNA, changing the structure temperature of the neighbour water by ions, by temperature change or by organic acceptors or second organic acceptors inside the DNA (like some antibiotics do in bacteria).

With our refined model of the water structure and electrolyte solutions we want to apply our spectroscopic tools on two questions: first we consider the fundamental band of water in different membranes (Fig. 12). Although the state of water in synthetic desalination membranes seems to be far from the topic under discussion in this paper, it may be of principal interest in order to understand transport mechanisms across biological membranes, too. If we compare the band of pure water with that of H_2O sorbed in cellulose acetate or polyimide at 98% relative humidity

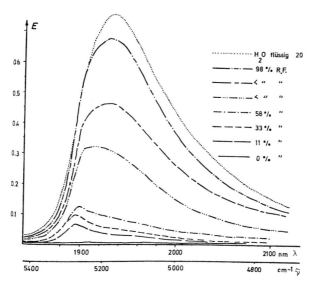

Fig. 13. Combination band of water sorbed in a glass membrane (pore diameter: 2.6 nm) at different relative humidies:

— . — . —	98%
— — — — —	< 98%
— ... — ...	< 98%
— . — . — .	58%
— — — —	33%
— —	11%
————	0%
....	control: liquid water

the differences are clearly seen (Fig. 12). We have a higher optical density at the short wavelength side of the OH water band. This means we have a considerably weaker H-bond formation inside the synthetic membranes. Especially in the acetylated cellulose membrane we may have water bonded with one OH group to a weak accepting site of the polymere, while other OH groups, giving rise to the absorbance at longer wavelengths, interact with other H_2O molecules. Water in these membranes, however, does not have the structure of pure water. On behalf of acceptor strength and sterical reasons, H-bonds of little strength may be formed predominantly. Thus only small amounts of salts may be solved inside these membranes, so that they may be used for desalination. If we look at the water inside a special glass membrane (Fig. 13) we see, that at low relative humidities water is weakly bonded giving rise to an absorbance at the short wavelength end of the band again. With increasing humidity - i.e. increasing water content of this membrane - the band becomes more and more similar to that of pure water (dotted line in the upper part of Fig. 13). This membrane with pores of 26 Å in diameter has a poor retention power for ions, since its water has a rather liquid-like structure and solves ions. The weak binding of water inside cellulose membranes should be advantageous, since it leaves the water more mobile and thus enhances water flow. These experiences are in agreement with studies on the two-phase formation of polyethyleneoxide derivatives with hydrophobic end groups (11). Inside the organic phase secondary hydrates with ~20 H_2O per ether group can exist. The outer hydrate layer has a similar structure as liquid water. Therefore it solves ions. But the primary hydrate layer of the ether groups seems to have a different structure which does not solve ions. The results of the two-phase formation (so-called "coacervation") can be described with a three-step model, too, of primary hydrate, secondary hydrate and normal water [2] and reminds to the parallelism between the interaction water-ion and water-organic acceptors.

As the last item we will discuss the system biopolymer-water. We found that biopolymers bind water more strongly than the membranes just discussed. With decreasing water content of collagen or gelatin we find a gradual shift of the maximal optical density of the H_2O combination band into the direction of higher wavelengths until - below 54% relative humidity - it remains constant at a structure temperature of -50°C. The comparison with the sorption isotherm (Fig. 17) shows, that (below 54% relative humidity) 24 g of H_2O are firmly bound per 100 g of protein dry weight. This corresponds to a H-bonding of 1.2 moles of water per aminoacid residue in a first hydration sphere [8].

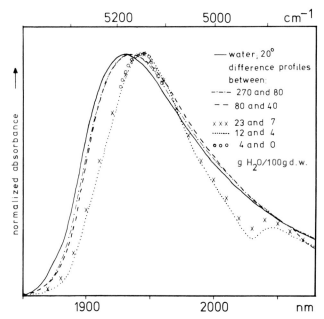

Fig. 14. Difference profiles of various amounts of water in gelatin (after normalization).

The difference spectra of gelatin (Fig. 14), which yield the same spectroscopic result as tendon collagen, show still more differences: below 23 g H_2O per 100 g of protein we always find the same difference profiles due to the first hydration sphere of this protein as just discussed. But at higher water content up to approximately 200 g/100 g dry weight the difference profiles are still different from that of pure water indicating a stronger H-bond balance of a second hydration sphere. Above ~200 g water per 100 g of protein the difference profiles are identical with that of pure water. So we can be sure that this additional water - the so-called "bulk" water - has the same structure as pure water. We can now understand the dependence of the melting point of gelatin gels on added salts (Fig. 15). Structure makers as ammonium sulphate enlarge the cluster size; therefore relatively more

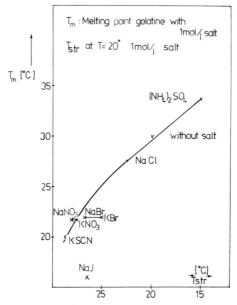

Fig. 15. Melting point (T_M) of gelatin gels in dependence of added salts. T_M plotted as function of the structure temperature (T_{str}) of the respective salts.

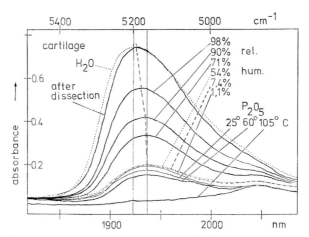

Fig. 16. Combination band of water sorbed in bovine nasal cartilage at different relative humidities.

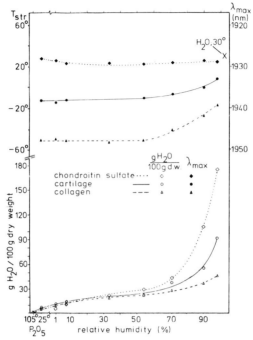

Fig. 17. Structure temperatures (———) and sorption isotherms (-----) of chondroitin sulfate, cartilage and collagen.

protein is available to stabilize the borders of one cluster. With structure breakers the opposite is true, so that the gelatin present in solution is not sufficient to form a gel by localizing the borders of the clusters present.

For total bovine nasal cartilage the spectroscopic result shows the same tendency as collagen (Fig. 16). We see that the water band in cartilage directly after dissection is already slightly different from that of pure water. With decreasing water content it further shifts to higher wavelengths until (below 54% relative humidity) it again keeps constant at a structure temperature of approximately -10°C (Fig. 17) [6]. The difference profiles give further insight into the different hydration spheres [7]: 5.5% of the total water present in fresh bovine nasal cartilage are strongly bound to the cartilage matrix with a structure temperature of -10°. Therefore, the difference profiles keep closely constant at water contents below this value. Another 20% of the total water in cartilage (up to 98% relative humidity) yields (at an experimental temperature of 37°C) a second set of difference profiles belonging to a second hydration sphere with a structure temperature of 10°C. The difference profile of cartilage after dissection and 98% relative humidity is still slightly different - with a structure temperature of approximately 32°C - from the band of pure water. It is, however, not clear whether these 75% of the total water in cartilage form a third separate hydration sphere or whether they are a mixture of water structures of the second sphere and bulk water. Since it is known, that this cartilage consists of approximately equal amounts of collagen and chondroitin sulphate, a sulphated glycosamin-glycan, which makes up to approximately 80% of the total dry weight, we may compare the spectroscopic results and sorption isotherms of the two pure substances and the tissue (Fig. 17). We recognize the shift which we have already discussed for collagen with its deep structure temperature in the first hydration sphere. The structure temperature of H_2O in chondroitin sulphate (Fig. 17, filled squares) hardly changes with water content and is nearly equal to the experimental temperature. This is a reflection of the corresponding water structures at the different acceptor sites of this polysaccharide. Cartilage finally shows a plot which lies between the two others. Something similar is true for the sorption isotherms. Both pure components build up the primary hydration sphere until 54% relative humidity. At higher humidities the water uptake of tendon collagen is only small, since the three-dimensional entanglement of the triple helices does not permit a further uptake like in gelatin. Cartilage takes up water amounts which lie between collagen and chondroitin sulfate, which itself shows a sharp increase of the sorption isotherm at higher

STRUCTURE OF AQUEOUS SYSTEMS 27

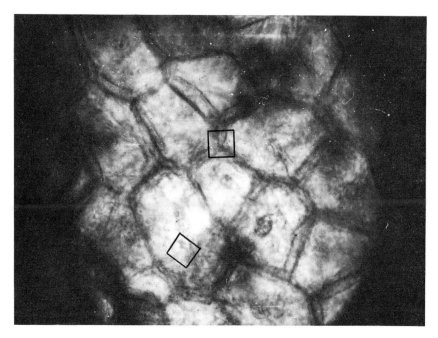

Fig. 18. Cell layer of *Rhododendron* flower leaves. Spectra (see Fig. 19) were recorded at the positions indicated by the squares.

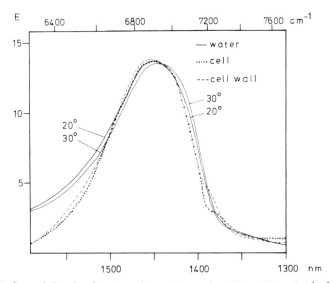

Fig. 19. Infrared band of water in cells and cell walls of *Rhododendron* flower leaves (details see text).

humidities. So one may think, that the hydration in cartilage may be explained by the combined action of the pure substances which it is composed of [8]. The comparison of the spectroscopic result obtained with cartilage and with that calculated from the pure components really seem to justify this assumption [8].

To get an answer to the question, whether the structure of water is uniform in different cellular regions, we looked at a monocellular layer of *Rhododendron* flower leaves (Fig. 18). We recorded the spectra inside the cell and in the region of the cell wall (indicated by squares in Fig. 18) at an experimental temperature of 24°C. Fig. 19 shows the result: while the band inside the cell fits quite well with the band of pure water at 20°C, that of the cell wall has an even deeper structure temperature. But until now we cannot say more than that in cellular structures the state of water is not unique and that it is stronger H-bonded than pure water at the same temperature. The abnormal isotherms of water in organic or biological systems, too, can be interpreted by the assumption of (at least) two different water "phases". Firstly we can assume bonded hydrate water in the innermost layers. Secondly the strong increase of water uptake above 50% - 60% relative humidity may be discussed as a decrease of the T_c value by two-phase formation of water (vapour-liquid) as it is also known in solutions. Thirdly in the range of 100% relative humidity a disturbed water structure like in solutions may coexist with some pure liquid water of normal structure.

References

1. Bernal, J.D. and Fowler, R.H. (1964). *J. Chem. Phys*. **1**, 516-548.
2. Bungenberg de Jong, H.D. (1949). In "Colloid Science" (H.R. Kruyd, ed.), Vol. II, 232-255. Elsevier Publishing Company, Amsterdam.
3. Hasted, J.B. (1974). In "Structure of Water and Aqueous Solutions" (W.A.P. Luck, ed.), 377-389. Verlag Chemie, Weinheim.
4. Hippel, P.H. von, and Schleich, T. (1969). In "Structure and Stability of Biological Macromolecules" (S.N. Timasheff and G.D. Gasman, eds), 417-574. Marcel Dekker, Inc., New York.

5. Hübner, G., Jung, K. and Winkler, E. (1970). "Die Rolle des Wassers in biologischen Systemen". Akademie-Verlag, Berlin.
6. Kleeberg, H. and Luck, W.A.P. (1977). *Naturwissenschaften* **64**, 223.
7. Kleeberg, H. and Luck, W.A.P. (1977). Submitted.
8. Kleeberg, H. and Luck, W.A.P. (1977). In preparation.
9. Lewin, S. and Williams, B.A. (1971). *Arch. Biochem. Biophys.* **144**, 1-5.
10. Luck, W.A.P. (1964). *Ber. Bunsenges. physik. Chem.* **68**, 895.
11. Luck, W.A.P. (1964). *Fortschr. Chem. Forsch.* **4**, 653-781.
12. Luck, W.A.P. (1965). *Ber. Bunsenges. physik. Chem.* **69**, 69-76.
13. Luck, W.A.P. (1965). *Ber. Bunsenges. physik. Chem.* **69**, 626-637.
14. Luck, W.A.P. (1965). *Naturwissenschaften* **52**, 49-52.
15. Luck, W.A.P. (1967). *Disc. Faraday Soc.* **43**, 115-127.
16. Luck, W.A.P. (1967). *Naturwissenschaften* **54**, 601-607.
17. Luck, W.A.P. (1970). *Informationsdienst Arbeitsgem. Pharm. Verfahrenstechn.* **16**, 127-159.
18. Luck, W.A.P. (1970). *Med. Welt* **21**, 87-101.
19. Luck, W.A.P. (1970). *Med. Welt* **21**, 857-870.
20. Luck, W.A.P. (1970). *Physik. Blätter* **26**, 133-137.
21. Luck, W.A.P. (1974). *In* "Structure of Water and Aqueous Solutions" (ed.). Verlag Chemie, Weinheim.
22. Luck, W.A.P. (1976). *In* "The Hydrogen Bond" (P. Schuster, G. Zundel and C. Sandorfy, eds), Vol. II, 527-562. North Holland Publishing Company.
23. Luck, W.A.P. (1976). *In* "The Hydrogen Bond" (P. Schuster, G. Zundel and C. Sandorfy, eds), Vol. III, 1367-1423. North Holland Publishing Company.
24. Luck, W.A.P. (1976). *Topics Curr. Chem.* **64**, 113-180.
25. Luck, W.A.P. (1976). *Naturwissenschaften* **63**, 39-40.
26. Luck, W.A.P. and Ditter, W. (1969). *Z. Naturforschg.* **24**b, 482-494.
27. Luck, W.A.P. and Ditter, W. (1970). *J. Phys. Chem.* **74**, 3687-3695.
28. Luck, W.A.P. and Ditter, W. (1971). *Tetrahedron* **27**, 201-220.
29. Luck, W.A.P. and Ditter, W. (1972). *Adv. Mol. Relaxation Proc.* **3**, 321-339.
30. Narten, A.H. (1974). *In* "Structure of Water and Aqueous Solutions" (W.A.P. Luck, ed.), 345-363. Verlag Chemie, Weinheim.
31. Némethy, G. and Scheraga, H.A. (1962). *J. Chem. Phys.* **36**, 3382-3400.
32. Samoilov, O.J. (1961). "Die Struktur wässriger Elektrolytlösungen". Verlag B.G. Teubner, Leipzig.
33. Schröder, W. (1969). *Z. Naturforschg.* **24**b, 500-508.
34. Walrafen, G.E. (1967). *J. Chem. Phys.* **47**, 114-126.
35. Walrafen, G.E. (1972). *In* "Water. A Comprehensive Treatise". (F. Frankles, ed.), Vol. I, 151-214. Plenum Press, New York.

REDUCTION DEGREE IN PHOTOREDOX SYSTEMS

G. CALZAFERRI

Institut für Anorganische und Physikalische Chemie der Universität, Bern (Schweiz)

Abstract

By means of both literature examples and data of own research work it will be explained how to succeed in understanding photochemical reactions by processing chemical, spectroscopic, and molecular-physical results. The description of light-induced photoredox reactions is suitable to show, how the discussion of the thermal equilibrium position can lead to important information on the behaviour of an irradiated system. In virtue of model reflections a possibility for the initiation and maintenance of an intramolecular charge separation will be referred to in the last chapter.

Introduction

Photochemistry studies the chemical properties of molecules in excited state. Since each molecule can attain several states, the variety of this chemistry is not smaller than that of the ground state. Without any help of quantum theory the chemistry of the ground state has practically succeeded in developing a simple formalism, which can settle a lot of experimental data and sometimes allows to make predictions. The chemistry of the excited state was from the very beginning confronted with quantum theory. Although many interesting observations were made long before the development of this theory. However, with our present knowledge it is no more reasonable to do photochemistry without the help of some results of quantum theory. Until now an exact quantum theoretical treatment of photochemistry is hardly possible. How extensive the difficulties are, demonstrates e.g. the detailed analysis of formaldehyde photolysis:

$$H_2CO \xrightarrow{h\nu(\sim 290 nm)} H + HCO$$
$$H_2CO \xrightarrow{h\nu(\sim 350 nm)} H_2 + CO$$
(1)

In spite of great efforts these relatively simple reactions are from the theoretical viewpoint only partly understood [26]. It is therefore our aim to find good models to work with. The particular difficulty - but at the same time the stimulus - of photochemistry consists in the necessity of processing chemical, spectroscopic and molecular-physical informations. This fact will be outlined by means of both literature examples and own research work. The arrangement chosen is for the most part incidental.

The insufficiently used possibility to have molecules reacting in a well-defined vibrational state - either by direct stimulation with infrared radiation or by the transfer of vibration energy - has gained remarkable interest with the development of infrared lasers![1]

The following examples show in an impressive manner, that the electromagnetic energy - even in the infrared region - possesses not only quantitative, but also important qualitative aspects, with regard to chemical reactions.

The pyrolysis of diborane leads to a product mixture, which does not contain icosaborane(16); normally tetra-

---

[1] By reason of the definition given above the chemistry with infrared radiation does not belong to photochemistry. This demonstrates a difficulty to be found, whenever one tries to describe a scientific discipline with the help of a simple definition.

B_2H_6 (200 torr) $\xrightarrow{h\nu(972.7 \text{ cm}^{-1})}$ $B_{20}H_{16}$ (main product) by chain reaction [4]   (2)

BCl_3 (150-200 torr) $\xrightarrow{h\nu(956 \text{ cm}^{-1})}$ $BCl_3^{\dagger}$

$3C_2Cl_4$ (20-80°C) $\xrightarrow{BCl_3^{\dagger}}$ C₆Cl₆ (main product) [5]   (3)

CH_3F (5 torr) $\xrightarrow{h\nu(9.6 \text{ μm})}$ $CH_3F^{\dagger}$ [19]   (4)

$$CH_3F^{\dagger} + H_3C-\underset{\underset{CH_3}{|}}{\overset{\overset{O-O}{|}}{C}}-\underset{\underset{CH_3}{|}}{\overset{}{C}}-CH_3 \xrightarrow{\text{(energy transfer)}} \left[CH_3-\underset{\underset{CH_3}{|}}{\overset{\overset{O-O}{|}}{C}}-\underset{\underset{CH_3}{|}}{\overset{}{C}}-CH_3\right]^{\dagger} + CH_3F$$

$$\longrightarrow 2\,\underset{CH_3\;\;CH_3}{\overset{\overset{O}{\|}}{C}} + h\nu \approx 410 \text{ nm}$$

$H^{35}Cl$ (ν=0)   no chemical reaction

$H^{35}Cl$ (ν=1) + $Br(^2P_{3/2})$ $\longrightarrow$ no chemical reaction [2]   (5)

$H^{35}Cl$ (ν=2) $HBr(\nu=0) + {}^{35}Cl$

chloroethylene cannot be converted to hexachlorobenzene below 700°C. The authors could show in both cases, that vibrational excitation is responsible for the declared course of reaction. This appears even more clearly in the sensitization of chemiluminescence of the tetramethyl-1,2-dioxetane by vibrationally excited methylfluoride. It likewise appears in the reaction behaviour of vibrationally excited hydrogenchloride, which confirms the postulated anticipations of microscopic reversibility. Vibrational excitation in the first excited state can also play a role in solutions. This was demonstrated by the example of 2,2-diethylchromene [6]. The fluorescence quantum yield of this compound decreases with increasing excitation energy, whereas the quantum yield of the photochemical reaction increases [29].

$$\text{chromene} \xrightarrow{h\nu} \text{product} \qquad (6)$$

A dependence of this kind can only be expected, if the photochemical reaction happens within one or only a few vibrations after excitation, i.e. within ~10^{-12} - 10^{-11} s. As we know by direct measurements [35,41], excess excitation energy is within this time dissipated by coupling of the molecules with their surroundings.

To describe a reaction, it would be an ideal situation to know the electronic (n), vibrational (ν) and rotational (j) state of the susbtrates A, B and the reaction products resulting from the first step.

$$A(n_A, \nu_A, j_A) + B(n_B, \nu_B, j_B) \longrightarrow R(n_R, \nu_R, j_R) \qquad (7)$$

The possibility of such a detailed analysis is restricted to the gas phase. The experimental and theoretical efforts are considerable. However, one approaches to information which is of great importance in connection with infrared lasers, isotope separation processes and research methods leading to a profound comprehension of chemical reactivity. The photodissociation of the cyano compounds XCN (with X = Cl, Br, I) was e.g. very thoroughly studied [3,39]. The distribution of the reaction energy on vibrational and rotational levels in the CN fragment is fairly well known.

$$XCN \xrightarrow{h\nu} X + CN(B^2\Sigma^+,\nu,j)$$

$$XCN \xrightarrow{h\nu} X + CN(A^2\Pi,\nu,j)$$

(8)

The potential curves of the cyanide radical for the three lowest electronic states together with the vibrational levels are given in Fig. 1 [40].

We would like to direct our attention only to the fact, that the states $CN(B^2\Sigma^+,\nu=0)$ and $CN(A^2\Pi,\nu=10)$ are accidentally degenerated, which - according to modern ideas - leads to a fast transfer between these two states [7,16, 18,30]. The experimental studies of this event - so-called "intersystem crossing" as function of the rotational states are in qualitative agreement with the theoretical ideas [3].

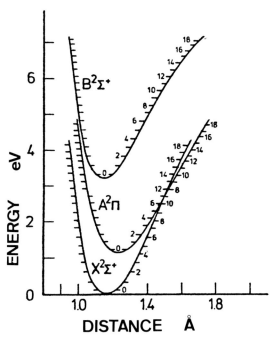

Fig. 1. Potential curves with vibration levels of the CN· radical in the lowest electronic states.

Intramolecular motions

In solutions rotational levels of molecules cannot be separated spectroscopically. Therefore they definitely play no role in the description of photochemistry in a condensed phase. However, the time during which a molecule can turn around its own axis - rotational diffusion - becomes an important figure. After excitation with polarized light a molecule emits polarized light only under the condition, that it cannot significantly rotate within the mean lifetime of the excited state. This circumstance has led to various methods to measure the rotational diffusion of electronically excited molecules [16,21,27,29,44]. We have found that the fluorescence quantum yield of methincyanines at room temperature rises considerably with increasing solvent viscosity [16]. According to the well-known principle, that for high fluorescence quantum yields the molecule skeleton has to be stiff [33], it was obvious to examine, whether this viscosity dependence is to be explained by hindrance of intramolecular movements with increasing viscosity. It was evident from former research work, that in general there exist no simple relations between the macroscopically measured viscosity and the viscosity as experienced by the single molecule [32]. The comparison already made by Th. Förster [21]

no fluorescence   fluorescence

X,Y = C-H, C-OH, C-Ph, N, N-Ph, O

has shown that this intramolecular movement is probably a rotation around the bond axis, which in Fig. 2 is marked by an arrow.

It is reasonable to assume, that in methincyanines a hindrance of the rotation of the whole molecule entails

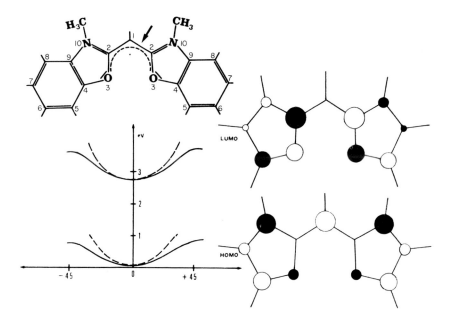

Fig. 2. Potential curves in the ground state and in the first excited state at twisting of the oxazole ligands. The straight lines give calculations with the EHT model [24]. The reason of the rotation barriers of ~1 eV are lower lying π orbitals. This is also the reason, why the barriers in the ground state and in the first excited state are nearly identical. The two hatched curves result, if the twisting of the two oxazole ligands is hindered by the viscosity of the solvent. The HOMO/LUMO behaviour as given in this figure and in Fig. 4 is realized, if no lower lying δ^* orbitals exist [12].

also a rotation aggravation around this bond axis. This idea could be verified by measuring the dynamic fluorescence depolarization [16,27].

Fig. 2 gives the highest occupied - HOMO - and the lowest unoccupied - LUMO - molecular orbitals as they are obtained by applying a one-electron model. The change of the charge distribution and of the binding forces connected with the transfer of the ground state to the first excited state can be estimated from these orbitals. The deficiencies of one-electron models are so well known, that this must not be stressed here. The justification to use these models nevertheless lies in their practical advantage. If carefully handled, it is possible to obtain reasonable pictures with very low calculation efforts. In Fig. 3 (a, b) results are demonstrated, which were obtained for an oxygen analogue of the thiathiophthene by applying EHT (extended Hückel; [24]) and PPP (Pariser, Parr, Popel; [34]) calculations.

The calculation results suggest the following interpretation of photochemical reactions of these compounds; they could be confirmed by careful experimental studies [13].

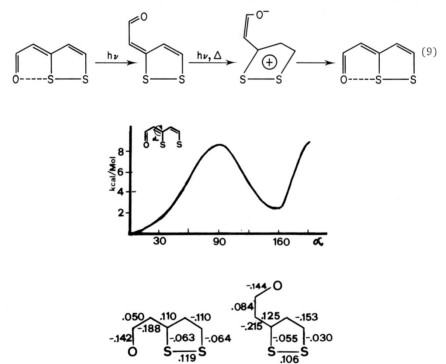

Fig. 3. (a) Energy profile in the ground state as function of the twist angle α, as obtained by an EHT calculation [13]. (b) Calculated change of the π bond order $\delta p_{\mu\nu}$ at the transition from the ground state to the electronically excited singlet state [13]:

$$\delta p_{\mu\nu} = p^*_{\mu\nu} - p^o_{\mu\nu}$$

$p^o_{\mu\nu}$ symbolizes the π bond order between the centers μ and ν in the ground state, $p^*_{\mu\nu}$ in the excited state.

Photoredox reactions of the thionine type

With thermodynamic reflections and the help of a simple orbital scheme a clear picture of photochemical redox reactions of the thionine type can be obtained [42]. These reactions are defined by the following minimal scheme [14, 15].

Photochemical redox reaction
$$A \longrightarrow A^*$$
$$A^* + Me^{n+} \longrightarrow R + Me^{(n+1)+} \tag{10}$$

dark reaction
$$A + Me^{n+} \rightleftharpoons R + Me^{(n+1)+} \quad ; K_{RM} \tag{11}$$

$$A + H \rightleftharpoons 2R \quad ; K_R$$

In the case of the thionine we symbolize A with TH^+ [36],

$$TH^+ = \underset{HN}{\overset{}{\diagup}}\!\!\!\diagdown\!\!\!\underset{S}{\diagdown}\!\!\!\diagup\!\!\!\underset{NH_2}{\overset{+}{\diagdown}}$$

R (the simple reduced form) according to the proton level with TH or TH^+ and H, the twice reduced form with TH^+. The energy diagram together with the relevant redox potentials of the thionine are shown in Fig. 4 (a,b).

For the photochemical redox reactions the difference

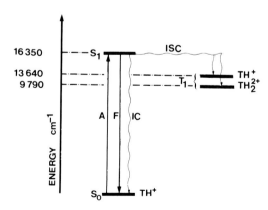

Fig. 4. (a) Energy level scheme of the thionine. A = absorption; F = fluorescence; IC = internal conversion; ISC = intersystem crossing; S_0 = ground state; S_1 = first excited state; T = triplet state [42].

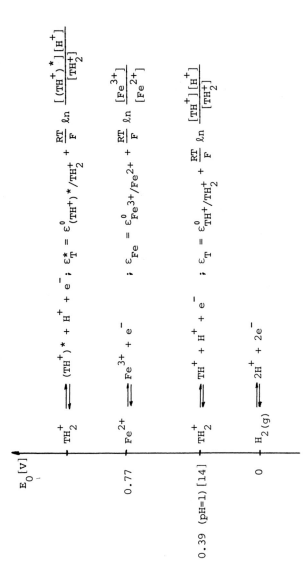

Fig. 4. (b) Redox potentials of the thionine in the ground state S_0 (ε_T) and the excited state triplet state T (ε_T^*) and of the iron ions (ε_{Fe}) [42].

$$\Delta^*_0 = \varepsilon^0{}_{(TH^+)/TH_2^+} - \varepsilon^0{}_{(TH^+)^*/TH_2^+} \qquad (12)$$

is significant. The fact, that an iron(II) oxidation takes place can be demonstrated in a one-electron scheme (Fig. 5).

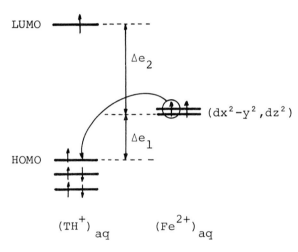

Fig. 5. One-electron scheme, demonstrating the oxidation of Fe(II) by electronically excited thionine. The orbital difference Δe_1 is given by $\Delta e_1 = e(dx^2-y^2, dz^2) - e_{HOMO}$

This scheme would gain valence, should it be possible to establish a quantitative connection between Δ^*_0 and Δe_1. To find this connection we first examine the energy balance given in Fig. 6.

With the help of this scheme it is easy to get a quantitative relation between Δ^*_0 and Δe.

$$\Delta^*_0 \cdot F = \Delta G^0{}_{TH^+/TH_2^+} - \Delta G^0{}_{(TH^+)^*/TH_2^+} = \qquad (13)$$

$$= \Delta H^0{}_{TH^+/TH_2^+} - \Delta H^0{}_{(TH^+)^*/TH_2^+} + T(\Delta S_{(TH^+)^*/TH_2^+} - \Delta S_{TH^+/TH_2^+})$$

It is reasonable to suppose, that the entropy difference

$$T \cdot \delta S = T(\Delta S_{(TH^+)^*/TH_2^+} - \Delta S_{TH^+/TH_2^+}) \qquad (14)$$

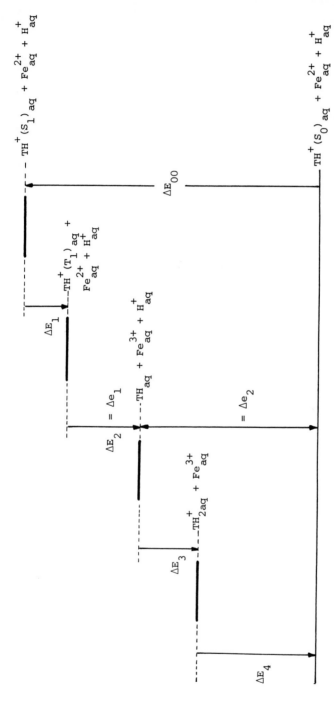

Fig. 6. Energy balance of the thionine/iron redox system [42]. Not considered is the disproportionation reaction, which splits ΔE_4 into $\Delta E_4(1)$ and $\Delta E_4(2)$. E_{00} = energy difference between the lowest vibrational level of the thionine in the ground state and in the first excited singlet state S_1, compare Fig. 4a. ΔE_1 = energy difference between the singlet state S_1 and the triplet state T_1, compare Fig. 4a; ΔE_2 = energy difference corresponding to Δe_1 as given in Fig. 5; ΔE_3 = protonation enthalpy ΔH_p^0

$\Delta G_p^0 = -RT \ln \frac{[H^+] \cdot [TH]}{[TH_2^+]} = \Delta H_p^0 - T\Delta S_p^0$. (Since the radical $TH\cdot$ is a strong base, it is immediately protonated.)

ΔE_4 = redox enthalpy ΔH_r: $\Delta G_r^0 = -RT \ln K_{RM} = \Delta H_r^0 - T\Delta S_r^0$

in general can be disregarded compared to the enthalpy difference. Under this condition there is

$$\Delta_0^* \cdot F \approx \Delta H^0_{TH^+/TH_2^{\pm}} - \Delta H^0_{(TH^+)^*/TH_2^{\pm}} = \Delta E_{00} + \Delta E_1 \qquad (15)$$
$$= \Delta e_1 + \Delta H_p + \Delta H^0_{TH^+/TH_2^{\pm}}$$

which gives the wanted connection between Δ_0^* and Δe:

$$\Delta_0^* \approx \frac{1}{F}(\Delta e_1 + \Delta H_p + \Delta H^0_{TH^+/TH_2^{\pm}}) \qquad (15.a)$$

It is interesting to see, that ΔE_1, ΔE_2 and ΔE_3 show deficit mechanisms referring to the finally stored energy of ΔE_{00}. The stored energy per mol is[1]

$$\frac{\Delta E_{00} + \Delta E_1 + \Delta E_2 + \Delta E_3}{\Delta E_{00}} = \frac{-\Delta E_4}{\Delta E_{00}} \qquad (16)$$

With this we come to a problem of primary interest in connection with all efforts to transform light energy into chemical or electrochemical energy. It is the question of the highest possible efficiency for such a transformation of a specific reaction.

Efficiency of a photoredox system

Calvert [8,9] has given the efficiency Q of such a transformation for monochromatic light of wavelength λ at the temperature T with

$$Q = \frac{\Delta G_T^0}{E_\lambda} \Phi \qquad (17)$$

E_λ is the light energy per Einsten at the wavelenth λ, Φ the quantum yield and ΔG_T^0 the change of free molar enthalpy between the substrates and the photochemically produced end products. For polychromatic light this

---
[1] Equation (16) does not yet contain the disproportionation reaction, which in the thionine example means another deficit mechanism.

equation can be changed accordingly referring to E_λ [1,15]. A careful analysis shows, that this change of the free molar enthalpy ΔG_T^0 is not always an appropriate measure for estimating Q. For this reason we regard the following photochemical redox reaction, which shall be denoted as the iodine type:

photochemical redox reaction
$$I_3^- \xrightarrow{h\nu} (I_3^-)^* \qquad (18)$$
$$(I_3^-)^* + 2Me^{n+} \longrightarrow 3I^- + 2Me^{(n+1)+}$$

dark reaction
$$3I^- + 2Me^{(n+1)+} \rightleftharpoons I_3^- + 2Me^{n+} \quad ; \quad K'_{IM}$$
$$I_3^- \rightleftharpoons I_2 + I^- \quad ; \quad K_I \qquad (19)$$

To describe this system, it is convenient to introduce a reduction degree r. It shall serve as measure for the number of the reduction equivalents, which are added in relation to $Me^{(n+1)+} = M_0$ and $3I_3^- = I_0$. With this M_0 is for the total concentration of metal ions and I_0 for the whole iodine amount:

$$M_0 := Me^{n+} + Me^{(n+1)+}$$
$$I_0 := I^- + 2I_2 + 3I_3^- \qquad (20)$$

It is useful to define the reduction degree as follows:

$$r := \frac{Me^{n+} + I^-}{M_0 + I_0} \quad ; \quad 0 \leq r \leq 1 \qquad (21)$$

With v the relation M_0 to I_0 will be indicated:

$$v := \frac{M_0}{I_0} \qquad (22)$$

To discuss the system (18/19) as a function of the three most important degrees of freedom r, v and M_0, it is reasonable to perform the following normalization of the concentrations.

$$M\emptyset := \frac{M_0}{e_c} := 1 \quad (23)$$

$$[e_c] := mol/l$$

For clearness sake we assume that the balance $I_3^- \rightleftharpoons I_2 + I^-$ lies so much on the left side that the concentration of I_2 is neglectable in all following considerations. This assumption does not restrict the validity of the considerations and can in special cases be discarded right off. By definition [23] follows

$$K'_{IM} = \frac{[Me^{n+}]^2[I_3^-]}{[Me^{(n+1)+}]^2[I^-]^3} = \frac{MN^2 \cdot e_c^2 \cdot I3 \cdot e_c}{MN1^2 \cdot e_c^2 \cdot I1^3 \cdot e_c^3} = \frac{K_{IM}}{e_c^2} \quad (24)$$

$$K_{IM} = K'_{IM} \cdot e_c^2 \quad (25)$$

If the normalized (equation 23) light-induced turnover of I_3^- is denoted with D, we gain for the photochemically produced change of the free Gibbs' enthalpy ΔG:

$$\Delta G = -RT \ln K_{IM} + RT \ln \frac{(MN-2D)^2(I3-D)}{(MN1+2D)^2(I1+3D)^3} \quad (26)$$

If two ideal electrodes would be available, one of which reversible for I_3^-/I^- only, the other one for $Me^{(n+1)+}/Me^{n+}$ only, the potential difference

$$\Delta E = \frac{\Delta G}{2F} \quad (27)$$

between these two electrodes could be measured. Only in the special case $\Delta G = -RT \cdot \ln K_{IM}$ equation (25) would be applicable to get Q. In the general case it is necessary to use a formulation as given in equation (26). Fig. 7a gives the normalized equilibrium concentrations of the reaction partners Me^{n+}, $Me^{(n+1)+}$, I^- and I_3^- together with the redox potential $E_{Me^{(n+1)+}/Me^{n+}}$ - for the equilibrium constant K_{IM} (equation (25)) and the ratio v = 1 - as function of the reduction degree. For the calculation of the curves we used a formerly described method [15,17]. If at a fixed reduction degree a sample is irradiated, the photo redox reaction - as given by equation (18) - shifts the individual concentrations from the equilibrium in directions marked by arrows. Since no reduction equivalents are supplied to the system, the reduction degree r remains unchanged.

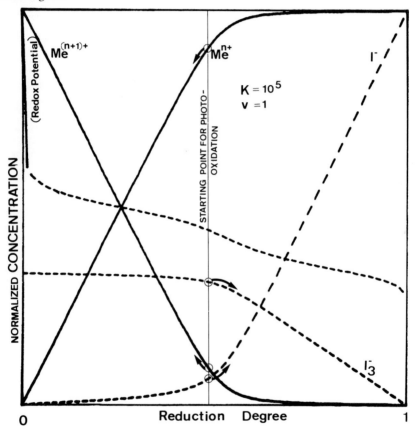

Fig. 7. (a) Normalized equilibrium concentrations of the redox partners Me^{n+}, $Me^{(n+1)+}$, I^-, I_3^- and the redox potential $E_{Me(n+1)+/Me^{n+}}$ (for the equilibrium constant $K_{IM} = 10^5$ and the ratio v = 1) as function of the reduction degree r. For the redox potential the total ordinate corresponds to 1 V (The same is valid for the Fig. 7b and 8).

Fig. 7b is equivalent to Fig. 7a with the only exception, that in addition the potential differences starting from three different reduction degrees - as calculated by equation (27) - are added, which are produced with the photochemical turnover rate ΔI_3^-. According to the Nernst equation it is possible to obtain potential differences of any size. In the sharply decreasing part of Fig. 7b, the potential differences are, however, not anymore connected to a metabolic turnover. For this reason in this potential region a photogalvanic cell will not show any remarkable electric current.

A region where currents can be gained over a constant potential difference is (in Fig. 7b) marked with (1/2,

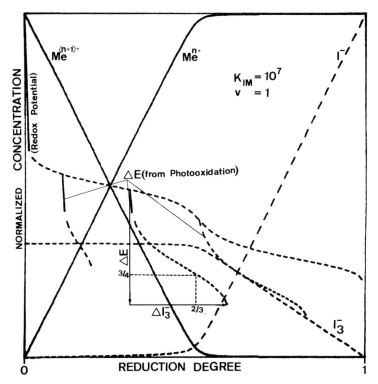

Fig. 7. (b) Like (a), but supplemented by the potential difference resulting from the photochemical turnover ΔI_3^-. The data were calculated by equation (26), starting from three different reduction degrees.

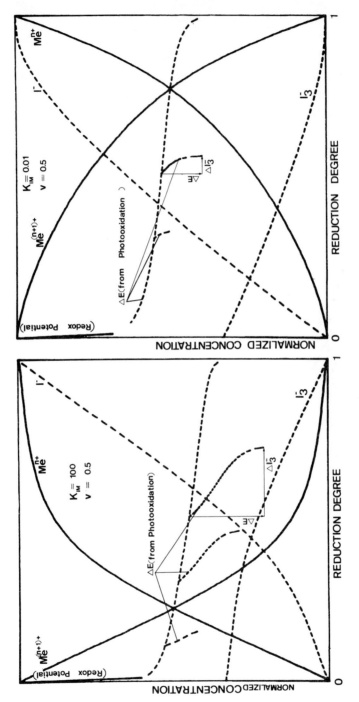

Fig. 8. For explanation see text and legends to Fig. 7 (a, b).

1/2). The electrode problems, which are to be solved for a photogalvanic cell of this kind were examined by Gerischer [22]. Since K_{IM} quadratically depends on the normalization constant e_c, its value will be changed with an alteration of the total concentration - e.g. for a factor of 100 by four orders of magnitude. The results of such a change for $K_{IM} = 10^2$, respectively $K_{IM} = 10^{-2}$ and $v = 0.5$ are demonstrated in Fig. 8. According to primary concentration and reduction degrees it is readily possible, that in a system which principally should be able to enter a photo redox reaction this cannot actually be observed due to unfavourable equilibrium conditions. With this it is shown, how important in some cases the discussion of the thermal equilibrium situation can be for the comprehension of the photochemical behaviour of chemical systems.

Perspectives

The intermolecular redox reactions shown in schemes (10) and (18) raise the question how far one succeeds by photochemical procedures in producing an intramolecular charge separation and maintaining this over a longer period. The complex chemistry knows a wide selection of charge transfer (CT) absorptions (see e.g. [11,20,28, 31, 37]). The life-time of such charge separations generally lies in the ns to µs region. CT transfers are known, in which an electron from the central atom specifically migrates to a ligand, so that it is justified to talk about an "intramolecular charge separation". There is e.g. the pyrazine-pentacyanoferrate complex $[Fe(CN)_5 \text{ pyrazine}]^{m-}$, m=2,3 [11,20]. In Fig. 9 the observed transfers in this compound and their attribution are demonstrated. The molecular orbitals received by an EHT calculation with quadratic charge iteration [11] serve to explain the charge displacement. In this connection the d-π* transfer marked with ① is of interest. It is evident, that with this transfer one electron primarily mainly localized on the central atom will be completely transferred to the pyrazine ligand. By this means both the basic character and the redox potential of the heterocycle are considerably changed.

In a binuclear complex it should be possible to stabilize such a charge displacement over a longer period against a potential gradient. Without going into details this statement shall be supported with the help of the following qualitative considerations. Here it is favourable to divide the most important one-electron orbitals of a common binuclear complex

$$(L_1)_n M_1 - B - M_2 (L_2)_m \tag{28}$$

into a part ϕ^{M_1}, ϕ^{M_2} for both centres M_1 and M_2, a part ϕ^{L_1}, ϕ^{L_2} for the ligands $(L_1)_n$ and $(L_2)_m$ and finally a part ϕ_B for the bridging ligand.

$$\Psi_i = (a_{1i}\phi_i^{M_1} + a_{2i}\phi_i^{M_2}) + (b_{1i}\phi_i^{L_1} + b_{2i}\phi_i^{L_2}) + c_i \phi_i^B \tag{29}$$

A specially simple example of a binuclear complex, where the importance of function (29) can be demonstrated, is the symmetrical binuclear (2,2)-pentacyanoferrate complex, which is bridged over a pyrazine. This compound possesses a "closed-shell" configuration. At the longest wavelengths the transfer with the (2,2)-complex shows definite MBCT (B = bridging ligand) character and may in very good approximation be described as d-π^* (pyrazine) [11,20]. The local symmetry of both Fe(CN)$_5$ fragments is C_{4v}. That means, the formal d_{xy}, d_{xz} and d_{yz} orbitals of these parts

Fig. 9. MO scheme for the [Fe(CN)$_5$ pyrazine]$^{m-}$, m=2,3, and demonstration of some molecular orbitals. In the absorption spectrum of the [Fe(CN)$_5$pyrazine]$^{3-}$ in water (at room temperature) the longest wavelength band ① (22 100 cm^{-1}) and the band ③ (38 000 cm^{-1}) can experimentally be identified as d-π^* and π-π^* transitions. The absorption bands denoted with ② probably correspond to one d-δ^* and one n-π^* transition.

REDUCTION DEGREE IN PHOTOREDOX SYSTEMS

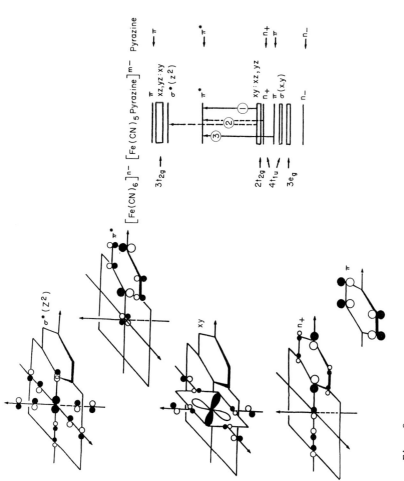

Fig. 9.

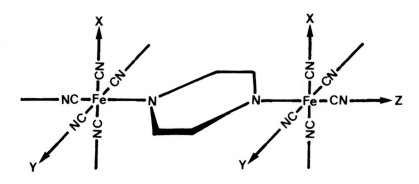

split into xy and (xz, yz). For symmetry reasons the xy orbitals of both centres cannot - neither directly nor over the bridging ligand - interact with each other. Against that a small mutual influence between the xz and yz function over the bridging ligands occurs, which in non-symmetrical complexes - e.g. (2,3)-complexes - leads to intervalence absorptions (see [11,20,23,25,31,37,38, 45]). This interaction decreases with comparable ligands with increasing distances.

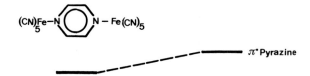

Fig. 10. Energy level scheme of the binuclear pentacyanoferrate compound as obtained by an EHT calculation [11].

The orbital scheme of the pyrazine-bridged compound as obtained by an EHT calculation is demonstrated in Fig. 10. The most important molecular orbitals in this connection are illustrated in Fig. 11. The splitting of these two b_{2g} orbitals from which the (in the z direction) polarized d-π^* transfer results, is so small, that in the already broad - experimentally determined [20] - CT band only one transfer can be recognized. From the viewpoint of clearness the wave function given in equation (29) is in this form not especially favourable. It is better to write linear combinations, which lead to the following expressions:

$$\Psi_i(+) = 2a_{1i}\Phi_i^{M_1} + 2b_{1i}\Phi_i^{L_1} + C_i(+)\Phi_i^B(+)$$

$$\Psi_i(-) = 2a_{1i}\Phi_i^{M_2} + 2b_{2i}\Phi_i^2 + C_i(-)\Phi_i^B(-)$$

(32)

Here $\Psi_i(+)$ is localized at the centre M_1, $\Psi_i(-)$ at the centre M_2. One can consider the normally small error, which has its origin in the formulation of these linear combinations, as being contained in $C_i(+)$ respectively $C_i(-)$. The degenerated molecule orbitals ($d_{xz}^{(1)}$, $d_{xz}^{(2)}$) as given in Fig. 11 correspond to the functions (32). The illustrated d-π^* transfer indicates, that during the lifetime of the π^* state one of the two iron centres possesses a higher oxidation degree, whereas the bridging ligand formally exists as B^-.

$$(L_1)_n M_1 - B - M_2(L_2)_m \xrightarrow[\text{quick}]{h\nu} (L_1)_n \overset{\oplus}{M_1} - \overset{\ominus}{B} - M_2(L_2)_m \qquad (33)$$

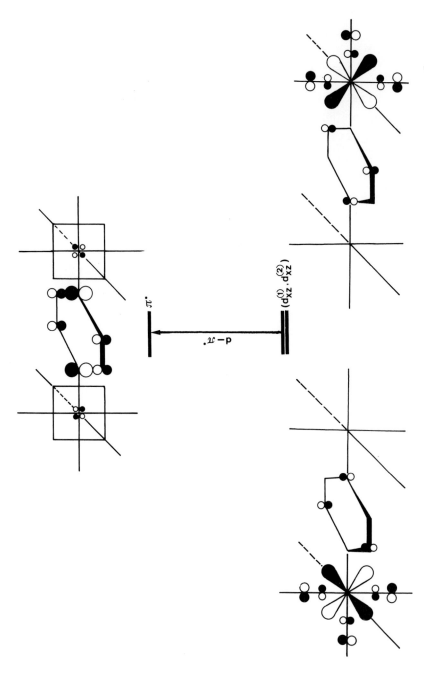

Fig. 11. Demonstration of the d-π^* transition in the binuclear pentacyanoferrate(II)-pyrazine complex with localized orbitals [11,28].

A charge separation of this kind can be maintained at most during some ns or μs. That depends whether the relaxation takes place under conservation or change of the spin multiplicity. However, we are interested in transfers, which lead to a charge separation with a much higher lifetime. One succeeds in this, when the electron can be kept in a potential trough. In a suitable complex the central atom M_2 can play the role of such a potential trough.

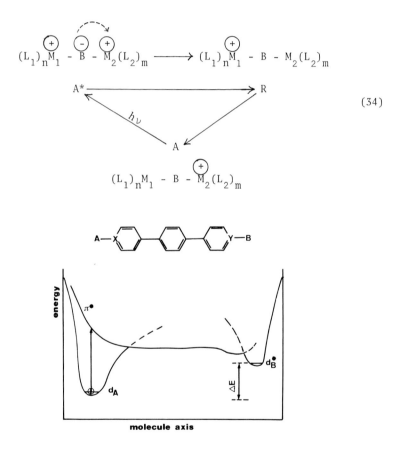

Fig. 12. One-electron model for the demonstration of a photochemically produced intramolecular charge separation in binuclear complexes. The figure gives the potential trough for an electron as a function of its position along the molecule axis [10].

The one-electron energy in function of the longitudinal axis of a model molecule is demonstrated in Fig. 12. By light absorption an electron which is located at the centre A will reach the π^* level of the bridging ligand. It possesses a certain probability - which should be as high as possible - to relax into the empty level d_{B^*} at the centre B, which lies about ΔE higher than the level A. From this level d_{B^*} there are various deactivation mechanisms back to d_A - not further discussed here [10] - which must be suppressed as far as possible. It is not difficult to realize, that there should be compounds, for which the life-time of the d_{B^*} state is considerably higher than that of the π^* state. In analogy one could denote such compounds as molecular photodiodes. They possibly represent ideal models for the study of multiphoton processes [10]. A somewhat different - quite recently successfully employed - possibility for the photochemical production of a charge separation depends on the exploitation of d-d transfers in two-dimensional crystal arrays [43].

Acknowledgements

This paper is part of project no. 2.475-0.75 of the Swiss National Foundation for Scientific Research.

I want to thank Mr H.R. Grüniger for numerous discussions and Prof. Dr E. Schumacher for support of this work.

References

1. Archer, M.D. (1975). *J. Appl. Electrochem.* **5**, 17-38.
2. Arnoldi, D. and Wolfrum, J. (1976). *Ber. Bunsenges. physik. Chem.* **80**, 892-902.
3. Ashfold, M.N.R. and Simons, J.P. (1977). *Chem. Phys. Lett.* **47**, 65-69.
4. Bachmann, H.R., Nöth, N., Rinck, R. and Kompa, K.L. (1974). *Chem. Phys. Lett.* **29**, 627-629.
5. Bachmann, H.R., Nöth, N., Rinck, R. and Kompa, K.L. (1977). *Chem. Phys. Lett.* **45**, 169-171.
6. Becker, R.S., Dolan, E. and Balke, D.E. (1969). *J. chem. Phys.* **50**, 239-245.
7. Bixon, M. and Jortner, J. (1968). *J. chem. Phys.* **48**, 715-726.
8. Calvert, J.G. (1953). *Ohio J. Sci.* **53**, 293-299.

9. Calvert, J.G. (1963). *In* "Introduction to the Utilization of Solar Energy" (A.M. Zarem and D.D. Erway, eds). McGraw-Hill, New York.
10. Calzaferri, G. (1977), (in press).
11. Calzaferri, G. and Felix, F. (1977). *Helv. Chim. Acta* **60**, 730-740.
12. Calzaferri, G. and Gleiter, R. (1975). *J. Chem. Soc. Perkin Transact.* II, 1975, 559-566.
13. Calzaferri, G. Gleiter, R., Knauer, K.-H., Rommerl, E., Schmidt, E. and Behringer, H. (1973). *Helv. Chim. Acta* **56**, 597-609.
14. Calzaferri, G. and Grüninger, H.R. (1977). *Chimia* **31**, 58-59.
15. Calzaferri, G. and Grüninger, H.R. (1977). *Z. Naturforschg.*, in press.
16. Calzaferri, G., Gugger, H. and Leutwyler, S. (1976). *Helv. Chim. Acta* **59**, 1969-1987.
17. Dubler, Th., Maissen, C. and Calzaferri, G. (1976). *Z. Naturforschg.* **31b**, 569-579.
18. Englman, R. and Jortner, J. (1970). *Mol. Physics* **18**, 145-164.
19. Farneth, W.E., Flynn, G., Slater, R. and Turro, N.J. (1976). *J. Am. Chem. Soc.* **98**, 7877-7878.
20. Felix, F. (1976). Thesis, University of Bern.
21. Förster, Th. (1951). "Fluoreszenz organischer Verbindungen". Verlag Vandenhoek & Ruprecht, Gottingen.
22. Gerischer, H. (1970). *In* "Physical Chemistry. An Advanced Treatise" (H. Eyring, ed.). Vol. IX A, pp. 463-540. Academic Press, New York, London.
23. Hay, P.J., Thibeault, J.C. and Hoffmann, R. (1975). *J. Am. Chem. Soc.* **97**, 4884-4899.
24. Hoffmann, R. (1963). *J. Chem. Phys.* **39**, 1397-1412.
25. Hush, N.S. (1967). *Progr. Inorg. Chem.* **8**, 391-444.
26. Jaffe, R.L. and Morokuma, K. (1976). *J. chem. Phys.* **64**, 4881-4886.
27. Jörg, U., Binkert, Th. and Calzaferri, G. (1977). *Z. angew. Math. Phys.* **28**, 363-366.
28. Jörgensen, C.K. (1969). "Oxidation Numbers and Oxidation States". Springer-Verlag, Berlin.
29. Labhart, H. (1977). *Chimia* **31**, 89-93.
30. Lin, S.H. and Bersohn, R. (1968). *J. chem. Phys.* **48**, 2732-2736.
31. Maissen, C. (1975). Thesis, University of Fribourg.
32. Oster, G. and Nishijima, Y. (1964). *Fortschr. Hochpolymer.-Forschg.* **3**, 313-331.
33. Parker, C.A. (1968). "Photoluminescence of Solutions". Elsevier Publ. Company, Amsterdam, Oxford.
34. Parr. R.G. (1963). "The Quantum Theory of Molecular Electronic Structure". Benjamin, New York.
35. Penzkofer, A., Falkenstein, W. and Kaiser, W. (1976). *Chem. Phys. Lett.* **44**, 82-87.
36. Rabinowitch, E. (1940). *J. chem. Phys.* **8**, 551-566.
37. Rieder, K., Hauser, U., Siegenthaler, H., Schmidt, E. and Ludi, A. (1975). *Inorg. Chem.* **14**, 1902-1907.
38. Robin, M.B. and Day, P. (1967). *Adv. Inorg. Chem. and Radiochem.* **10**, 247-422.
39. Sabety-Dzvonik, M. and Cody, R. (1976). *J. chem. Phys.* **64**, 4794-4796.

40. Smith, I.W. (1975). *In* "The Excited State in Chemical Physics", (J.W. McGowan, ed.). Vol. XXIII, pp. 32-64. John Wiley, Interscience, New York.
41. Spanner, K., Lauberau, A. and Kaiser, W. (1976). *Chem. Phys. Lett.* **44**, 88-92.
42. Sulzberger, B. (1977). Diploma thesis, University of Bern.
43. Tributsch, H. (1977). *Ber. Bunsenges. physik. Chem.* **81**, 361-369.
44. Tschanz, H.P., Binkert, Th. and Zinsli, P.E. (1974). *Z. angew. Math. Phys.* **25**, 117-120.
45. Ying-Nan Chin (1976). *J. phys. Chem.* **80**, 993-996.

OXYGEN EVOLUTION AS ENERGETIC PROBLEM

H. METZNER

*Institut für Chemische Pflanzenphysiologie
der Universität, Tübingen (F.R.G.)*

Introduction

The modern way to deal with the light-induced charge transfer reactions in photosynthesis is a combination of photophysical and electrochemical considerations. The photophysical part consists in the excitation of sensitizer molecules and various ways of energy transfer from antenna pigments to reaction centres. Here the excitation energy is used for an electron exchange between two redox couples. It is exactly this step, with which the electrochemical part of the whole sequence starts.

Many authors believe that the first electrochemical reaction at the PS II centre is an electron exchange between an excited chlorophyll molecule and an as yet unknown acceptor, probably a quinone. To describe this step energetically, it has become common use to compare the redox potentials of donor and acceptor. It seems doubtful, whether this is a reasonable way, when dealing with photochemical reactions. At least in all relevant physiological processes we have no information on the relative shares of the oxidized and reduced states of the participating systems. We can therefore only compare the so-called standard values (E_o), which are defined for a [Red] : [Ox] ratio of 1 : 1. It is very seldom possible to correct these data for the actual experimental conditions. Even then they are restricted to *free* - not to adsorbed - redox couples.

Whenever the adsorption energies of the two oxidation states differ - and this will be the most probable case -, there must be a significant potential shift.

It would certainly be much better to replace redox potentials by ionization energies and electron affinities. Unfortunately these data are not yet available for the biochemically most important compounds. For this reason physiologists still use the conventional redox scale and refer their data to the - for pH 7.0 and 25°C corrected - E_0' values. This habit explains why the level of the oxygen electrode, characterized by E_0' = 0.81 V, is chosen as starting point of the Hill-Bendall diagram.

Water Electrolysis

To decompose water takes - under reversible conditions - the same energy as can be gained by burning hydrogen, *e.g.* in a suitable fuel cell. This energy requirement can be given by the total enthalpy (ΔH_0). The value for the reaction

$$2 H_2 + O_2 \rightarrow 2 H_2O \tag{1}$$

is ~-570 kJ·mol^{-1}. Since the reaction product - liquid water - has a lower entropy than the two reacting gases, the free energy is less negative; it may be given with ΔG_0 = -475 kJ·mol^{-1} = 4.92 eV. If we realize that the reaction between hydrogen and oxygen includes the exchange of *four* electrons, we can predict an energy requirement of 4.92/4 = 1.23 eV per charge transfer step. This is exactly the value, which characterizes the distance between the hydrogen and the oxygen electrode; it is this value, with which the physiologists use to calculate. It is, however, the potential of a completely reversible system (E_{rev}). If a system - like a fuel cell - performs work, *i.e.* delivers an electric current, its terminal potential (E_T) is definitely lower. The difference η = E_T - E_{rev}, the so-called overvoltage, characterizes the distance of the system from the real equilibrium; it is a measure for the necessary activation energies. In fact, the potential of the hydrogen fuel cell is not higher than 1.06 V [19].

The same is valid for the reverse reaction: the electrolytic water decomposition. The calculated E_{rev} of 1.23 V characterizes a completely reversible reaction. This value does not regard the activation barriers of the process, which depend on the electrode material used. In fact, water electrolysis - even at very low current densities - requires at least 1.65 V [20]. The higher the current density, the higher the decomposition potential.

For photosynthesis the comparison with the electrolytic water decomposition is much more interesting than the reverse process of hydrogen oxidation. In electrochemistry we have to assume that the rate-limiting reactions are electron exchanges between the electrodes and movable charge carriers, normally cations (like H_3O^+) and anions (like OH^-). Besides this there might be an exchange reaction with neutral water molecules. In any case the real reactant will, however, be an *adsorbed* species. At the anode we observe the release of molecular oxygen. This reaction must include - whatever its mechanism might be - the breakage of an oxygen-hydrogen bond. This is an extremely endergonic step, which requires ~462 kJ·mol^{-1} ≈4.8 eV. It is not easy to understand, how this energetically expensive partial reaction is embedded in the sequence of reactions, which altogether need only ~1.6 V. Since the adsorption energy of the primary reaction product - OH· or H· radicals - may be definitely higher than that of water, the missing energy may perhaps - at least partly - result from the preceding adsorption step.

The dissociation of water molecules into hydrogen and hydroxyl ions - *i.e.* the proton exchange between neighbour water molecules - requires a free energy of ~0.8 eV. This is low enough to allow a certain "dissociation" at room temperature already. Even if by this step only 2 out of 1 billion H_2O molecules are ionized, it provides aqueous systems with a sufficient number of charge carriers. So we observe primarily a reaction between the electrode surface and adsorbed ions of water. We have to discriminate between the processes at the two opposite electrodes. On the cathode we have the transfer

$$\text{Electr}^- + [\,H_3O^+\,]^{(a)} \longrightarrow \text{Electr}^o + [\,H\cdot\,] + [\,H_2O\,] \qquad (2)$$

on the anodic side

$$\text{Electr}^+ + [\,OH^-\,] \longrightarrow \text{Electr}^o + [\,OH\cdot\,] \qquad (3)$$

For a comparison with PS II reactions we may concentrate on the "oxygen side". The first step seems to be the discharge of OH^- anions (equation 3). Unfortunately we have to realize, that the electrochemists do not yet understand the sequence of secondary reactions, which ultimately results in the release of molecular oxygen. There is apparently no formation of H_2O_2 [8,54]. So there remain two possible reactions to follow:

$$(a) \quad [\,OH\cdot\,] + [\,OH\cdot\,] \longrightarrow [\,H_2O\,] + [\,O\cdot\,] \qquad (4)$$

$$(b) \quad [\,OH\cdot\,] + \text{Electr}^+ \longrightarrow [\,OH^+\,] + \text{Electr}^o \qquad (5)$$

$$[\,OH^+\,] + [\,H_2O\,] \longrightarrow [\,O\cdot\,] + [\,H_3O^+\,] \qquad (6)$$

In both cases we gain adsorbed oxygen atoms, which can freely diffuse in the electrode surface. They finally become united to adsorbed molecules

$$[\,O\cdot\,] + [\,O\cdot\,] \longrightarrow [\,O_2\,] \qquad (7)$$

(a) The brackets symbolize adsorbed species

which might need additional energy to become desorbed

$$[O_2] \longrightarrow O_2 \qquad (8)$$

If we add mineral acids to the electrolytic cell, we measure practically the same energy requirement. This means that most of the anions ask for the same energy for their charge transfer to the electrode. On the other hand, there are considerable differences between various electrode materials.

The two sequences given differ in the number of exchange steps. In the first case (equation 4) we assume only *one* electron exchange, followed by a hydrogen transfer. Electrochemically this could be characterized by the potential $OH^-/OH\cdot$, which is $E_0' = 2.4$ V [4]. If this rate-limiting step is nevertheless possible with a voltage of ~1.6 V, we have to realize not only differences in adsorption energies, but also the fact that at the electrode surface $[OH^-]/[OH\cdot]$, *i.e.* the actual E value is less positive than the E_0' value. If we postulate a two-step mechanism (equations 5,6), we have to realize that the second step needs much less energy than the first one.

Measurements during water electrolysis have demonstrated, that near the cathode hydrated electrons (H_2O^-) are formed [53]. This can only be explained, if we assume an electron exchange between neutral (adsorbed) water molecules and the electrode material. In this case we would observe at the cathode

$$Electr^- + [H_2O] \longrightarrow Electr^o + [H_2O^-] \qquad (9)$$

On the anodic side we should then expect

$$Electr^+ + [H_2O] \longrightarrow Electr^o + [H_2O^+] \qquad (10)$$

The two resulting radicals are well-known from radiochemical experiments. The extent of this charge transfer seems to be very small.

Water Radiolysis

It has long been known, that water cannot only be decomposed by electrolysis, but also by irradiation in one of its UV absorption bands. Unfortunately we are still missing well-documented data on the secondary processes of excited water molecules [11]. We use to assume that the reactions start with an electron exchange between two identical molecules

$$H_2O + H_2O \xrightarrow{h\nu} H_2O^- + H_2O^+ \qquad (11)$$

The reduced species - the so-called hydrated electron - has a rather long half-lifetime of ~300 μs [11]; its redox potential may be given with E'_o = -3.0 V [11]. The oxidized species has a half-lifetime of only $1.6 \cdot 10^{-14}$ s [30]. During this short time interval it exchanges a proton with a neighbour water molecule:

$$H_2O^+ + H_2O \longrightarrow H_3O^+ + OH\cdot \qquad (12)$$

The resulting OH· radicals do not form molecular oxygen. The necessary steps between OH· and O_2 seem to be limited to surfaces.

Photosystem II reactions and model systems

The situation on the oxidative site of PS II may be compared to reactions on charged electrode surfaces. If we replace the positive anode ($Electr^+$) by the oxidized donor (Don^+), we have at first sight a very similar situation. Quite different from water electrolysis, however, we may not assume that the reaction starts from OH^- anions. There is good evidence, that the so-called "water decomposition" must be localized inside the thylakoids. These vesicles enclose a very small volume of only ~$5 \cdot 10^{-4}$ μm^3. The probability to find a hydroxyl ion inside a thylakoid is therefore less than 1 : 1 million. In fact, modern schemes of the photosynthetic electron transport avoid a reaction between the oxidized donor and the OH^- ion. Instead they postulate the energetically more expensive oxidation of neutral water molecules. Nevertheless the plant physiologists believe that it is justified to start the Hill-Bendall diagram from the level of the oxygen electrode. Its potential characterizes, however, a *two*-electron transfer ($O^{--}/O \cdot$). But the postulated H_2O oxidation is - like in radiochemistry - a *one*-electron reaction

$$H_2O + Don^+ \longrightarrow H_2O^+ + Don$$

How can we calculate its standard potential? For the energetic consideration it would certainly be much better to avoid the application of electrochemical potential values at all. It seems more appropriate to deal with the energy requirement of the elementary steps. The radiochemical data inform us, that the radiolysis of water starts in the region ≤ 190 nm. Quanta in this frequency range have an energy content of ~6.5 eV. This energy is apparently sufficient for the charge exchange between neighbour molecules (equation 12), *i.e.* $E_{o(T)} = 6.5$ eV is necessary to create the couple H_2O^-/H_2O^+. From the electrochemical point of view we are dealing with two additive potentials $E_{o(1)}$ (H_2O/H_2O^-) and $E_{o(2)}$ (H_2O/H_2O^+). To obtain the wanted potential $E_{o(2)}$ (H_2O/H_2O^+) we need the difference $E_{o(T)} - E_{o(1)} = 6.5 - 3.0 = 3.5$ V. This is certainly a very rough estimate, but fortunately this potential has in the meantime been experimentally determined to 3.0 $\pm$ 0.5 V [14].

Some data favour the assumption, that the primary donor of PS II might be a chlorophyll *a* dimer at the reaction centre. The redox potential of chlorophyll has been repeatedly determined; dependent on the solvent it was found between 0.4 V [28] and 0.8 V [49]. Unfortunately these *in vitro* data are not relevant for the situation *in vivo*. On the other hand we can roughly estimate the redox potential, which the primary donor should have. If we really postulate, that the light-induced oxidation of the donor leads to an exergonic oxidation of water, thermodynamic calculations force us to assume that the E_o' value of the primary donor must exceed 3.0 V (Fig. 1). Even if we correct this value for realistic experimental conditions - regarding that $[H_2O] \gg [H_2O^+]$ - we obtain a very high energy requirement, which makes an H_2O oxidation step extremely unlikely.

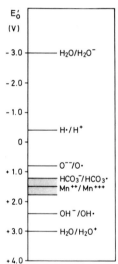

Fig. 1. Potential diagram (E_o' values) for relevant redox couples. If the E_o' values of the primary electron acceptor of PS II is near zero, the potential of the primary donor may be expected in the shaded region between 1.2 and 1.8 V (details see text).

To understand the situation at the PS II reaction centre, we need the energy content of the absorbed photons. The electromagnetic energy of light quanta can without any loss be transformed into the excitation energy of the absorbing molecule. The question is, to what extent this excitation energy can become converted into chemical energy. Photons absorbed in the long-wavelength band of chlorophyll *a* have an energy content of 1.8 eV. The energy of photons, which are emitted as delayed light, is >1.75 eV. This comparison favours the assumption, that the primary charge separation transforms the excitation energy to ~98%. The redox potential difference between the primary donor and the primary acceptor ($\Delta E_0'$) does, on the other hand, not represent the total energy change of the transfer step. There remain the shares of electrostatic forces, which are partly reflected in hydration energy changes of the redox couple (Fig. 2). It seems very doubtful, whether thermodynamic calculations based on the simple application of the Second Principle [2,12,23] are justified. If they were correct, the free energy of the red light quanta should be near 1.8 x 0.7 ≈ 1.2 - 1.3 eV.

If we postulate that the primary electron acceptor of PS II has a redox potential near zero, the primary donor should have an E_0' value ~1.8 eV. This would exclude, that

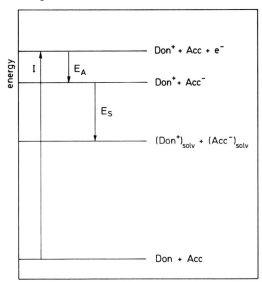

Fig. 2. Energy diagram for a charge transfer reaction between donor and acceptor molecules (after [7]). I = ionization energy of the donor molecule; E_A = electron affinity of the acceptor molecule (distance between ionization level and highest unoccupied orbital); E_S = solvation energy of the two charged species.

the oxidized donor could regain the missing electron from water molecules. If we, on the other hand, would discuss the possibility of two successive photoreactions in PS II [34], we could not explain the high quantum efficiency of photosystem II reactions as determined by Sun and Sauer [51].

This dilemma made several authors take their refuge to experiments with easily preparable inorganic models. There are meanwhile many data on the decomposition of water on irradiated semiconductor electrodes [15, 42] or in contact with illuminated powders like TiO_2, ZnO [25,26] or AgCl [36,39]. Our own experiments with aqueous silver chloride crystal suspensions demonstrated, that the quantum yield of the O_2 evolution clearly depends on the kind and concentration of added anions [40]. Especially bicarbonate anions increased the reaction rate by orders of magnitude. This observation - together with the well-known influence of bicarbonate anions on the Hill reaction [35,50,55] - encouraged new experiments on the possible precursor of the photosynthetic oxygen.

Isotope discrimination in photosynthesis

Our present-day hypothesis of a light-induced "water decomposition" is mainly based on mass spectrometric data. It may be unnecessary to repeat the severe objections against the validity of most data given in the literature and the far-reaching conclusions [37]. In spite of all pitfalls of the available methods there nevertheless remain two different ways to come to reliable results. Both use the isotope discrimination in chemical reactions. Since the zero point energy of a chemical bond depends on the mass of the connected nuclei, there is a difference in reaction rates between bonds like ^{18}O-H and ^{16}O-H. Statistical mechanics tell us, that the bond with the heavier isotope reacts somewhat slower; the difference can be observed as "kinetic" isotope effect. If we decompose natural water samples - which always contain ~0.2% ^{18}O - we have to expect that the released oxygen possesses a lower $^{18}O/^{16}O$ ratio. In electrolytic experiments this is definitely the case [1]. This observation initiated mass spectrometric measurements with AgCl suspensions supplied with ^{18}O-enriched sodium bicarbonate. These experiments together with control experiments with ^{18}O-labelled suspension water demonstrated that the released oxygen does not come from water but - at least predominantly - from the added bicarbonate anions [41]. Quite different experiments on the oxygen evolution in irradiated duroquinone solutions containing carbonate anions showed that here, too, the released oxygen had its source in the CO_3^{--} ions [17].

There is a second way to study isotope discriminations. If we introduce a heavy oxygen isotope (^{17}O or ^{18}O) into an equilibrium system like

$$CO_2 + 2 H_2O \rightleftharpoons HCO_3^- + H_3O^+ \quad (14)$$

the oxygen isotopes are unevenly distributed between the reactants; the heavier isotope becomes enriched in the heaviest species [52]. In the case of the H_2O-CO_2 system this gives the data of Table 1. If we irradiate this system in contact with thylakoids or whole cells, we start with reaction partners which differ in their $^{18}O/^{16}O$ ratios. If the photosynthetic oxygen were the result of a

TABLE 1

^{18}O content of the different species in the natural H_2O-CO_2 system (after [5] and [52])

Compound	^{18}O content (%)	^{18}O content (relative)
H_2O	1.981	1.000
CO_2	2.062	1.041
HCO_3^-	2.033	1.026

water decomposition we should - as in electrolysis - expect that the ^{18}O content in the released O_2 is somewhat lower than in the water molecules. This is definitely not the case (Table 2). If we consider the ^{18}O content of the

TABLE 2

^{18}O *content of oxygen samples evolved from fresh water (after* [37]*)*

	^{18}O content (%)	^{18}O content (relative)
Fresh water (control value):	1.981	1.000
Electrolytic oxygen:	1.961	0.990
Photosynthetic oxygen:		
a) *Elodea* leaves	2.003	1.011
b) Land plants	1.992	1.006
c) Fresh water algae	1.991	1.005

atmospheric oxygen, which we regard as the result of photosynthesis, we get even more pronounced ^{18}O enrichments (Table 3). One explanation of this unexpected result

TABLE 3

^{18}O *content of different constituents of the natural* H_2O-CO_2 *system (after* [37]*)*

Compound	^{18}O content (%)	^{18}O content (relative)
Fresh water	1.981	1.000
Ocean water	1.995	1.007
CO_2 in fresh water	2.067	1.043
CO_2 in ocean water	2.078	1.049
Atmospheric oxygen	2.042	1.031

might offer the discrimination of the oxygen isotopes in respiratory processes, which should remove a higher percentage of the lighter isotope from the medium. The actually observed data (see [37]) are too small to explain the discrepancies. Unfortunately we have not yet reliable data on the isotope discrimination in thylakoid suspensions to exclude the principal possibility, that the isotope shift might be caused by an unusually high isotope discrimination by photorespiration.

Since the real precursor of the photosynthetic oxygen must have a higher ^{18}O content than the released O_2 molecules, the mass spectrometric data speak against a water decomposition. There remains, however, the argument, that the situation might be quite different, if we start from a kind of *bound* water. Unfortunately we have only very incomplete data on the properties of "bound" water molecules. It therefore remains a mere speculation to assume, that the replacement of bulk water against surface-bound water would explain the observed data. There are nevertheless several models which postulate the decomposition of an as yet unknown type of bound water [29], in some cases in a very detailed connection with manganese complexes [45]. Indeed, hydration water has a somewhat increased $^{18}O/^{16}O$ ratio [33], but there remain some more data which speak against any breakage of an oxygen-hydrogen bond.

If there is really a discrimination between ^{18}O-H and ^{16}O-H bonds, there should be a much more pronounced difference in the decomposition rate of H_2O and D_2O molecules. In fact, the photosynthesis rate of *Chlorella* cells in D_2O is by a factor of ~3 lower than in H_2O [10,44]. Experiments with light-dark cycles demonstrate, however, that this discrimination cannot be ascribed to the fast reactions leading to oxygen release [38,43]. If short flashes are separated by sufficiently long dark periods (in the order of some ms) the isotope discrimination disappears completely [43] or - under different experimental conditions - does at least not exceed a factor of 1.3 [38]. This observation is in accordance with data, which Sinclair *et al.* have obtained in their experiments with *Chlorella* cells [48] and chloroplasts [3].

We are not yet well-informed on hydrogen isotope effects

in unimolecular decomposition reactions [32], but we have many reliable observations on the discrimination between normal and heavy water both in electrochemical and radiochemical experiments. Whereas in electrolysis separation factors up to 20 have been described [24], several comparisons between the radiolysis of H_2O and D_2O gave discrimination factors of 3.5 ± 0.3 [32]. These observations point in the same direction as the experiments with ^{18}O-labelled reactants.

The Problem of the Oxygen Precursor

There cannot be any doubt that the electrons used in the CO_2 reduction reactions of photosynthesis *finally* must come from water. But this does not mean, that H_2O - free or "bound" - is the *immediate* precursor of O_2 molecules. Since all experimental data cannot be taken as evidence in favour of a water oxidation, we have to look for possible alternatives. One possible explanation could be the assumption, that the oxygen precursor is a more complicated "photolyte", in whose synthesis CO_2 takes part. From the energetic point of view it would seem more reasonable to postulate a compound, which can exergonically replace the electron(s) which the primary donor has lost in the light reaction. This primary donor should - as we have seen - have a redox potential between 1.2 and 1.8 V. This is the region in which manganese complexes might work (Fig. 1). The electron removal from H_2O molecules requires nearly twice this potential, whereas the potential of the $HCO_3^-/HCO_3\cdot$ couple may be given with 1.2 V [18]. This value, calculated from pulse radiolysis experiments, would be in perfect agreement with the hypothesis, that the oxidized donor regains its missing electron(s) from an anion instead of water molecules [35].

This agreement alone is, of course, no argument in favour of a "CO_2 hypothesis". The CO_2 effect on the Hill reaction might possibly find another explanation [16]. We

know that carbon dioxide has a strong influence on other
reaction sequences, too. All what these data say is, that
the energy of a single quantum of red light is sufficient
to remove an electron from a bicarbonate anion.

If during PS II reactions bicarbonate radicals would
appear, what would be their fate? From the energetic point
of view it is unlikely that they decompose to CO_2 and OH·
radicals. Electrolytic experiments with deeply cooled bicarbonate solutions demonstrate that - at sufficiently
high current densities - the radicals combine to dimers
(Fig. 3); they may be characterized as peroxidicarbonic
acid (see also [27]). Similar dimer formation has been
observed in the xanthine oxidase reaction [21]. This
intermediate product has been observed a long time ago
already [9]; its sparingly soluble potassium salt can
easily be isolated. Peroxy acids of this kind are known
to cause epoxidations of conjugated double bonds [31].
It could, in fact, be demonstrated that the electrolyti-
cally prepared peroxidicarbonic acid transfers oxygen
atoms to carotenoids [13], producing a series of different epoxides (see also [5]).

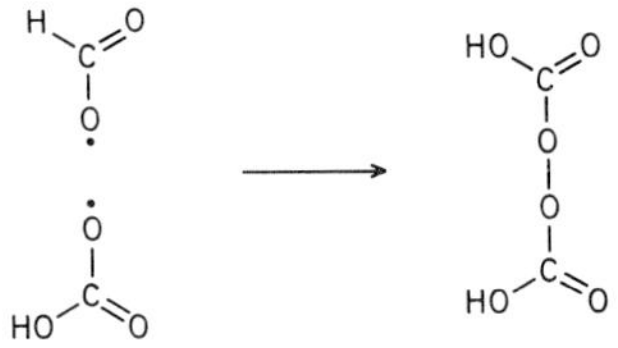

Fig. 3. Dimerization of two bicarbonate radicals to peroxidicarbonic acid.

The replacement of water molecules by bicarbonate
anions would lead to the following reaction sequence

$$2\ Don^+ + 2\ HCO_3^- \longrightarrow 2\ Don + 2\ HCO_3\cdot \qquad (15)$$

$$2\ HCO_3\cdot \longrightarrow H_2C_2O_6 \qquad (16)$$

$$H_2C_2O_6 \longrightarrow 2\ CO_2 + H_2O + O\cdot \qquad (17)$$

$$2\ CO_2 + 4\ H_2O \longrightarrow 2\ HCO_3^- + 2\ H_3O^+ \qquad (18)$$

$$\overline{2\ Don^+ + 3\ H_2O \longrightarrow 2\ Don + 2\ H_3O^+ + O\cdot} \qquad (19)$$

Since CO_2 is always regained and - in reaction with water - regenerated to bicarbonate, we understand the fast isotope equilibration process, which has obscured all the old mass spectrometric determinations on $^{18}O/^{16}O$ ratios. This is the reason, why only very fast measurements can give reliable data.

Since the proposal (equations 15-19) only replaces the H_2O molecules as reaction partners of the oxidized PS II donor by HCO_3^-, all considerations on the primary processes including oscillations may remain unchanged. The main modification of our present-day scheme is the inclusion of a new cycle, which continuously regenerates the necessary electron donor (HCO_3^-). It might well be, that the carbonic anhydrase plays a decisive role in this dark reaction.

References

1. Anbar, M. and Taube, H. (1956). *J. Am. Chem. Soc.* **78**, 3252-3255.
2. Archer, M.D. (1975). Book of Abstracts of the VIII. International Conference on Photochemistry, Edmonton, p. J2.
3. Arnason, T. and Sinclair, J. (1976). *Biochim. Biophys. Acta* **430**, 517-523.
4. Baxendale, J.H. (1964). *Rad. Res. Suppl.* **4**, 139-140.
5. Bodea, C. and Florescu, M. (1956). *Rev. de Chim. Roum.* **1**, 105-114.
6. Bottinga, Y. and Craig, H. (1969). *Earth Planet. Sci. Lett.* **5**, 285-295.
7. Briegleb, G. (1961). "Elektronen-Donator-Acceptor-Komplexe". Springer-Verlag, Berlin, Göttingen, Heidelberg.
8. Butler, J.A.V. and Leslie, W.M. (1936). *Trans. Faraday Soc.* **32**, 435-444.
9. Constam, E.J. and Hansen, A. von (1896). *Z. Elektrochem.* **3**, 137-144.
10. Curry, J. and Trelease, S.F. (1935). *Science* **82**, 18.
11. Draganić, I.G. and Draganić, Z.D. (1971). "The Radiation Chemistry of Water". Academic Press, New York.
12. Duysens, L.N.M. (1958). Brookhaven Symp. in Biol. **11**, 18-20.
13. Flügger, H.W. (1977). Diploma thesis, University of Tubingen.
14. Frank, A.J., Grätzel, M. and Henglein, A. (1976). *Ber. Bunsenges. physik. Chem.* **80**, 593-602.
15. Fujishima, A. and Honda, K. (1972). *Nature* **238**, 37-38.
16. Govindjee and Khanna, R. (1977), this volume.
17. Grätzel, M., Henglein, A., Scheerer, R. and Toffel, P. (1976). *Angew. Chem.* **88**, 690-691.

18. Henglein, A., personal communication.
19. Hoare, J.P. (1962). *J. Electrochem. Soc.* **109**, 858-865.
20. Hoare, J.P. (1968). "The Electrochemistry of Oxygen". Interscience Publishers, New York, London, Sidney, Toronto.
21. Hodgson, E.K. and Fridovich, J. (1976). *Arch. Biochem. Biophys.* **172**, 202-205.
22. Johannin-Gilles, A. and Vodar, B. (1954). *J. Phys. Radium* **15**, 223-224.
23. Knox, R.S. (1969). *Biophys. J.* **9**, 1351-1362.
24. Kortum, G. (1948). "Lehrbuch der Elektrochemie". Dietrich'sche Verlagsbuchhandlung, Wiesbaden.
25. Krasnovsky, A.A. (1977), this volume.
26. Krasnovsky, A.A. and Brin, G.P. (1961). *Dokl. Akad. Nauk SSSR.* **139**, 142-145.
27. Kreutz, W. (1974). *In* "Proceedings of the VI. International Congress on Photobiology" (G.O. Schenck, ed.). Nr. 029, Deutsche Gesellschaft fur Lichtforschung, Frankfurt.
28. Kutyurin, V.M. (1970). *Biologiya* **4**, 569-580.
29. Kutyurin, V.M., Nazarov, N.M. and Semeniuk, K.G. (1966). *Dokl. Akad. Nauk SSSR.* **171**, 215-217.
30. Lampe, F.W., Field, F.H. and Franklin, J.L. (1957). *J. Am. Chem. Soc.* **79**, 6132-6135.
31. Lawesson, S.-O. and Schroll, G. (1969). *In* "The Chemistry of Carboxylic Acids and Esters", (S. Patai, ed.), pp. 670-703. Interscience Publishers, London, New York,
32. Lifshitz, Ch. and Stein, G. (1964). *Israel J. Chem.* **2**, 337-353.
33. Maiwald, B. and Heinzinger, K. (1972). *Z. Naturforschg.* **27a**, 819-826.
34. Malkin, R. (1971). *Biochim. Biophys. Acta* **253**, 421-427.
35. Metzner, H. (1966). *Naturwissenschaften* **53**, 141-150.
36. Metzner, H. (1968). *Hoppe-Seyler's Z. physiol. Chem.* **349**, 1586-1588.
37. Metzner, H. (1975). *J. Theor. Biol.* **51**, 201-231.
38. Metzner, H. (1977), unpublished.
39. Metzner, H. and Fischer, K. (1974). *Photosynthetica* **8**, 257-262.
40. Metzner, H., Fischer, K. and Lupp, G. (1975). *Photosynthetica* **9**, 327-330.
41. Metzner, H. and Gerster, R. (1976). *Photosynthetica* **10**, 302-306.
42. Nakato, Y., Abe, K. and Tsubomura, H. (1976). *Ber. Bunsenges. physik. Chem.* **80**, 1002-1007.
43. Pratt, R. and Trelease, S.F. (1938). *Am. J. Bot.* **25**, 133-139.
44. Reitz, O. and Bonhoeffer, K.F. (1935). *Z. physik. Chem. A* **172**, 369-388
45. Renger, G. (1977), this volume.
46. Scheerer, R., Grätzel, M. (1976). *Ber. Bunsenges. physik. Chem.* **80**, 979-982.
47. Seely, G.R. (1966). *In* "The Chlorophylls", (L.P. Vernon and G.R. Seely, eds), pp. 523-568, Academic Press, New York, London.
48. Sinclair, J. and Arnason, T. (1974). *Biochim. Biophys. Acta* **368**, 393-400.
49. Stanienda, A. (1963). *Naturwissenschaften* **50**, 731-732.
50. Stemler, A. and Govindjee (1973). *Plant Physiol.* **52**, 119-123.
51. Sun, A.S. and Sauer, K. (1971). *Biochim. Biophys. Acta* **234**, 399-414.

52. Urey, H.C. (1947). *J. Chem. Soc.* 562-581
53. Walker, D.C. (1967). *Canad. J. Chem.* **45**, 807-811.
54. Walker, O.J. and Weiss, J. (1935). *Trans. Faraday Soc.* **31**, 1011-1017.
55. Warburg, O. and Krippahl, G. (1960). *Z. Naturforschg.* **15b**, 367-369.

ON THE STRUCTURAL ARRANGEMENT OF LIGHT REACTION CENTRES I AND II IN THE PHOTOSYNTHETIC MEMBRANE

W. KREUTZ

Institut für Biophysik und Strahlenbiologie der Universität Freiburg im Breisgau, Freiburg i. Br., (BRD)

The first electronmicroscopic information regarding defined surface structures was presented in 1963 by a thylakoid surface picture by Park and Pon [17]. Also in 1963, the evaluation of X-ray diffraction diagrams of chloroplasts of *Antirrhinum majus* supplied indications on particle arrays in the thylakoid membrane planes [7]. In the following chapters an attempt is made to co-ordinate the structural data gained since then by both methods, in order to obtain a three-dimensional concept of the thylakoid membrane.

In order to set up a three-dimensional model on the basis of electronmicroscopic data, pictures of the outer and inner surface of the thylakoid membrane, of the inner fracture faces, as well as projection pictures with sufficient structural resolution must be available. In the course of the last 15 years it was possible to make these picture types sufficiently available. A choice of these types is represented in the Figs 1 to 6. Fig. 1 by Park and Biggins shows the inner surface of a thylakoid of spinach [16]. This structure of the inner surface, obtained by shadowing isolated thylakoids with Pt/C, was confirmed by deep-etching pictures during the last few years [18]. The deep-freeze-etching pictures further show that the highly ordered state of the particles, apparently consisting of 4 subunits, does not represent the normal

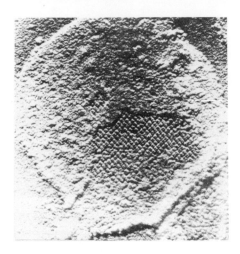

Fig. 1. Electronmicroscopic view of the inner surface of a thylakoid membrane (after [16]).

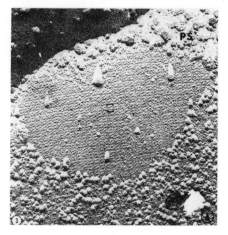

Fig. 2. Plane view of the outer surface of the thylakoid membrane of higher plants (after [13]).

case, but is only realized partly or spontaneously in the membrane.

The deep-freeze technique also supplied indications in regard to the structure of the outer surface [13]. Fig. 2 demonstrates that in this surface extremely well ordered structures may partly occur, too. The structural characteristics of these particles are different from those

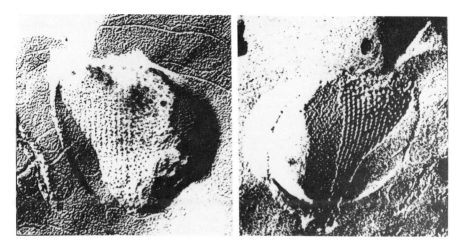

Fig. 3. Replica pictures of the two inner fracture faces of the thylakoid membrane of higher plants (after [15]).

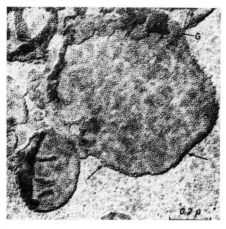

Fig. 4. Outer surface view of bacterial thylakoids (after [3]).

visible on the inner surface of the thylakoid. They rather show a chain-like arrangement and their dimensions are considerably different from those on the inner surface. The chain distance amounts to only about 83Å.

The particles, detectable on the outer surface, appear to be correlated with the particles visible on the inner surface, however, not in a direct but in an indirect manner. This is demonstrated by freeze-fracture pictures, e.g. by the replica pictures [15] in Fig. 3. On the inner

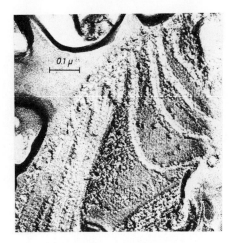

Fig. 5. Freeze fracture picture of bacterial thylakoids (after [3]).

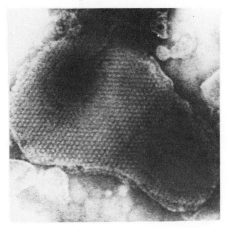

Fig. 6. Negative staining picture of bacterial thylakoids (after [11]).

fracture face of the membrane (Fig. 3), no particles with 4 subunits in square order and dimensions of about 160 Å are visible, but uniform particles of about 80 Å diameter, which protrude from the inner membrane leaflet and must, therefore, penetrate into the outer membrane leaflet. But, if these particles are arranged lattice-like as in Fig. 3, they form a square lattice with dimensions of about 170 x 190 Å on the inner surface. On the other hand, electronmicroscopic negative staining pictures representing

Fig. 7. X-ray diffraction pattern of orientated bacterial thylakoids. Vertical reflection orders = meridional orders, horizontal = equatorial. (After [4]).

the mass projection of all negative stained membrane parts, show a regular arrangement of 90 Å particles, consisting of subunits in an orthogonal lattice with lattice periods of 170 Å and 190 Å length. These molecules are regarded as coupling factor or as carboxydismutase molecule arrays on the outer surface of the membrane [6]. This interpretation is supported by electronmicroscopic positive staining views of the cross-section of the thylakoid membrane, from which particles project on the outer surface [14]. Finally, we have to remember negative staining pictures of the cross-section of lipid extracted thylakoids. A more or less continuous "membrane" of double layer nature with a total thickness of about 40 Å remains.

In the case of photosynthetic bacteria three electronmicroscopic pictures are presented. One picture (Fig. 4) gives the outer surface of thylakoids of *Rhodopseudomonas viridis*, the second (Fig. 5) shows the two inner fracture faces of the thylakoid membrane [3], while the third (Fig. 6) depicts a negative staining picture [11]. Similar as in thylakoids of higher plants, paracrystalline molecule distributions are observed on the outer surface (Fig. 4) possessing two preferential view directions. The particle strands in these view directions have a centre-to-centre distance of about 85 Å. The freeze fracture picture (Fig. 5) clearly shows that in the layer situated under the paracrystalline outer layer particles are embedded. On the negative staining picture (Fig. 6) an almost perfect hexagonal arrangement of about 80 Å particles is presented. On account of the projection view picture it cannot be decided, whether these particles are situated on the surface or within the membrane. From X-ray diffraction data, however, it may be assumed that these particles lie in the membrane: the thylakoid period in *Rhodopseudomonas viridis* can shrink down to 140 Å if dehydrated carefully. Even if the particles would only possess 50 Å thickness, they would require a thickness of 100 Å on both sides of the thylakoid. For the membrane thickness itself only 20 respectively 35 Å would remain, a fact not compatible with the cross-section profile of this membrane as determined by X-ray diffraction.

The evaluation of X-ray diffraction patterns of 100% hydrated isolated thylakoids with a stacking period of 170 Å supplies the electron density cross-section profile

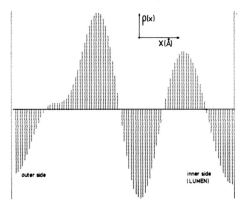

Fig. 8. Electron density distribution of the cross-section of the bacterial thylakoids. The abscissa denotes the level of the average electron density of the thylakoid stack. The modulation above and below the abscissa indicates higher or lower electron densities than the average density. Distance between vertical lines = 1 Å. (After [4]).

shown in Fig. 8 [4]. This evaluation was obtained by the
so-called Q-function (=Fourier transform of the intensity
function), under consideration of distance distortions in
the stacking period and in the centre-to-centre distance
of the two thylakoid membranes as well as possible undula-
tions in the membrane place. This evaluation, for the
first time taking into account such distortion factors,
which might influence the scattering of thylakoid stacks,
was provided by a new evaluation theory, developed during
the last 2 years [19].

The electron density distribution shown in Fig. 8 gives
the deviation of the electron density from the mean electron density of the thylakoid membranes; positive peaks
signify higher, negative peaks lower electron density than
the mean electron density. Consequently the highest electron density exists in the outermost half of the thylakoid
membrane, the lowest density is positioned in the centre
of the membrane. On the inner side, the electron density
again increases. This electron density was calculated from
the meridional X-ray reflections (vertical reflection
series) of Fig. 7, consisting of 5 orders of a period of
170 Å [4]. In addition to this reflection series, the diffraction pattern of Fig. 7 shows a second series orientated orthogonally to the meridional series, i.e. a reflection series, to be attributed to a plane lattice. These
equatorial reflections correspond to a hexagonal plane
lattice of particles, as directly seen on the electronmicrograph of Fig. 6. In agreement with the electronmicroscopic lattice pattern the centre-to-centre distance of
the particles amounts to 130 Å obtained from the X-ray
data, corresponding to the largest hexagonal lattice distance of 112 Å.

A further interesting viewpoint is that the width of
the equatorial reflections (extension vertical to scattering direction) is more than twice as large as that of the
meridional series. From the absolute width of the

equatorial reflections a maximal particle diameter parallel to the stacking axis of about 80 Å can be estimated.

For thylakoids of higher plants *in vivo*, already in 1964, the electron density distribution of the cross-section was calculated [8]. This distribution has been confirmed by the recently developed evaluation techniques. The electron density profiles indicate that on the outer side considerably more proteins are arranged than on the inner side of the thylakoids. Furthermore, the profiles show that the outer protein layer has a thickness of about 40 Å. In the centre of the membrane the lowest electron density exists and the electron density becomes positive again on the inner side. The width of the inner positive peak amounts to about 20 Å.

The main difference between the electron density cross-section profiles of thylakoids of higher plants and those of bacteria consists in the width and height of the inner positive peak. In higher plants this peak is broader by 5-10 Å. This is to indicate that in thylakoids of higher plants more protein is settled within the inner membrane half than in the bacterial thylakoids (compare Fig. 11 and Fig. 13).

Apart from the calculations of the cross-section profile, investigations were carried out with intrinsic proteins of chloroplasts isolated by 100% formic acid [5] and chlorophyll-containing complex I and complex II proteins, which were isolated by detergents [1]. On the basis of these investigations a two-dimensional lattice in the outer layer of the thylakoid membrane was proposed [1,9]. In a first approximation a square lattice was assumed, the major mass of the particles had to be directed along the diagonal of the lattice cells in order to explain the observed interferences. The Bragg-spacing of the 10-reflection of the square lattice was found to be 42 Å. In the diagonal direction, containing the major mass, the lattice period was found to be about 60 Å. First calculations of electron density projection planes and electron density layer distributions [5,8] with 15 Å resolution, supplied indication for the presence of peptide strands, running horizontally and vertically along the periphery of the 42 Å lattice cell, thus forming a kind of hollow cylinder which can best be recognized by the electron density projection onto the membrane cross-section plane (Fig. 9).

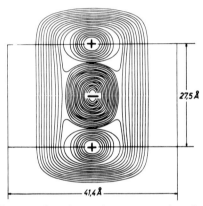

Fig. 9. Cross-section projection of electron density distribution of a chloroprotein particle. The region signified by a minus sign means the hollow part of the particle (after [5]).

Recently Fenna and Mathews [2] carried out an electron density calculation of a water soluble chlorophyll protein, extracted from bacteria with a 2.8 Å resolution. They find a hollow ellipsoid with dimensions of about 65 Å x 43 Å. The hollow space amounts to 15 x 35 x 57 Å. The walls of the hollow ellipsoid are wrapped by β-peptide chains and partly by short α-helical pieces.

Up to now, we assumed that those particles actually forming the plane lattice, possess dimension of 42 x 42 Å or 42 x 84 Å, to explain the orthogonal super structures, occurring in the electronmicrograph in Fig. 1.

Should the lattice forming protein particle show dimensions of 42 x 65 Å instead of 42 and 42 x 84 Å, an oblique planar lattice should exist, to which an orthogonal

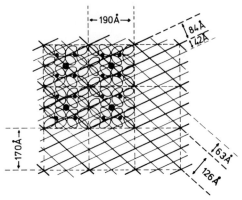

Fig. 10. Structure proposal for a combined planar lattice in the thylakoid of higher plants, obtained by the superposition of an oblique and an orthogonal lattice. Full circles mark the position of the membrane "spanning particles" and the 4-subunit particles in the inner lipid leaflet.

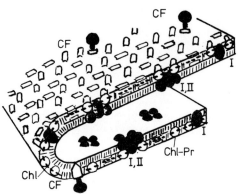

Fig. 11. Three-dimensional model of the thylakoid membrane of higher plants. CF = coupling factor + ATPases, I, II = enzymatic protein of reaction complex of light reaction I and II, other abbreviations see Fig. 12.

lattice with dimensions of about 170 x 190 Å can be superimposed as shown in Fig. 10. This lattice combination between an oblique lattice, formed of 42 x 65 Å particles and a superimposed orthogonal super lattice (≈square lattice), formed by additional associated particles can also explain the X-ray diffraction data. The co-ordination sites (binding sites) for the additional particles, forming the super lattice, are designated by full circles. The special

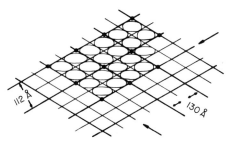

Fig. 12. Structure proposal for the hexagonal planar lattice in the bacterial thylakoids. Full circles designate the position of the membrane spanning particle.

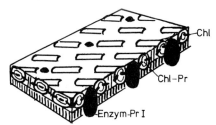

Fig. 13. Three-dimensional model view of a membrane section of bacterial thylakoids. Enzym-Pr I = enzymatic protein part of light reaction centre I, Chl = chlorophyll, Chl-Pr = chloroprotein.

protein particle arrangement within an area of 170 x 190 Å is to explain the repetitive nature of this structure unit. It is provided by the introduction of a two-fold rotation axis [10]. With the same particle type a second oblique lattice can be spanned, allowing the superposition of a hexagonal superlattice, as required for the plane structure of bacterial thylakoids (Fig. 12).

In our opinion a class of chlorophyll-bearing lattice forming proteins exists, which are very similar in their tertiary structure, however, showing deviations in their primary and secondary structure, thus being able to settle special chlorophyll populations. Further on we shall designate this protein class particles as "chloroproteins" and depict them as hollow cylinders for the sake of simplicity.

In the case of bacterial thylakoids it appears that to the binding sites, provided by the chloroprotein lattice, additional particles are coupled in a hexagonal superlattice within the membrane of the thylakoid. These particles span the inner lipid monolayer and will penetrate into the thylakoid lumen (Fig. 13). There is only very little space left for lipids in the chloroprotein arrangement, which results in a very asymmetrical lipid distribution within the cross-section of the thylakoid membrane. Roughly estimated at a maximum only 20% of the lipids can be situated in the outer monolayer. Therefore, it is hardly possible to speak of a lipid bilayer. It appears probable that the chlorophyll molecules are sterically arranged in a similar manner as in the water soluble chloroprotein described by Fenna and Mathews.

The structural situation of thylakoids of higher plants in this respect appears more complicated. The particles visible on the inner surface of the membrane apparently consisting of 4-subunits, stick only into the inner lipid monolayer, since they are not visible on the inner fracture surface of the membrane. From this fracture plane, however, single particles protrude forming the same lattice type (170 x 190 Å) as the subunit particles on the inner surface, which have to penetrate into the chloroprotein layer. These particles will, more or less, span the whole membrane, thus co-ordinating the oblique chloroprotein arrangement with the orthogonal lattice of the 4-subunit particles. In these membrane regions the structural situation demonstrated in Fig. 11 should be observed. In these complexes of the membrane the particles are so densely packed that there is no space available for the coupling of additional particles from the outside, for instance for coupling factor particles. Outside coupling may only occur onto such membrane parts, not occupied by 4-subunit particles on the inner side of the membrane (Fig. 11).

High order lattice regions seem to occur less frequently than lattice distorted membrane parts, i.e. complex regions consisting of chloroprotein and enzymatic protein mostly associate more or less loosely within the membrane.

Very probably the 4-subunit particles represent the enzymatic part of the light reaction centres II, the spanning particles the enzymatic protein of light reaction centres I. Light reaction I and II would, accordingly, be correlated in a very compact manner in the same complex. This structural situation would also explain the breakdown of light reaction II by treatment with detergents, while light reaction I remains intact. Detergents probably decouple the chloroproteins of system II from the enzymatic 4-subunit proteins, which only contact in the centre part of the membrane. Enzyme protein I (spanning protein) is, however, anchored within the chloroprotein lattice and therefore is disintegrated from the membrane together with the chloroproteins I. In such a complex the two light reactions should be coupled as a ratio 1:2. This concept regarding the arrangement of the light reaction centres is supported by the fact that some thylakoid membranes are solely made of such complexes in a nearly perfect crystalline arrangement (compare Figs 1 and 2). We further suspect that in this way two chlorophyll levels are constituted in the reaction complexes, one made by the chloroproteins in the outer half of the membrane and the other by the enzymatic proteins I and II in the inner half of the membrane.

The ratio of light-harvesting chlorophylls and active chlorophylls P_{700} and P_{680} reaches a minimum when the reaction centres constitute a perfect crystalline arrangement covering the whole membrane. In this situation the chlorophyll ratio is strictly defined. The portion of the light-harvesting chlorophylls increases the more, the more chloroproteins are incorporated between the highly organized reaction complexes. In the paracrystalline state, 16 chloroproteins with 7 chlorophylls each, belong to one reaction complex I,II. Further chlorophylls may be added by the chlorophyll equipment of the reaction centres in the enzymatic proteins.

In the case of photosynthetic bacteria 4 chloroproteins, probably also containing 7 chlorophylls each, have to be attributed to each reaction centre I.

The share of the crystalline and statistically organized membrane parts possibly depends on the intensity and duration of the incident light.

References

1. Bailey, J.L. and Kreutz, W. (1969). *In* "Progress in Photosynthesis Research" (H. Metzner ed.). Vol. I, pp. 149-158. International Union of Biological Sciences, Tübingen.
2. Fenna, R.E. and Mathews, B.W. (1975). *Nature* **258**, 573-577.
3. Giesbrecht, P. and Drews, G. (1966). *Arch. Mikrobiol.* **54**, 297-330.
4. Hodapp, N. and Kreutz, W. (1977). Unpublished.
5. Hosemann, R. and Kreutz, W. (1966). *Naturwissenschaften* **53**, 298-304.
6. Howell, S.H. and Moudrianakis, E. (1967). *J. Mol. Biol.* **27**, 323-333.
7. Kreutz, W. (1963). *Z. Naturforschg.* **18**b, 1098-1104.
8. Kreutz, W. (1964). *Z. Naturforschg.* **19**b, 441-446.
9. Kreutz, W. (1970). *Adv. Bot. Res.* **3**, 53-169.
10. Kreutz, W. (1972). *Angew. Chem.* **84**, 597-614.
11. Kreutz, W., Hofmann, K.P. and Uhl, R. (1974). *In* "25. Colloquium Ges. Biol. Chem", pp. 312-328, Mosbach.
12. Menke, W. (1967). *Arbeitgem. f. Forschg. Nordrhein-Westfalen,* No. 171.
13. Miller, K.E. (1976). *J. Ultrastruct. Res.* **54**, 159-167.
14. Miller, K.E. and Staehelin, L.A. (1976). *J. Cell Biol.* **68**, 30-47.
15. Mühlethaler, K. and Wehrli, E. (1977). Unpublished.
16. Park, R.B. and Biggins, J. (1964). *Science* **144**, 1009-1011.
17. Park, R.B. and Pon, N.G. (1963). *J. Mol. Biol.* **6**, 136-158.
18. Staehelin, L.A. (1976). *J. Cell Biol.* **71**, 136-158.
19. Welte, W. and Kreutz, W. (1977). Unpublished.

**CHARACTERIZATION OF THE OXYGEN-EVOLVING SIDE
OF PHOTOSYSTEM II BY ANTISERA TO
POLYPEPTIDES AND CAROTENOIDS**

G.H. SCHMID, A. RADUNZ, F. KOENIG and W. MENKE

*Max-Planck-Institut für Züchtungsforschung
(Erwin-Baur-Institut), Abteilung Menke, Köln-Vogelsang (F.R.G.)*

Introduction

Despite the efforts of many laboratories the knowledge of the biochemistry on the oxygen-evolving side of photosystem II has not much advanced in the last years. The present state of knowledge is still that manganese [1] and chloride ions [4] play a role in oxygen evolution. The involvement of bicarbonate in reactions on this side is uncertain [5,14]. Among the possible factors at least one proteinaceous component is assumed between the site of oxygen-evolution and the site of the primary photochemistry of photosystem II [8,9,29]. More recently, a soluble manganese-containing factor from *Phormidium luridum* was isolated which restores the Hill activity that has been lost in hypotonic washing [25]. By means of serological methods we were able to show the participation of further thylakoid components in reactions on the water-splitting side of photosystem II [16-19,22]. It should be borne in mind that the serological reactions can only occur if antigens and antibodies come into contact. If a serological reaction with stroma-free chloroplasts is observed, it is concluded that the antigen is localized at the outer surface of the thylakoid membrane. It is also obvious that antibodies cannot be adsorbed onto the surface of the thylakoid membrane where the thylakoids touch i.e. in the grana. If the antibodies react bifunctionally agglutination is observed. If, however, for sterical reasons, the antibody molecules react only monofunctionally no agglutination is seen. In this case, agglutination occurs after addition of the soluble antigen or of anti-γ-globulin (Coombs test). The sterical hindrance of the bifunctional reaction might be explained by the possibility that the antigenic determinants are situated in gaps or clefts. Antigens which are located at the inner surface or inside the thylakoid membrane are made accessible by ultrasonication. In comparison to other methods the advantage of our method is that it gives very specific answers concerning the localization and function of components in the thylakoid membrane.

In these investigations, the main difficulty in preparing the protein antigens is due to their insolubility in water. For our purpose, sodium dodecyl sulfate solutions which contain β-mercaptoethanol are found to be the best way to solubilize the thylakoid membrane. From the mixture obtained we have isolated the individual polypeptides by gel filtration and other methods. Despite the fact that after the removal of the detergent the secondary and tertiary structure of the polypeptides is greatly altered [13], the antisera to these polypeptides agglutinate suspensions of stroma-freed chloroplasts and affect photosynthetic electron transport. Consequently, the antisera contain antibodies which are directed towards native antigenic determinants of the thylakoid membrane.

In the following the effect on electron transport reactions of two protein antisera which are directed to polypeptides of the apparent molecular weight 66 000 and 11 000 (the molecular weight of this polypeptide was determined to be 6300 by ultracentrifugation and diffusion measurements [2]) is described. Moreover we report on the effect of the antisera to the carotenoids lutein and neoxanthin.

Protein Antisera

An antiserum to the polypeptide 11 000 agglutinates stroma-freed chloroplasts from *Antirrhinum* and *Nicotiana*. The antiserum also inhibits photosynthetic electron transport in stroma-free swellable chloroplasts of tobacco. Consequently, antigenic determinants towards which the antiserum is directed are located in the outer surface of the thylakoid membrane. In Fig. 1 the dependence of the degree of inhibition on the amount of added antiserum is plotted for the optimal pH 7.4. The curve shape is sigmoidal which hints at a co-operative effect. The calculation of the Hill interaction coefficient yields 10. It

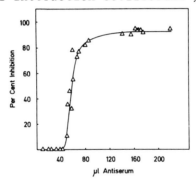

Fig. 1. Dependence of the inhibition degree on the amount of added antiserum to polypeptide 11 000 at pH 7.4. Electron transport system tetramethyl benzidine/ascorbate → anthraquinone-2-sulfonate.

should be noted that sigmoidal curve shapes are also obtained by the antisera to polypeptide 33 000 [12] to plastocyanin [20] and to cytochrome f [21]. At the optimal pH and in the presence of saturating amounts of antiserum the anthraquinone-2-sulfonate Hill reaction is inhibited by the antiserum to polypeptide 11 000 by 90 per cent. Addition of tetramethyl benzidine which is an artificial electron donor to photosystem II [6] does not relieve the inhibition. However, addition of diphenyl carbazide, another artificial donor to photosystem II [27], restores

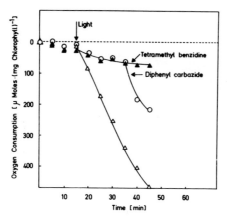

Fig. 2. Photosynthetic electron transport measured as oxygen consumption in the anthraquinone-2-sulfonate mediated Mehler reaction in wild tobacco chloroplasts. (Δ) Hill reaction in the presence of control serum; (▲) Hill reaction in the presence of antiserum with indication of tetramethylbenzidine addition; (O) Hill reaction in the presence of antiserum; diphenylcarbazide restores electron transport.

electron transport (Fig. 2). Typical photosystem I reactions are not affected by the antiserum. From this it follows that the inhibition site is on the oxygen-evolving side of photosystem II between the sites of electron donation of tetramethyl benzidine and diphenyl carbazide (Fig. 3).

The inhibitory action of the antiserum on electron transport requires light. Despite the fact that agglutination of the chloroplasts occurs immediately upon addition of the antiserum, the inhibition of electron transport becomes only apparent in the course of the illumination period.

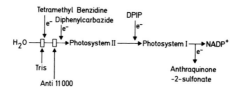

Fig. 3. Inhibition site of antiserum to polypeptide 11 000 in the electron transport scheme.

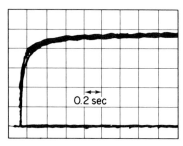

Fig. 4(a). Fluorescence rise curve of tobacco chloroplasts in the assay system tetramethyl benzidine/ascorbate → anthraquinone-2-sulfonate. One assay in the presence of antiserum to polypeptide 11 000, the other in the presence of control serum. Excitation wavelength 440 nm, excitation slit 40 nm; emission wavelength 685 nm, emission slit 34 nm.

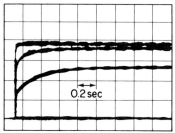

Fig. 4(b). Fluorescence rise of the same assay as in (a). The assay was shaken in the Warburg apparatus at 30 000 erg·cm^{-2}·s^{-1} of red light 580 nm <λ< 700 nm until the inhibition caused by the antiserum was 93%. Then the assay was kept in the dark for 20 min prior to the fluorescence measurement. Lower curve in the presence of antiserum; middle curve in the presence of control serum; upper curve in the presence of control serum plus 10 µM DCMU.

The localization of the inhibition site was also attempted by investigating the effect of the antiserum on the fluorescence. In these measurements the influence of light on the inhibitory action of the antiserum is especially obvious. The assay system contained tetramethyl benzidine/ascorbate as the electron donor and anthraquinone-2-sulfonate as the acceptor. A dark-adapted sample shows no influence of the antiserum on the steady state level of fluorescence (Fig. 4a). If, however, the reaction mixture has been illuminated until the inhibitory effect was established, then the fluorescence level was lower in the presence of antiserum than in the presence of control serum (Fig. 4b). This shows, that the inhibition is on the water-splitting side of photosystem II. In this condition quencher Q is not or only slowly reduced because electrons are faster removed by photosystem I than they are supplied by photosystem II. If the drain of electrons through photosystem I is prevented by the addition of DCMU, fluorescence is immediately high because the quencher stays in the reduced state.

TRIS-washing of chloroplasts abolishes according to the literature the capacity for oxygen evolution [30]. Addition of the artificial electron donor couple tetramethyl benzidine/ascorbate restores electron transport to $NADP^+$ or to a suitable artificial electron acceptor such as anthraquinone-2-sulfonate. Fig. 5 shows that also in TRIS-washed chloroplasts the antiserum exerts its inhibitory effect. It appears, however, that the relative degree of inhibition is lower than with unwashed chloroplasts. If instead of tetramethyl benzidine the artificial donor diphenyl carbazide is added then the antiserum does not

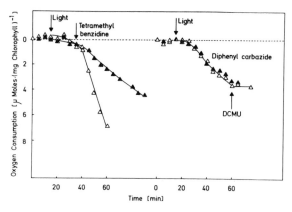

Fig. 5. Effect of the antiserum to polypeptide 11 000 on TRIS-buffer washed tobacco chloroplasts. The electron transport system is tetramethyl benzidine/ascorbate ⟶ anthraquinone-2-sulfonate or diphenyl carbazide ⟶ anthraquinone-2-sulfonate. ▲ in the presence of antiserum, △ in the presence of control serum. DCMU is 2.10^{-5} M in the assay where indicated.

influence the electron transport rate. Again, it shows that the inhibition site is between the site of electron donation of tetramethyl benzidine and that of diphenyl carbazide. The inhibition by the antiserum reaches apparently somewhat into the TRIS-block.

Application of the repetitive flash spectroscopy permits the direct measurements of the photoreaction of the reaction centre chlorophyll a_{II} [3] and that of the primary electron acceptor X_{320} [24]. It appears that the initial amplitudes of the absorption change of Chl a_{II} at 690 nm and that of the primary electron acceptor X_{320} are both diminished by the antiserum [22].

The effect of the antiserum on low temperature fluorescence emission shows also that the antiserum affects the functionability of photosystem II reaction centres in that it enhances the energy spill-over to photosystem I. The effect of the antiserum is opposite to Murata's effect of the Mg^{2+} ion induced inhibition of energy spill-over from photosystem II to photosystem I [15].

The data presented mean that the polypeptide 11 000 or the protein to which it belongs is associated with photosystem II and somehow participates in a reaction which occurs on the water-splitting side of photosystem II. The polypeptide is located in the outer surface of the thylakoid membrane. However, our result does not permit any conclusion as to where the splitting of the water molecule itself occurs.

For comparison we describe in the following an antiserum which affects the reaction centre of photosystem II. The antigen which served for immunization exhibited an apparent molecular weight of 66 000 (in order to distinguish this antiserum from antisera to other polypeptides of the same molecular weight we designate this antiserum 66 000 PS II-42) and was obtained by repeated gel chromatography of a starting preparation using an alkaline elution medium [10]. The difference to the antiserum to polypeptide 11 000 is that the antiserum also inhibits the photooxidation of diphenyl carbazide in TRIS-washed chloroplasts (Fig. 6). From this it follows that the site of inhibition lies behind the site of electron donation of diphenyl carbazide. As with the antiserum to polypeptide 11 000 the dependence of the degree of inhibition of electron transport on the amount of antiserum added yields a sigmoidal curve shape (Fig. 7). From the analysis of the Kautsky effect in the presence of this antiserum a further difference to the antiserum to polypeptide 11 000 emerges, because the antiserum causes no effect on fluorescence. Repeated illuminations do not change the fluorescence level (Fig. 8). As the variable portion of fluorescence depends on the ratio of Q_{red}/Q_{ox} this result infers that this ratio is unchanged also under conditions in which electron flow is inhibited by more than 90%. Consequently, the adsorption of antibodies influences the oxidation as well as the reduction of the quencher to the same extent. Provided the reaction of the antiserum is monospecific, unchanged fluorescence together with blocked electron transport can be explained if the antigen is a component of reaction centre II and if intermolecular interactions exist between the antigen and the natural electron donor and acceptor. DCMU addition does not induce a difference between the fluorescence rise curves.

According to the results described it appears that the two polypeptides play a role at different sites of the electron transport system. The two antigens are either electron carriers or are in contact *via* intermolecular interactions with antigen molecules which are directly involved in electron transport as carriers. In both cases the effect of the antibody binding can be due to a conformational change in the protein molecule which in turn decreases the reaction rate. Only if the first possibility applies an effect at the active site of the carrier itself can be visualized.

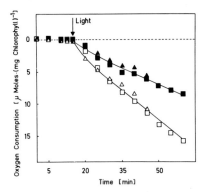

Fig. 6. Influence of the antiserum 66 000 PS II-42 on TRIS-washed tobacco chloroplasts. □ Photooxidation of tetramethyl benzidine with anthraquinone-2-sulfonate as the electron acceptor in the presence of control serum; ■ reaction as □ in the presence of antiserum; △ photooxidation of diphenyl carbazide with the same electron acceptor in the presence of control serum; ▲ same reaction as △ in the presence of antiserum.

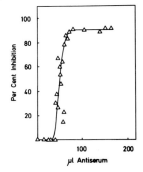

Fig. 7. Dependence of the inhibition degree on the amount of added antiserum to polypeptide 66 000; electron transport system tetramethyl benzidine/ascorbate → anthraquinone-2-sulfonate.

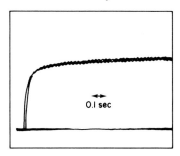

Fig. 8. Influence of the antiserum 66 000 PS II-42 on the fluorescence in tobacco chloroplasts. Experimental conditions as indicated in Fig. 4.

Carotenoid Antisera

As already published we have prepared antisera to the carotenoids lutein and neoxanthin, which all agglutinate stroma-free tobacco chloroplasts and inhibit on the donor side of photosystem II [16-18]. We can show that the sites of inhibition are distinctly different from each other in electron transport of higher plants. A peculiarity of the reaction of the antisera to lutein and neoxanthin with morphologically well preserved stroma-free chloroplasts of higher plants e.g. tobacco is a low degree of inhibition of 10-20%. In swollen chloroplasts higher degrees of inhibition are obtained. We have attributed this relatively low inhibition partly to thylakoid stacking which is reduced by swelling. From the literature it is known that the lamellar system of blue-green algae consists of single unstacked thylakoids [11]. Therefore with isolated thylakoids of blue-green algae higher degrees of inhibition should be observed which was indeed the case. In fact with thylakoids of *Nostoc muscorum* and *Oscillatoria chalybea* inhibition degrees of 50 to 60% were observed [18]. Despite the notion that neoxanthin is not occurring in blue-green algae [28] we observe an effect by our antiserum. From this we infer that the antiserum to neoxanthin cross-reacts with a carotenoid of similar configuration. As, in addition, the antisera agglutinate the thylakoid preparations, lutein as well as the carotenoid, reacting with the antiserum to neoxanthin are located in the outer surface of the thylakoid membrane. The inhibition sites of both antisera are identical in *Nostoc muscorum* and are located between the sites of electron donation of the artificial donors tetramethyl benzidine and diphenyl carbazide (Table 1). However, in the case of *Oscillatoria* the inhibition sites of both antisera differ (Table 2). Whereas the inhibition of the antiserum to neoxanthin lies again between the sites of electron donation of tetramethyl benzidine

TABLE 1

Effect of the antiserum to neoxanthin and lutein on photosynthetic electron transport in thylakoids of Nostoc muscorum

Electron transport system	Antiserum to neoxanthin		Antiserum to lutein	
	Control rate μmoles acceptor reduced/mg chlorophyll/h	% Inhibition	Control rate μmoles acceptor reduced/mg chlorophyll/h	% Inhibition
$H_2O \to$ A-2-Sulf	12	60	16	50
TMB/asc $\to$ A-2-Sulf	30	65	40	40
DPC $\to$ A-2-Sulf	50	0	40	0
$H_2O \to$ DCPIP	35	60	6	35
DPC $\to$ DCPIP	20	10	30	10
DCPIP/asc — A-2-Sulf	190	0	254	0

TABLE 2

Effect of the antiserum to neoxanthin and lutein on photosynthetic electron transport in thylakoids of Oscillatoria chalybea.

Electron transport system	Antiserum to neoxanthin		Antiserum to lutein	
	Control rate μmoles acceptor reduced/mg chlorophyll/h	% Inhibition	Control rate μmoles acceptor reduced/mg chlorophyll/h	% Inhibition
$H_2O \rightarrow$ A-2-Sulf	20	30	40	48
TMB/asc $\rightarrow$ A-2-Sulf	70	40	60	20
DPC $\rightarrow$ A-2-Sulf	60	0	80	0
$H_2O \rightarrow$ DCPIP	15	25	20	37
DPC $\rightarrow$ DCPIP	25	10	20	13
DCPIP/asc $\rightarrow$ A-2-Sulf	240	0	200	0

and diphenyl carbazide, the inhibition of the antiserum to lutein appears to be situated at least partially beyond the site of electron donation of tetramethyl benzidine. The situation with *Oscillatoria* thylakoids is the same as in tobacco chloroplasts [17]. The results presented show that at least part of the carotenoids lutein and neoxanthin are involved in photosystem II activity and that binding of antibodies to these carotenoids leads to an inhibition of photosynthetic electron transport on the donor side of photosystem II. In this context it should be mentioned that Sewe and Reich [23] have found that lutein forms preferably a complex with chlorophyll *b* which according to the literature is mainly located in those regions of the membrane which belong to photosystem II [7,26].

Summarizing we would like to say that with the help of antisera one can demonstrate that the polypeptide 11 000 and the two carotenoids lutein and neoxanthin are surface-exposed in the thylakoid membrane and are somehow involved in reactions on the oxygen-evolving side of photosystem II whereas the polypeptide 66 000 PS II-42 seems to belong to reaction centre II.

References

1. Cheniae, G.M. (1970). *Ann. Rev. Plant Physiol.* **21**, 467-498.
2. Craubner, H., Koenig, F. and Schmid, G.H. (1977). *Z. Naturforschg.* **32c**, 384-391.
3. Glaser, M., Wolff, Ch., Buchwald, H.E. and Witt, H.T. (1974). *FEBS Lett.* **42**, 81-85.
4. Gorham, P.R. and Clendenning, K.A. (1952). *Arch. Biochem. Biophys.* **37**, 199-223.
5. Govindjee and Khanna, R.K. (1977), this volume0.
6. Harth, E., Oettmeier, W. and Trebst, A. (1974). *FEBS Lett.* **43** 231-234.
7. Hind, G. and Olson, J.M. (1968). *Ann. Rev. Plant Physiol.* **19**, 249-282.
8. Jones, L.W. and Kok, B. (1966). *Plant Physiol.* **41**, 1044-1049.
9. Katoh, S. and San Pietro, A. (1967). *Arch. Biochem. Biophys.* **122**, 144-152.
10. Koenig, F., Menke, W., Radunz, A. and Schmid, G.H. (1977). *Z. Naturforschg.*, in press.
11. Menke, W. (1961). *Z. Naturforschg.* **16b**, 543-546.

12. Menke, W., Koenig, F., Radunz, A. and Schmid, G.H. (1975). *FEBS Lett.* **49**, 372-375.
13. Menke, W., Radunz, A., Schmid, G.H., Koenig, F. and Hirtz, R.-I. (1976). *Z. Naturforschg.* **31c**, 436-444.
14. Metzner, H. (1977), this volume.
15. Murata, N. (1969). *Biochim. Biophys. Acta* 189, 171-181.
16. Radunz, A. and Schmid, G.H. (1973). *Z. Naturforschg.* **28c**, 37-44.
17. Radunz, A. and Schmid, G.H. (1975). *Z. Naturforschg.* **30c**, 622-627 627.
18. Schmid, G.H., List, H. and Radunz, A. (1977). *Z. Naturforschg.* **32c**, 118-124.
19. Schmid, G.H., Menke, W., Koenig, F. and Radunz, A. (1976). *Z. Naturforschg.* **31c**, 304-311.
20. Schmid, G.H., Radunz, A. and Menke, W. (1975). *Z. Naturforschg.* **30c**, 201-212.
21. Schmid, G.H., Radunz, A. and Menke, W. (1977). *Z. Naturforsch.* **32c**, 271-280.
22. Schmid, G.H., Renger, G., Gläser, M., Koenig, F., Radunz, A. and Menke, W. (1976). *Z. Naturforschg.* **31c**, 594-600.
23. Sewe, K.-U. and Reich, R. (1977). *Z. Naturforschg.* **32c**, 161-171.
24. Stiehl, H.H. and Witt, H.T. (1968). *Z. Naturforschg.* **23b**, 220-224.
25. Tel-Or, E. and Avron, M. (1975). "Proceedings of the 3rd International Congress on Photosynthesis" (M. Avron, ed.). Vol. I, pp. 569-578, Elsevier Scientific Publishing Company, Amsterdam, Oxford, New York.
26. Thomas, J.B. (1972). *In* "Proceedings of the 2nd International Congress on Photosynthesis Research" (G. Forti, M. Avron and A. Melandri, eds). Vol. II, pp. 1509-1513, Dr W. Junk N.V., The Hague.
27. Vernon, L.P. and Shaw, E.R. (1969). *Plant Physiol.* **44**, 1645-1649.
28. Weedon, B.C.L. (1971). "Carotenoids" (O. Isler, ed.), pp. 267-323. Birkhauser Verlag, Basel, Stuttgart.
29. Yamashita, T. and Butler, W.L. (1968). *Plant Physiol.* **43**, 1978-1986.
30. Yamashita, T. and Butler, W.L. (1969). *Plant Physiol.* **44**, 435-438.

PREPARATION OF THYLAKOID MEMBRANES ACTIVE IN OXYGEN EVOLUTION AT HIGH TEMPERATURE FROM A THERMOPHILIC BLUE-GREEN ALGA

T. YAMAOKA, K. SATOH and S. KATOH

Department of Pure and Applied Sciences,
College of General Education, University of Tokyo, Tokyo (Japan)

Summary

The thylakoid membranes isolated from a thermophilic blue-green alga were active in oxygen evolution coupled with a stoichiometric ferricyanide reduction. The activity was low at room temperature but increased steeply as temperature was raised above 30°C to reach a maximum at temperatures as high as 50-55°C. An irreversible inactivation occurred above this range of temperature. The maximum rate of oxygen evolution in the membranes was comparable to that of photosynthetic oxygen evolution in the intact cells. The reaction was mediated by photosystem II alone, since it was sensitive to DCMU but not to DBMIB, and methylviologen could not replace ferricyanide as electron acceptor.

A spin labelling experiment indicated that the transition of the physical phase of the membrane lipids appeared between 30 and 40°C and at about 10°C. The high transition temperatures could be related with the fatty acid composition of the membrane lipids, which contained palmitic, oleic and palmitoleic acids as major fatty acids but very low concentration of poly-unsaturated fatty acids. The Arrhenius plot of the reaction rates showed a transition of the activation energy at 14°C but not between 30 and 40°C. It is suggested that the solid state of the membrane lipids does not support normal function of the oxygen evolving system. However, the oxygen evolving system is located in an environment which is particularly abundant in lipids with low melting point so that it stays in the liquid-crystalline state during the lateral phase separation.

Introduction

Thermophilic blue-green algae are the most heat-resistant among photosynthetic organisms. They grow in hot springs at temperatures as high as 50-75°C [4]. The algae showed the maximum photosynthetic efficiency at the temperature of their habitats [3].

The optimal functioning of the photosynthetic apparatus at such a high range of temperature premises an extreme thermostability of the thylakoid membrane and individual enzymes involved in photosynthetic processes. It is expected, therefore, that the algae provide us a unique and excellent opportunity of studying the significance of the membrane lipids in supporting photochemical activities such as electron transport or photophosphorylation. The recent view of the biological membrane envisages that proper function of the membrane enzymes requires the liquid-crystalline state of the membrane lipids [9,10,12-14]. However, it would be difficult with ordinary higher plants and algae to relate an activity of the thylakoid membrane to the physical state of lipids, since due to their high linolenic acid contents the phase transition would appear at a low range of temperature where the thylakoid membrane ceases to function. On the contrary, it is highly possible that the thylakoid membrane of thermophilic algae exhibits the phase transition at room temperature, thus enabling us to study photochemical activities of the membrane and the physical state of the membrane lipids at the same range of temperature.

In the present work, thylakoid membranes having a high

activity of oxygen evolution were isolated from a thermophilic alga. The activity was maximal at 50-55°C but very low at room temperature. The phase transition and fatty acid composition of the membrane lipids were determined to relate them to the thermophilic character of the activity. The results suggest a unique environment for the oxygen evolving system with respect to lipid composition.

Material and Methods

A unicellular blue-green alga, which had been isolated from the Beppu Hot Springs, Kyushu, and tentatively identified as a species of *Synechococcus*, was employed as the source of thylakoid membranes. The alga was grown in an inorganic medium of Gafford and Craft [6] at 55°C with continuous bubbling with air containing 5% CO_2. The continuous illumination of 5,000 lux was provided from a bank of tungsten lamps.

Protoplasts were prepared according to the method of Biggins [1]. The cells were incubated for 1 h at 40°C in a medium containing 0.4 M mannitol, 40 mM phosphate buffer, pH 6.8, and 0.025% lysozyme. The protoplasts were collected by centrifugation, resuspended in the preparation medium containing 30% polyethylene glycol 4000, 0.4 M sucrose, 50 mM phosphate, pH 7.0 and 10 mM NaCl and passed through a French press at 400 kg/cm^2. The homogenate was centrifuged at 20,000 g for 15 min and the supernatant obtained was again centrifuged at 80,000 g for 1 hr. The thylakoid membranes or their fragments were suspended in the preparation medium and kept at -20°C.

Ferricyanide photoreduction was determined by following absorbance decrease at 420 nm with a Hitachi 356 spectrophotometer. The sample was illuminated with a broad band red light (600-800 nm) of 1.2×10^5 erg/cm^2 s. Oxygen evolution was determined with a Clark-type oxygen electrode. The reaction mixture contained 30% polyethylene glycol 4000, 50 mM 2-(N-morpholino)ethanesulfonic acid·NaOH, pH

5.5, 10 mM MgCl^2 and 1 mM ferricyanide.

Results and Discussion

The alga employed in the present work grows well at a high range of temperature up to 60°C, showing the maximum growth rate at 57°C. No measurable growth occurred at temperatures above 65°C. Fig. 1 shows effects of temperature on photosynthetic oxygen evolution in intact cells. The maximum rate was obtained at about 55°C. The activity showed a steeper decrease at temperatures above than below the optimum.

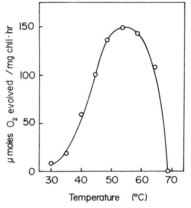

Fig. 1. Effects of temperature on photosynthetic oxygen evolution in intact cells. The reaction mixture contained 50 mM phosphate, pH 7.5, 10 mM NaCl, 5 mM NaHCO$_3$ and the algal cells containing 10 μg chlorophyll/ml. The samples were illuminated with white light of a saturating intensity (20,000 lux).

Dependence of photosynthesis of the thermophilic alga on the extremely high range of temperature, where the photosynthetic apparatus of ordinary plants is readily destroyed, indicates that the thylakoid membranes or functional proteins are more thermostable in the thermophilic alga than in mesophilic photosynthetic organisms. In fact, the thylakoid membranes isolated from the alga were found to be photochemically active at high temperature. The membranes could produce molecular oxygen in the light in the presence of ferricyanide as electron acceptor. Amounts of oxygen produced were stoichiometric to those of ferricyanide reduced. The activity was, therefore, determined either by measuring oxygen evolution polarographically or by following ferricyanide reduction spectrophotometrically.

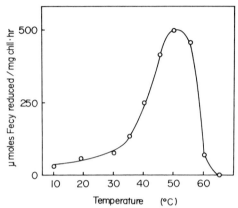

Fig. 2. Effects of temperature on ferricyanide photoreduction in the thylakoid membranes. Chlorophyll concentration, 11 µg/ml.

Fig. 2 shows temperature-dependence of ferricyanide photoreduction. Rates of ferricyanide reduction were very low at room temperature, but increased steeply as temperature was raised above 30°C to reach a maximum at 50-55°C. At this optimum range of temperature, the membranes could produce oxygen at rates of 100-150 µmoles oxygen evolved/mg chlorophyll·h, which are comparable to those of photosynthetic oxygen evolution in intact cells. This temperature profile clearly indicates that the thylakoid membranes are thermophilic, i.e., they are not only thermostable but also require high temperature to function normally.

The oxygen evolving system was irreversibly inactivated by incubating the membranes at temperatures higher than 55°C. Thus, the temperature-dependent changes in the activity at high temperature region are mainly determined by a thermal inactivation of some essential component of the electron transport such as a manganese-protein complex. Note, however, that the upper temperature limit for the activity was significantly higher in the algal membranes than in chloroplasts of ordinary plants. It has been shown previously that the oxygen evolving system was inactivated by incubating chloroplasts for a few minutes at temperature above 40°C [7].

The membranes could also photoreduce DCPIP at the high range of temperature, although rates of the dye reduction were always lower than those of ferricyanide photoreduction. On the other hand, methylviologen was ineffective as electron acceptor. This suggests that either photosystem I or the electron transport chain connecting photosystems I and II is inactive in the membranes. Compatible with this view were the effects of electron transport inhibitors (Fig. 3). Ferricyanide photoreduction was suppressed by DCMU, although about one half of the activity was relatively resistant to the poison (see ref. [5]). On the other hand, DBMIB which blocks electron transport at the oxidizing site of plastoquinone [2], was without effect on the activity at the concentration range where it is usually employed as an inhibitor of electron transport. Furthermore, spectrophotometric titrations of the membranes with redox reagents revealed that the membranes were largely devoid of cytochrome of f-type. It appears, therefore, that the electron transport between photosystems I and II is interrupted by a loss of the cytochrome

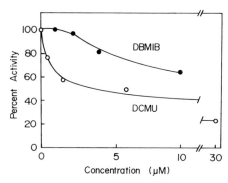

Fig. 3. Effect of DCMU and DBMIB on ferricyanide photoreduction in the thylakoid membranes. Temperature, 50°C. Chlorophyll concentration, 11 and 5 μg/ml for experiments with DCMU and DBMIB, respectively.

during the preparation of the membranes.

The above observations indicate that the oxygen evolving system under investigation involves a span of electron transport chain from the water-splitting enzyme to plastquinone and is driven by photosystem II alone. These electron carriers and the reaction center of photosystem II are tightly associated with the lipo-protein matrix of the thylakoid membranes. Temperature-dependence of such a multi-enzyme-lipid system is determined not only by the activation energies and thermostabilities of the membrane enzymes but also by the physical state of the membrane lipids [9,10,12-14].

The transition of the solid to the liquid-crystalline state occurs over a wide range of temperatures in biological membranes which are highly heterogenous in their lipid compositions. The phase transition in such a case consists of a lateral phase separation of the solid and the liquid-crystalline states. Schimishick and McConnell showed that the upper and lower boundaries of the transition could be detected by the spin labelling study [14].

The fluidity of the membrane lipids was studied with a spin probe, N-oxyl-4',4'-dimethyloxazolidine derivative of 5-ketostearic acid (5-SAL). When added to a suspension of the membranes, the radicals partitioned predominantly into the membranes and showed restricted isotropic motion. The hyperfine splitting, $2T_{II}$, was determined as a parameter of the mobility of the radicals in the membranes [10,13]. A plot of this parameter against the reciprocal of the absolute temperature shows a distinct discontinuity point between 30 and 40°C. Another break at about 10°C was also repeatedly observed. The two points correspond to T_f and T_s in the nomenclature of ref.[14]. The discontinuity point between 30 and 40°C, T_f, is the temperature of the transition between the liquid-crystalline state and the liquid crystal-solid state, and the point at 10°C, T_s, the temperature of the transition between the liquid/crystal-solid state and the solid state.

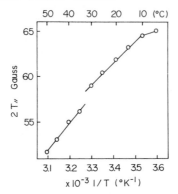

Fig. 4. Temperature dependence of the maximum hyperfine splitting of EPR spectra, $2T_{II}$. Concentrations of 5-SAL and the membranes were 50 μM and 143 μg chlorophyll/ml, respectively.

The phase transition temperature depends, among others, on the degree of unsaturation of fatty acids [8]. T_f and T_s would be far below room temperature in the thylakoid membranes of mesophilic plants, since they contain high concentration of poly-unsaturated fatty acids. With *Anacystis* thylakoid membranes which contain no linolenic acid, Murata *et al.* observed T_f, but not T_s, at room temperature [10]. Thus, appearance of the two phase transitions at temperatures above 0°C is a unique feature of the thylakoid membrane of the thermophilic alga, which can be related to the fatty acid composition, on the one hand, and to the thermophilic character of the oxygen evolving activity, on the other.

TABLE 1

Major fatty acids of the thylakoid membranes

Carbon atoms : double bonds	16:0	16:1	Unidentified	18:0	18:1	18:3
Percent of total fatty acid	42.8	17.4	1.8	3.1	33.5	1.8

Percent fatty acid composition of the lipids in the

thylakoid membranes determined by gas-liquid chromatography is shown in Table 1. The most abundant was palmitic acid (16:0), followed by two mono-unsaturated fatty acids, oleic acid (18:1) and palmitoleic acid (16:1). As expected from the high phase transition temperatures, the membranes contained only small quantities of poly-unsaturated fatty acid. Linolenic acid (18:3) and an unidentified fatty acid, which might be a di- or tri-unsaturated acid, comprise less than 2% of the total fatty acid present, respectively.

In order to compare with the temperature-dependent changes in the physical state of the membrane lipids, the rate of ferricyanide photoreduction is given in the Arrhenius plot in Fig. 5. A transition of the activation

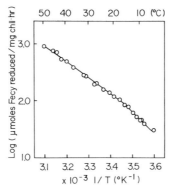

Fig. 5. Arrhenius plot of ferricyanide photoreduction in the thylakoid membranes. Chlorophyll, 10 µg/ml.

energy appeared at 14°C. Unexpectedly, however, no break was observed between 30 and 40°C, where a distinct discontinuity in the mobility of the membrane lipids appeared. The activation energy was 22.6 kJ/mol between 14 and 50°C and 35.2 kJ/mol below 14°C. Addition of methylamine or gramicidin J had no effect on the rate of ferricyanide photoreduction or its temperature dependence (c.f. ref. [11]). That the reaction was not limited by a diffusion process was confirmed by an experiment in which effects of ferricyanide concentration were examined.

In contrast to ferricyanide photoreduction, the entire electron transport system, which was monitored by methylviologen photoreduction in the intact cells, showed a transition of the activation energy at 32°C. Thus, the phase transition at T_f affects some other part of electron transport.

The absence of any change in the activation energy of ferricyanide photoreduction at T_f indicates that this part of electron transport is not influenced by the lateral phase separation of the solid and the liquid-crystalline states. This suggests that the oxygen evolving system does not depend on the physical state of the membrane lipids. However, the results obtained showed that the membrane lipids in the rigid gel state do not support normal function of the oxygen evolving system: the deviation of the activation energy from a linear Arrhenius plot at T_s indicates that the reaction system is affected when all the membrane lipids are solidifed. Between T_f and T_s, domains of the solid state and the liquid-crystalline state exist simultaneously. It is suggested that the oxygen evolving system is located predominantly in the domains of the liquid-crystalline state during the lateral phase separation. It may be that the environment surrounding the oxygen evolving system is particularly abundant in lipids with low melting point so that it resists solidification until temperature is lowered to reach T_s.

Acknowledgements

The authors wish to thank Dr R. Hirasawa for the EPR measurement and Mr J. Ohnishi and Dr M. Yamada for the determination of fatty acids, respectively. This work was supported in part by grants from the Ministry of Education, Japan.

References

1. Biggins, J. (1967). *Plant Physiol.* **42**, 1442-1446.

2. Böhme, H., Reimer, S. and Trebst, A. (1971). *Z. Naturforschg.* **26b**, 341-352.
3. Brock, T.D. (1967). *Nature* **214**, 882-885.
4. Castenholz, R.W. (1969). *Bacteriol. Rev.* **33**, 476-504.
5. Fujita, Y. and Suzuki, R. (1973). *Plant Cell Physiol.* **14**, 261-273.
6. Gafford, R.D. and Craft, C.E. (1959). USAF School of Aviation Medicine Rep. 58-124.
7. Katoh, S. and San Pietro, A. (1967). *Arch. Biochem. Biophys.* **122**, 144-152.
8. Ladbrook, B.D. and Chapman, D. (1969). *Chem. Phys. Lipids* **3**, 304-359.
9. McElhaney, R.N. (1976). *In* "Extreme Environments. Mechanisms of Microbial Adaptation" (M.R. Heinrich, ed.), pp. 255-281, Academic Press, New York.
10. Murata, N., Troughton, J.H. and Fork, D.C. (1975). *Plant Physiol.* **56**, 508-517.
11. Nolan, W.G. and Smillie, R.M. (1976). *Biochim. Biophys. Acta* **440**, 461-475.
12. Raison, J.K. and McMurchie, E.J. (1974). *Biochim. Biophys. Acta* **363**, 135-140.
13. Sackmann, E., Träuble, H., Galla, H.-J. and Overath, P. (1973). *Biochemistry* **12**, 5360-5369.
14. Schimishick, E.J. and McConnell, H.M. (1973). *Biochemistry* **12**, 2351-2360.

COMPARISON OF THE O_2 EVOLVING SYSTEM IN GRANAL AND AGRANAL CHLOROPLASTS OF MAIZE

G. HORVÁTH, M. DROPPA, A. JÁNOSSY*,
K. CSATORDAY and Á. FALUDI-DÁNIEL

*Institute of Plant Physiology and *Institute of Biophysics,
Biological Research Center, Hungarian Academy of Sciences,
Szeged (Hungary)*

Abstract

Chloroplasts of very high purity and intactness were prepared from enzymatically separated mesophyll protoplasts and bundle sheath cells.

O_2-evolving capacity and Hill-activity was practically absent from the bundle sheath chloroplasts, at the same time as Fe and Mn contents of the thylakoids were high, indicating the presence of the electron transport chain and the water splitting enzyme. The amount and proportion of the Fe and Mn associated with lipids was much lower in the bundle sheath than in the mesophyll thylakoids. This suggests that in bundle sheath thylakoids photosystem II is present but with an irregular structure and/or environment. The same is indicated by the fluorescence spectra of bundle sheath chloroplasts at room temperature by revealing the absence of F-685.

Introduction

Ecological and economic importance of the malate-type C_4 plants as well as the current interest in the relationship between the structure and function of chloroplast constituents incited a great number of works investigating the characteristics of granal and agranal chloroplasts [10]. The results reported in the literature are often conflicting which can be due most likely to the various degrees of cross contamination and intactness in chloroplast preparations.

A procedure developed in our laboratory allowed to isolate mesophyll and bundle sheath chloroplasts of maize with a very high purity and intactness. In such preparations the O_2 evolving system of these two types of chloroplasts could be characterized more reliably than earlier.

In this report data on O_2 evolving capacity and Hill-activity are given. Metal contents of thylakoids were determined after water-washing and lipid extractions, and fluorescence spectra of intact chloroplasts were recorded.

Material and Methods

Isolation of intact granal chloroplasts started from mesophyll protoplasts of maize (*Zea mays* L., convar KSC 360). Seedlings were grown in the greenhouse for 8-10 days. The leaves were subjected to enzymic digestion [12]. The medium contained 0.6 M D-sorbitol, 0.1 M K-phosphate, 0.5% Macerozyme R-10, 2.0% cellulase (Onozuka R-10) and 0.5% K-dextran sulfate, pH 5.8. Incubation was for 90 min at 37°C. Mesophyll protoplasts were washed in a medium of 0.6 M D-sorbitol, 10 mM NaCl, 5 mM $MgCl_2$, 1 mM $MnCl_2$, 2 mM EDTA, 0.4% bovine serum albumine and 5 mM HEPES, pH 7.5 [13]. The protoplasts were forced through a hypodermic needle of 0.5 mm diameter. The cell debris and chloroplasts were separated and the chloroplasts were collected by low speed centrifugations.

Isolation of intact agranal chloroplasts started from bundle sheath cells with tapered cell walls [2]. Bundle sheath strands were digested in a medium containing 0.4 M D-sorbitol, 0.1 M K-phosphate, 0.5% Macerozyme R-10, 1.0% cellulase and 1.0% helicase pH 5.8. (Detailed description see [6]).

The measurement of O_2 evolution was carried out by using a Clark-type electrode [4] in a temperature controlled cuvette (Rank-Brothers, Cambridge, U.K.) in a medium containing 0.1 M D-sorbitol, 10 mM K_2HPO_4, 20 mM NaCl, 4.0 mM $MgCl_2$, 2.0 mM EDTA, 20 mM HEPES, pH 7.5, 2.0 mM $K_3[Fe(CN)_6]$ and chloroplast suspension containing 50 µg chlorophyll. Final volume was 3.0 ml. Samples were illuminated by white light of saturating intensity, and the temperature was maintained at 30°C.

Hill-activity was measured by the method of Vernon and Shaw (14) with an Aminco DW-2/UV-VIS spectrophotometer equipped with a side illumination. The reaction was carried out in a medium containing 0.2 M sucrose, 30 mM KH_2PO_4, 10 mM dichlorophenol indophenol, pH 6.5, chloroplast suspension with 10 µM chlorophyll content and in some cases 0.5 mM 2.5-diphenyl carbazide. Actinic red light (> 630 nm) was of saturating intensity.

Metal content was determined in thylakoids washed thoroughly with distilled water and in thylakoids fixed in 0.1 M glutaraldehyde washed in 50 mM phosphate buffer pH, 7.2, and extracted with increasing concentrations of ethanol, ethanol/ethyl ether mixtures and ethyl ether. Both washed and extracted thylakoids were treated with 16% perchloric acid for 24 hours at 25°C [5]. The insoluble material was sedimented at 10 000 g and the supernatant was analysed for Mg, Mn, Fe and Ca content in a Varian 1000 atomic absorption spectrometer.

Fluorescence spectra were measured at room temperature with a Perkin-Elmer MPF - 3 spectrofluorimeter. Excitation wavelength was 435 nm. The spectra were corrected for reabsorption and the spectral sensitivity of the apparatus.

Results and Discussion

Chloroplast preparations from enzymatically separated protoplasts and bundle sheath cells with tapered cell walls had the advantage that cross-contamination was reduced to 5% and the high degree of envelope retention

(90-95%) protected the soluble membrane components from solubilization.

The yield was variable, depending on seasonal changes but regular checks under the electron microscope have shown that purity and intactness were fairly reproducible.

TABLE 1

O_2 evolution and Hill activity of intact mesophyll and bundle sheath chloroplasts (O_2 μmol/mg Chl.h; $4e^-$/mg Chl.h)

	O_2 evolution		Hill activity	
	-MA	+MA	-DPC	+DPC
M	49	150	172	172
BS	0	0	0	10

MA = 10 mM methylamine, DPC = 0.5 mM 2,5-diphenyl carbazide, M = mesophyll chloroplast, BS = bundle sheath chloroplast.

Photosynthetic O_2 evolution of mesophyll chloroplasts was around 50 μmoles O_2/mg Chl. h, which is in the range observed with intact spinach chloroplasts [11]. The O_2 output could be increased considerably by methylamine uncoupling (Table 1), which indicates that O_2 evolution was coupled with photophosphorylation prior methylamine treatment. Under similar conditions bundle sheath chloroplasts were completely incapable of evolving O_2 (Fig. 1). As shown in earlier studies [7] such chloroplasts exhibit 8-12% stacking *in situ*, about one-fifth of the stacking observed in mesophyll chloroplasts. Thus, the amount of stacked regions would be enough to provide for a detectable O_2 evolution. The complete lack of the O_2 evolving capacity means that stacked regions of bundle sheath chloroplasts are not functional analogues of grana.

Hill-activity ($H_2O \longrightarrow DCPIP$) of mesophyll chloroplasts was comparable to the methylamine uncoupled O_2 evolution, and was not stimulated by DPC. Bundle sheath chloroplasts exhibited a very low Hill-activity and only in the presence of DPC. This Hill-activity (never exceeding 5 per cent of that of mesophyll chloroplasts) can never be attributed to trace contamination by mesophyll chloroplast fragments, and it is at the same level as observed by Ku et al. [9].

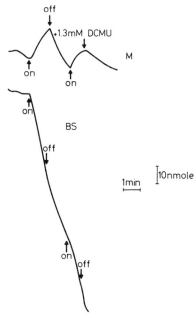

Fig. 1. O_2 evolution by mesophyll protoplasts and bundle sheath cells. Reaction conditions as given in "Methods". DCMU dissolved in ethanol was added in the reaction mixture to a final concentration of 1.3 mM DCMU and 0.3 per cent ethanol. The ethanol alone had no appreciable effect. M = mesophyll protoplast, BS = bundle sheath cells.

The reason why bundle sheath chloroplasts do not evolve O_2 has not been unravelled so far. Electron transport chain is not missing as shown in a classical work of Bazzaz and Govindjee [1].

In a previous X-ray microanalytical study we could show that in bundle sheath chloroplasts appreciable amounts of bound iron and manganese were present [8].

This observation is in accordance with the metal content of thylakoids as shown by atomic absorption analyses. As demonstrated by Table 2, iron content expressed in the number of atoms per mole of chlorophyll was even higher in the thylakoids of bundle sheath thylakoids, than in the thylakoids of mesophyll chloroplasts. The same holds for the manganese content of the membranes which can be an indication for the presence of some form of the water splitting enzyme. However, it has to be noted that the metal content is only a circumstantial evidence for the presence of cytochromes, P_{430} and water splitting enzyme. It is a completely open question whether these metal-containing proteins *per se* are active or present only in non-active rudimentary forms.

TABLE 2

Magnesium, iron, manganese and calcium content in thylakoids of mesophyll and bundle sheath chloroplasts (atom/mol chlorophyll)

		Mg	Fe	Mn	Ca	Protein-bound $\frac{Mg + Ca}{Fe + Mn}$
	Washed membrane	1.64	0.126	0.027	1.21	–
M	after extraction	0.53	0.102	0.013	0.97	13
	% with lipids	68(17)*	19	52	20	
	Washed membrane	2.19	0.266	0.040	1.74	–
BS	after extraction	0.92	0.260	0.034	1.04	7
	% with lipids	58(22)*	2	15	40	

*Mg apart from chlorophyll

Relative amounts of the metals extractable by lipid solvents from mesophyll and bundle sheath thylakoids are considerably different. In the thylakoid membranes of mesophyll chloroplasts much higher amounts of iron and manganese are in lipid environment than in the membranes of bundle sheath chloroplasts. This demonstrates that electron transport chain and the water splitting enzyme of bundle sheath thylakoids are in a molecular environment different from that provided by the membranes of mesophyll chloroplasts.

Magnesium and calcium content of bundle sheath thylakoids was higher than that of the mesophyll thylakoids. Magnesium (not belonging to the chlorophyll) and calcium of the chloroplast membrane indicates the coupling factor and ATP-ase content. The observation that in this respect bundle sheath membranes are superior to mesophyll membranes points to a higher photophosphorylative potential of the former.

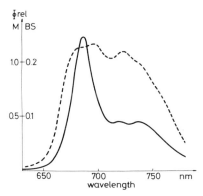

Fig. 2. Fluorescence spectra of intact mesophyll (M) and bundle sheath (BS) chloroplasts measured at room temperature. M = solid line, BS = dashed line.

Room temperature fluorescence spectra of intact mesophyll and bundle sheath chloroplasts (Fig. 2) show the well-known fact, that bundle sheath chloroplasts contain more chlorophyll absorbing at longer wavelength, characteristic to a surplus of photosystem I, and *vice versa* to a deficiency of photosystem II. New information from these curves is the trough at 685 nm in the fluorescence spectra of bundle sheath chloroplasts. The lack in F 685 is indicative to a photosystem II deficiency in the light harvesting complex [3].

The main conclusion from the data presented above is that the lack of the O_2 evolving capacity in bundle sheath chloroplasts might be due to an aberrant molecular structure of the thylakoid membrane.

Acknowledgements

The authors are indebted to Dr Mihály Varju, at the Central Institute for Chemical Research in Budapest for the atomic absorption measurements, to Mrs. Katalin Nyerges and Miss Zsuzsanna Toth for the skilful technical assistance.

References

1. Bazzaz, M.B. and Govindjee, R. (1973). *Plant Physiol.* **52**, 257-262.
2. Bialek, G.E., Horváth, G., Garab, Gy.I., Mustárdy, L.A. and Faludi-Dániel, Á. (1977). *Proc. Natl. Acad. Sci. Washington* **74**, 1455-1457.
3. Butler, W.L. and Kitajima, M. (1975). *Biochim. Biophys. Acta* **396**, 72-85.
4. Delieu, T. and Walker, D.A. (1972). *New Phytol.* **71**, 201-255.
5. Gross, E. (1972). *Arch. Biochem. Biophys.* **150**, 324-329.
6. Horváth, G., Droppa, M., Mustárdy, L.A. and Faludi-Dániel, Á. (1977), in preparation.
7. Horváth, G., Garab, Gy.I., Mustárdy, L.A., Halász, N. and Faludi-Dániel, Á. (1975). *Plant Sci. Lett.* **5**, 239-244.
8. Jánossy, A.G.S., Mustárdy, L.A. and Faludi-Dániel, Á. (1977). *Acta Histochem.* **58**, 317-323.
9. Ku, S.B., Gutierrez, M., Kanai, R. and Edwards, G.E. (1974). *Z. Pflanzenphysiol.* **72**, 320-337.
10. Laetsch, W.M. (1974). *Ann. Rev. Plant Physiol.* **25**, 27-52.
11. Nishimura, M. and Akazawa, T. (1975). *Plant Physiol.* **55**, 712-716.
12. Rathnam, C.K.M. and Edwards, G.E. (1976). *Plant Cell Physiol.* **17**, 177-186.
13. Reeves, S.G. and Hall, D.O. (1973). *Biochim. Biophys. Acta* **314**, 66-78.
14. Vernon, L.P. and Shaw, E.R. (1969). *Biochem. Biophys. Res. Comm.* **36**, 878-884.

STUDIES OF THE DONOR SIDE OF PHOTOSYSTEM II BY ABSORPTION SPECTROSCOPY

P. MATHIS

Service de Biophysique, Département de Biologie, Centre d'Etudes Nucléaires de Saclay, GIF-sur-YVETTE (France)

Introduction

In green plants, the electron transfer from the complex leading to oxygen evolution to the terminal stable electron acceptor ($NADP^+$) involves two photochemical electron transfer reactions. These reactions take place at two reaction centers which are associated with a segment of the electron transfer chain and with some (more or less specific) light-harvesting pigments, thus constituting two "photosystems", named photosystem I and photosystem II.

The primary photochemical reaction of photosystem II is an electron transfer from the excited primary donor P^+ to a primary acceptor (this defines the acceptor side of PS II); P^+ is normally reduced by a first secondary donor (located on the so-called donor side, directly connected to the oxygen-evolving complex). In the last few years, a large amount of informations has been gained on the acceptor side of PS II [reviewed in 16 and 26; newer work is cited in 21]. In brief, an electron is transferred from P to the primary stable acceptor A_1, probably a specialized plastoquinone molecule (often called Q). In the next step A_1^- reduces the secondary acceptor A_2 (probably also a plastoquinone). Two successive photoacts have to provide two electrons at the level of A_2 in order to lead to the next step, which involves the full reduction of one plastoquinone molecule in a large pool of this carrier

($PQ + 2e^- + 2H^+ \longrightarrow PQH_2$). This formally simple scheme does not account for a few important features which are poorly understood (effect of various inhibitors, of H^+, etc... References may be found in 20,21) and for the possible occurrence of a transitory step between P and A_1, as in bacteria. Such a step has been proposed by Van Best and Duysens [25].

$$\text{Acceptor side} \qquad\qquad \text{Donor side}$$
$$\text{pool of PQ}\ldots A_2\ldots A_1^-\ldots P^+\ldots D_1\ldots \ldots \ldots O_2 \text{ evolution}$$
$$\vdots$$
$$\text{(Accessory donors)}$$

The questions are much less simple as regards with the donor side, for which it is even difficult to draw a reasonably accurate scheme. In this article we will discuss a few properties of P and of its immediate donor(s) and present several data that we obtained by flash absorption spectroscopy.

The Primary Donor P

The primary donor P of PS II had revealed to be much more difficult to study than the primary donor in photosynthetic bacteria (P_{870} or equivalents) or in PS I (P_{700}), which in both cases is made of several interacting (bacterio-)chlorophylls. The identification of P as one or several molecules of chlorophyll a has been obtained by flash absorption spectroscopy. The flash leads to the disappearance of the absorption band of chlorophyll a [9-11] and to the appearance of an absorption band at 820 nm [12, 20,23], due to the radical-cation of Chl a [4,5]. The recovery is usually very fast (this aspect will be discussed later). Flash EPR experiments performed at low temperature also allowed to observe P^+ and to attribute it to the radical-cation of Chl a. The narrowing of the EPR spectrum, in comparison with Chl a *in vitro*, has been attributed to the delocalization of the lone electron over several Chl a molecules [18,34]. Similar observations have also been made upon continuous illumination of chloroplasts or of chloroplast particles [15,25,31,32], under conditions where the normal donors are inhibited. It is possible, however, that in these experiments the spectra include some contribution from Chl a molecules other than P^+ since an oxidation of some Chl a by P^+ has been observed at low temperature under oxidizing conditions [18,34].

All available evidence thus points to P being a specialized Chl a complex. In this complex chlorophyll must be in a very special environment since the potentials of the (Chl a / Chl a^+) couple is 0.52 volt in solution[28], whereas it has to be over 0.82 volt in order to allow for a 4-step oxidation of water into O_2 in the photosynthetic reactions.

The First Secondary Donor

Physiological conditions

The chemical nature of the first secondary donor D_1 has not been determined. The rate at which D_1 reduces P^+ is still a subject of controversy. Den Haan *et al.* [8] concluded to a reduction in less than 1 µs on the basis of an analysis of the fluorescence increase following a flash whereas Jursinic and Govindjee [14] concluded to a reduction in 6 µs from similar experiments in chloroplasts. Gläser *et al.* [11] proposed that P^+ was reduced in a biphasic manner, with $t_{\frac{1}{2}}$ = 35 and 200 µs. In a recent work the same group interpreted their saturation curves by assuming that a submicrosecond phase was not observed in their experiments [27]. In our experiments we measured P^+ formed by a flash in means of its absorption band at 820 nm; we concluded that P^+ is reduced by D_1 in less than 5 µs [20] (recent experiments by J.A. Van Best and P. Mathis indicate a major reduction in less than 0.4 µs). At 820 nm the absorption change induced by a flash (Fig.1) is attributed mostly to $P\text{-}700^+$. In the presence of ferricyanide the signal is smaller and recovers more slowly, as expected for $P\text{-}700^+$. If the chloroplast suspension is illuminated by a background of far-red light, some signal remains, whose fast phase ($t_{\frac{1}{2}}$ = 36 μs) cannot be due to P-700. This phase is of small amplitude (about 25% of P^+). It probably corresponds to the phase observed by Gläser et al. [11]. Some of the discrepancies can be explained on the basis of recent experiments by Renger et al. [27], who found that the magnitude of the 35 μs phase of decay of P^+ increases at a lower pH inside the thylakoid.

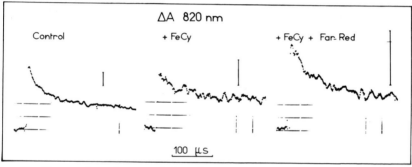

Fig. 1. Absorption changes induced at 820 nm in a suspension of spinach chloroplasts (osmotically broken and resuspended in a buffer 0.4 M sucrose - 50 mM Tricine, pH 7.6 - 10 mM KCl - 2 mM $MgCl_2$). Left trace: no addition. Median trace: + 5 mM potassium ferricyanide. Right trace: + 5 mM ferricyanide, with a background of far-red light. Average of 5 experiments, on a dark-adapted sample (3-10 min. at 9°C). The vertical bar represents 1.0×10^{-4} absorbance unit. Chlorophyll concentration: 34 μM. 1 cm path in the cuvette. Excitation at ≃ 600 nm by a 1 μs dye laser flash.

Low temperature

At low temperature oxygen evolution is inhibited, but some primary photoreactions can still occur in PS II. The primary acceptor is photoreduced; it has often been studied by an absorption shift around 550 nm (C-550). A parallel photooxidation of cytochrome b_{559} has been extensively studied [6,16]. Studies by flash absorption spectroscopy [13,22,23] performed at 77K led to a full support to the scheme (see [6]):

$$A_1 - P - \text{Cyt } b_{559} \underset{4 \text{ ms}}{\overset{}{\rightleftarrows}} A_1^- - P^+ - \text{Cyt } b_{559} \xrightarrow{15\text{-}20 \text{ ms}} A_1 - P - \text{Cyt } b^+_{559}$$

It has often been implied that the physiological donor D_1 was not operative at 77K and was replaced, although inefficiently, by cytochrome b_{559}. That view has been questioned by Visser [34], who proposed that the donor D_1 (named Z) was still operative and that cytochrome b_{559} was donating to D_1^+. When cytochrome b_{559} is chemically oxidized before the illumination at 77K it is apparently replaced by an unknown donor ($E_{\frac{1}{2}} \simeq 480$ mV) and, under strongly oxidizing conditions, the terminal donor is a chlorophyll molecule [18,34]. At less extreme temperature (-50°C) the chemical identity of the final electron donor depends upon the state of the O_2-evolving system: cytochrome b_{559} is the donor in the states S_2 and S_3 and another molecule is operative in the states S_0 and S_1 [33].

The study of PS II reactions at low temperature led to propose that cytochrome b_{559} was located on the donor side of PS II. Its direct involvement in the electron transfer steps leading to oxygen evolution is, however, highly questionable (see [7] for a review).

Chemical inhibition of oxygen evolution

Several chemical treatments lead to a more or less severe and irreversible damaging of the donor side of PS II. We have used three of them: low pH (4.0), high concentration of TRIS (0.2 M, at pH 9.0) or of hydroxylamine (5 mM). After the treatment at a low pH, flash excitation of chloroplasts leads to the observation of P^+ (at 820 nm) and A_1^- (around 320 nm), which decay nearly in parallel, in a back reaction with a half-time of about 150 μs [12,20]. An inefficient (unknown) donor has been inferred to contribute to the reduction of P^+ with a half-time of about 0.8 ms [12,30].

After treatment of chloroplasts or *Chlorella* cells with

3-10 mM hydroxylamine, oxygen evolution is blocked and P^+ can be detected by flash absorption spectroscopy ([20,21] and Fig. 2). It decays mostly with a half-time of 20-40 μs, and partly with a half-time of 100-200 μs, in parallel with a rise in the fluorescence yield of chlorophyll [8]. The donor D'_1 which is operating is again chemically unknown. It is not known whether it is chemicall different from D_1, or a less efficient state of D_1.

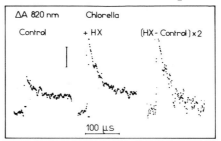

Fig. 2. Absorption changes induced in a suspension of *Chlorella* cells growth medium. Left trace: no addition (the signal is attributed to P_{700}^+ alone). Median trace: + 10 mM hydroxylamine. Right trace: difference between the two previous signals, with a twice expanded vertical scale (signal attributed to P^+). Average of 20 experiments per series. The vertical bar represents $1 \cdot 10^{-4}$ absorbance unit.

In case of treatment with TRIS we found that P^+ could be detected and decayed by a back-reaction with A_1^- ($t_{\frac{1}{2}}$ about 120 μs) provided that some positive charge(s) were stored on the donor side of PS II [12]. A flash given to dark-adapted TRIS-treated chloroplasts led to a reduction of P^+ in 30-40 μs, probably by the same donor D'_1 as after hydroxylamine treatment [21]. In similar conditions, cytochrome b_{559} is photooxidized by PS II continuous light [17] but this reaction is probably inefficient and D'_1 is probably not cytochrome b_{559}

The Oxygen-evolving System

Recently some flash-induced absorption changes have been observed in the ultraviolet, whose magnitude displays a 4-fold periodic behaviour [19,24]. The changes are stable for hundreds of milliseconds and their difference spectra are not typical, but they vary from one flash to the other. They have been attributed to transitions amongst the S states of the donor side of PS II, $i.e.$ variations in the oxidation state of the O_2-evolving complex. In Fig. 3 is presented an example of absorption changes induced by a sequence of short flashes in dark-adapted spinach chloroplasts. In this particular experiment the fast phase (due to the reoxidation of A_1^-, of A_2^-, and may be also to the transition from S_4 to S_0) is not fully resolved. At the present time it is not possible to derive a chemical information on the S states from the shape of the difference spectra [19,24] although this might occur in the future if more extended difference spectra can be measured.

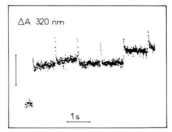

Fig. 3. Absorption changes at 320 nm induced in a suspension of spinach chloroplasts (same conditions as in Fig. 1) by a train of 6 short xenon flashes. Addition of 40 μm potassium ferricyanide and 10^{-7} M gramicidin D. Average of 100 experiments. The vertical bar represents $2 \cdot 10^{-4}$ absorbance units. Chlorophyll concentration: 14 μM. Cuvette: 1 cm.

Conclusion

A rather large amount of information has been gained on the donor side of PS II by absorption spectroscopy. It is, however, still impossible to draw a unique reaction scheme for that part of the electron transfer chain. A major problem is to know whether the two donors which have been inferred (the "fast" donor D_1 and the "slow" donor D'_1) are chemically distinct or correspond to two configurations of a unique molecule or even of the reaction center itself. Previous experiments at low temperature led indeed to the proposal of two configurations for the reaction center of PS II [33]. It is possible to write the following two schemes for PS II at physiological temperatures:

"Fast donation" state:

$$A_1 \underset{}{\overset{< 0.4\mu s}{—\!\!—}} P —\!\!— (D_1) —\!\!— D_2 —\!\!— (O_2 \text{ evolving complex})$$
$$\text{(EPR II vf)}$$

"Slow donation" state:

$$A_1 \underset{}{\overset{35 \mu s}{—\!\!—}} P —\!\!— D'_1 —\!\!— D'_2$$

In the first case the secondary donor D_2 may give rise, in its oxidized state, to the EPR signal IIvf which is probably located between P and the oxygen-evolving complex [1] and is probably not donating directly to P^+ [3]. A similar EPR species, names signal II_f, has been studied in TRIS-treated chloroplasts [2]. It is not clear whether it corresponds to D'^+_1 or D'^+_2, and it is also not known whether it originates from the same chemical species as signal IIvf.

An important point which comes out of kinetic studies is the rapidity of the reactions on the donor side of PS II. This rapidity is needed to efficiently compete with the fast back reaction between P^+ and A^-_1 (half-time 120-150 μs). If a 35-μs phase of reduction of P^+ were dominant in physiological conditions, it would correspond to a large number of "misses" (about 20%) because the back reaction would annihilate the primary charge separation in the same fraction of the reaction centers.

Acknowledgements

Thanks are due to Dr. J. Haveman and to Dr. Van Best who collaborated in the experiments by flash absorption spectroscopy.

References

1. Babcock, G.T., Blankenship, R.E. and Sauer, K. (1976). *FEBS Lett.* **61**, 286-289.
2. Babcock, G.T. and Sauer, K. (1975). *Biochim. Biophys. Acta* **376**, 329-344.
3. Blankenship, R.E., McGuire, A. and Sauer, K. (1977). *Biochim. Biophys. Acta* **459**, 617-619.
4. Bobrovskii, A.P. and Kholmogorov, V.A. (1971). *Optics and Spectroscopy* **30**, 17-20.
5. Borg, D.C., Fajer, J., Felton, R.H. and Dolphin, D. (1972). *Proc. Natl. Acad. Sci. Washington* **67**, 813-820.
6. Butler, W.L. (1973). *Acc. Chem. Res.* **6**, 177-184.
7. Cramer, W.A. and Whitmarsh, J. (1977). *Ann. Rev. Plant Physiol.* **28**, 133-172.
8. Den Haan, G.A., Duysens, L.N.M. and Egberts, D.J.N. (1974). *Biochim. Biophys. Acta* **368**, 409-421.
9. Döring, G., Renger, G., Vater, J. and Witt, H.T. (1969). *Z. Naturforschg.* **24b**, 1139-1143.
10. Döring, G., Stiehl, H.H. and Witt, H.T. (1967). *Z. Naturforschg.* **22b**, 639-644.
11. Gläser, M., Wolff, C., Buchwald, H.E. and Witt, H.T. (1974). *FEBS Lett.* **42**, 81-85.
12. Haveman, J. and Mathis, P. (1976). *Biochim. Biophys. Acta* **440**, 346-355.
13. Haveman, J., Mathis, P. and Vermeglio, A. (1975). *FEBS Lett.* **58**, 259-261.
14. Jursinic, P. and Govindjee (1977). *Biochim. Biophys. Acta* **461**, 253-267.
15. Ke, B., Sahu, S., Shaw, E. and Beinert, H. (1974). *Biochim. Biophys. Acta* **347**, 36-48.
16. Knaff, D.B. (1977). *Photochem. Photobiol.* in press.
17. Knaff, D.B. and Arnon, D.I. (1969). *Proc. Natl. Acad. Sci. Washington* **64**, 715-722.
18. Malkin, R. and Bearden, A.J. (1975). *Biochim. Biophys. Acta* **396**, 250-259.
19. Mathis, P. and Haveman, J. (1977). *Biochim. Biophys. Acta* **461**, 167-181.
20. Mathis, P., Haveman, J. and Yates, M. (1976). Brookhaven Symp. in Biol., **28**, 267-277.
21. Mathis, P. and Van Best, J. (1977). Proceedings of the IV. International Photosynthesis Congress, Reading, in press.
22. Mathis, P. and Vermeglio, A. (1974). *Biochim. Biophys. Acta* **368**, 130-134.
23. Mathis, P. and Vermeglio, A. (1975). *Biochim. Biophys. Acta* **396**, 371-381.
24. Pulles, M.P.J., Van Gorkom, H.J. and Willemsen, J.G. (1976). *Biochim. Biophys. Acta* **449**, 536-540.
25. Pulles, M.P.J., Van Gorkom, H.J. and Verschoor, G.A.M. (1976). *Biochim. Biophys. Acta* **440**, 98-106.
26. Radmer, R. and Kok, B. (1975). *Ann. Rev. Biochem.* **44**, 409-433.
27. Renger, G., Gläser, M. and Buchwald, H.E. (1977). *Biochim. Biophys. Acta* in press.
28. Stanienda, A. (1965). *Z. Phys. Chem.* **229**, 257-272.

29. Van Best, J.A. and Duysens, L.N.M. (1977). *Biochim. Biophys. Acta* **459**, 187-206.
30. Van Gorkom, H.J., Pulles, M.P.J., Havemann, J. and Den Haan, G.A. (1976). *Biochim. Biophys. Acta* **423**, 217-226.
31. Van Gorkom, H.J., Pulles, M.P.J. and Wessels, J.S.C. (1975). *Biochim. Biophys. Acta* **408**, 331-339.
32. Van Gorkom, H.J., Tamminga, J.J. and Haveman, J. (1974). *Biochim. Biophys. Acta* **347**, 417-438.
33. Vermeglio, A. and Mathis, P. (1973). *Biochim. Biophys. Acta* **314**, 57-65.
34. Visser, J.W.M. (1975). Thesis, State University, Leiden.

FLUORESCENCE AND ABSORBANCE CHANGES IN TRIS-WASHED CHLOROPLASTS

H.J. VAN GORKOM, M.P.J. PULLES and A.-L. ETIENNE

Department of Biophysics, Huygens Laboratory of the State University, Leiden (The Netherlands) and Laboratoire de Photosynthèse du CNRS, Gif-sur-Yvette (France)

Introduction

The kinetics of photosystem II electron transport are relatively well known, thanks to many detailed studies on oxygen evolution and on prompt and delayed chlorophyll fluorescence. At least ten different electron transfer reactions are involved, namely the primary charge separation, the reduction of the oxidized donor, the two-step reduction of the secondary electron acceptor and its reoxidation by the plastoquinone pool, the four-step oxidation of the oxygen evolving enzyme and its discharge. We now feel rather confident that a probably dimeric chlorophyll a molecule acts as the primary electron donor, P 680, and that the primary acceptor, Q, is a plastoquinone molecule bound in such a way that it can be reduced only to the semiquinone anion [11]. Absorbance changes in the near ultra-violet which are correlated with the charge accumulation both on the secondary acceptor, R, and on the oxygen evolving enzyme, M, have been described [8,9]. R^- appeared to be a plastosemiquinone anion, and by lack of evidence to the contrary we assume that R^{2-} is a plastohydroquinone, but the protonation step is still an open question. Our main interest now, however, is to identify the molecular species involved in electron transport at the oxidizing site of the photosystem II reaction centre.

All known components of the photosynthetic electron transport chain cause absorbance changes in the near ultraviolet. The period four oscillations of absorbance and also some indications of possible absorbance changes caused by signal IIvf suggest that the oxygen evolving apparatus is no exception. Events at the acceptor site dominate absorption difference spectra in this wavelength region. Therefore we started out to try whether the contribution by absorbance changes of Q could be determined on the basis of the chlorophyll fluorescence yield [3].

The Φ/Q-relation

We succeeded in obtaining both a fluorescence variation of reasonable amplitude and an absorbance difference spectrum with very little contribution of other components than Q under the following conditions: TRIS-washed chloroplasts with ferricyanide to oxidize P 700, DCMU to prevent reduction of R, and as an electron donor tetraphenylboron, because it can be used in the presence of ferricyanide. The light-induced difference spectrum obtained under these conditions is shown in Fig. 1, together with the differ-

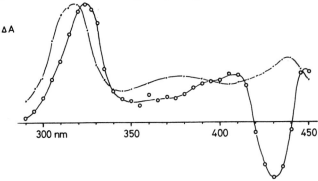

Fig. 1. Open circles: Flash-induced absorbance changes in TRIS-washed chloroplasts in the presence of 2.5 mM ferricyanide, 5 μM DCMU and 50 μM tetraphenylboron; chlorophyll concentration 250 μM. Corrected for particle flattening. Dots: Plastosemiquinone anion minus plastoquinone, from ref. [2].

ence spectrum of an ionic plastosemiquinone minus plastoquinone measured *in vitro* [2]. As observed earlier in PS II particles [11], the difference spectrum is somewhat shifted to longer wavelengths and there is an unexplained negative band around 430 nm which may be caused by C 550 or P 680. Also in the green and red regions absorbance changes unrelated to Q were found to be almost negligible. The difference spectrum of tetraphenylboron oxidation is essentially zero at wavelengths longer than 290 nm [11].

Then we measured fluorescence yield at 680 nm and absorbance at 320 nm simultaneously. As shown in Fig. 2,

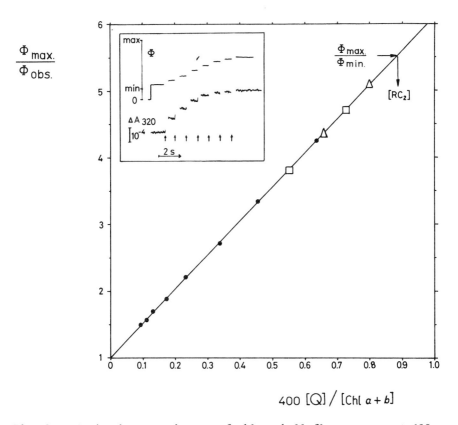

Fig. 2. Relation between changes of chlorophyll fluorescence at 680 nm and absorbance at 320 nm induced by a series of non-saturating flashes. Conditions as for Fig. 1. Different symbols refer to different experiments.

a linear relation was found between the reciprocal of fluorescence and the concentration of Q. If we assume 1) that all Q is oxidized at the minimum fluorescence level obtained in the presence of ferricyanide, 2) that [Q] is zero at the maximum fluorescence yield obtained in the presence of dithionite and 3) a differential extinction coefficient of 13 mM^{-1} cm^{-1} at 320 nm for the reduction of Q [2], a PS II reaction centre concentration of 1 per 400 chlorophyll $a + b$ is obtained[a]. This value seems very reasonable in view of the experimental error limit of at least 10% involved in the absolute calibration. Relative values of [Q] may be estimated with much better precision. As pointed out by Vredenberg and Duysens [12] the linearity of 1/F *versus* quencher concentration may be expected if excitation trapping by the reaction centres is a purely statistical process without limitations imposed by structural inhomogeneity in the fluorescing chlorophyll antenna. In this case the well-known Stern-Volmer equation is obtained:

$$\frac{F_{obs}}{F_{max}} = \frac{1}{1 + kQ}$$

with k, for normalized [Q], being equal to

$$\frac{F_{max}}{F_{min}} - 1 \quad (b)$$

This interpretation implies that the rate of exciton trapping by Q, which is proportional to $F_{max} - F_{obs}$ [7] should follow the relation

$$\frac{dQ}{dt} = \frac{kQ}{1 + kQ} \cdot cI$$

where cI indicates proportionality to light intensity. Under the conditions of our experiment, however, the fluorescence does not rise to the maximum level. Apparently a fraction α of the reaction centres is unable to stabilize the primary charge separation and remains in the quenching state, unless Q is reduced chemically with dithionite. The kinetics of the fluorescence rise then should be determined by the relation

$$-\frac{dQ}{dT} = \frac{Q - \alpha}{Q} \cdot \frac{kQ}{1 + kQ} \cdot cI$$

The coincidence of the experimental points with the theoretical curve in Fig. 3 shows that these kinetics are indeed observed.

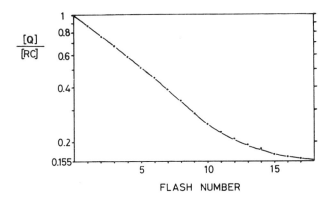

Fig. 3. Measured (dots) and predicted (line, see text) kinetics of the fluorescence rise, measured as in Fig. 2.

We conclude that under the conditions used exciton trapping by the PS II centres is indistinguishable from a random process in a statistical pigment bed; there is no indication of inhomogeneity or separation between photosystem II "units". Pigments which cannot transfer excitation energy to PS II reaction centres apparently do not contribute to fluorescence at 680 nm. In addition it seems that, except for the fraction α, the PS II centres are homogeneous with respect to the quantum yield of the formation of the non-trapping state P 680 Q^-. These conclusions, the last one in particular, may not apply to intact chloroplasts.

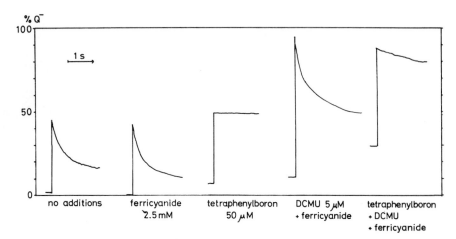

Fig. 4. Flash-induced redox changes of Q calculated from fluorescence changes according to Fig. 2. Additions as indicated.

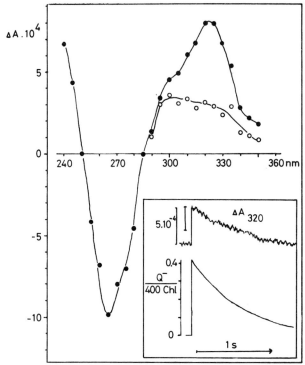

Fig. 5. Closed circles: Spectrum of the slowly (0.5 s) decaying flash-induced absorbance changes, corrected for particle flattening. Open circles: The same measurements after subtraction of absorbance changes caused by Q, see text.

Absorbance changes

Thus making use of fluorescence measurements we are now trying to unravel the absorbance changes in the near ultra-violet in TRIS-washed chloroplasts.

After illumination with a saturating flash 50% of the reaction centres exhibit a slow reoxidation of Q^- with half-times of 0.7 and several seconds (Fig. 4). The 0.7 s decay time is similar to the one reported for the so-called signal IIf [1]. Addition of 2.5 mM ferricyanide did not accelerate this decay dramatically, indicating that the decay is not limited by a lack of acceptor. The decay may largely be caused by back reaction, since addition of a donor (tetraphenylboron) inhibited the decay strongly. In the other half of the reaction centres Q^- apparently is too short-living to be observed in this experiment, and is seen only after addition of DCMU to slow down its reoxidation.

Fig. 5 shows the absorbance changes correlated with the slow fluorescence decay in the presence of ferricyanide. From the fluorescence change one may calculate that only 2/3 of the absorbance change at 320 nm is caused by the slow reoxidation of Q^- (indicated by the bar near the absorbance kinetics). Subtracting the absorbance changes caused by this amount of Q reduction (spectrum of Fig. 1) yields a spectrum which is definitely different from the one obtained upon reduction of plastoquinone to its semi-quinone anion, and may at least in part be attributed to the species responsible for EPR signal IIf on the basis of its decay time.

Illumination by repetitive flashes spaced at a time too short to permit the slow phase of Q^- reoxidation to proceed to a significant extent, induced the absorbance changes shown in Fig. 6. This result confirms earlier observations [4,10]. The absence of concomitant changes of fluorescence yield and the shape of the difference spectrum suggest that the changes are caused by the

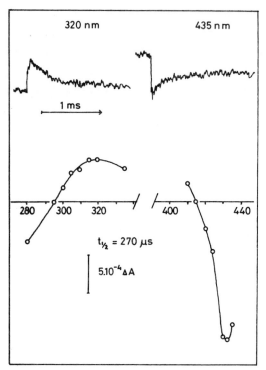

Fig. 6. Rapidly decaying absorbance changes observed with repetitive flashes at a frequency of 2 Hz. No additions. Spectrum corrected for particle flattening.

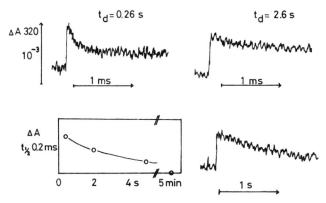

Fig. 7. Dark time dependence of the fast decaying absorbance changes. Conditions as in Fig. 6.

reaction P 680 Q → P 680$^+$ Q$^-$ and the similarity of the decay kinetics at 320 and 430 nm suggest that these kinetics reflect the back reaction. The decay time of these changes varied between 150 and 300 μs. The amplitude of the absorbance changes suggested that P 680$^+$ Q$^-$ was observed in at most 50% of the reaction centres.

With increasing dark time between the flashes the amplitude of these absorbance changes decreased to zero, reaching one half of the maximum amplitude at about 3 s dark time (Fig. 7). Since a rapid fluorescence rise then is observed, as reported earlier by Jursinic and Govindjee [16], we conclude that the secondary donor Z is still operative in these reaction centres but remains oxidized for 3 s. Fluorescence does decay within a few milliseconds, without a concomitant absorbance decrease at 320 nm, indicating that Q$^-$ is reoxidized by R. So after a few milliseconds the state Z$^+$ P 680 Q R$^-$ is obtained and decays to Z P 680 Q R with a half-time of 3 s. Jursinic and Govindjee [16] and Haveman and Mathis [4] reported that the 200 μs recombination of P 680$^+$ Q$^-$ is not observed after the first two flashes in dark-adapted chloroplasts, whereas we needed only one flash preillumination. The discrepancy may perhaps indicate that the chloroplasts used by these authors had much less of the slow and stable forms of signal II.

Taken together, our data indicate that in TRIS-washed chloroplasts the photosystem II reaction centres occur in two different types, present in about equal amounts. One type is inactive because the oxidized secondary electron donor, Z$^+$, is not reduced. Upon addition of artificial donors electron transport is restored in these reaction centres. The other type is inactive because the reduced primary acceptor, Q$^-$, is not reoxidized. Addition of ferricyanide does not relieve the inhibition. Addition of an electron donor stabilizes Q$^-$ even further, probably by preventing the back reaction. No absorbance changes of the secondary acceptor R were observed in these centres, which clearly do not participate in the net electron flow and oscillations of R occurring in TRIS-washed chloroplasts with an artificial electron donor. Signal IIf probably appears in these centres only, since its reported decay time agrees well with the reoxidation time of Q^- in this type of centres, while the only available secondary electron donor in the other type of centres remains oxidized four times longer.

We do not know whether the two types of centres are interconvertible states of the same structure or intrinsically different structures, which might even exist as such in intact chloroplasts.

Footnotes

(a) An uncertainty is introduced by the fact that only 85 to 95% of Q is photoreduced. The missing fraction cannot actually be shown to follow the linear relation with 1/F.
(b) The same relation is obtained from Joliot's equation [5] if p, the probability of energy transfer out of a closed unit, is $1-(F_{min}/F_{max})$.

Acknowledgements

Dolf Santbulte and Guido Swart participated in some experiments. This investigation was supported by the Foundation for Biophysics, financed by the Netherlands Organization for the Advancement of Pure Research.

References

1. Babcock, G.T. and Sauer, K. (1975). *Biochim. Biophys. Acta* **376**, 329-344.
2. Bensasson, R. and Land, E.J. (1973). *Biochim. Biophys. Acta* **325**, 175-181.
3. Duysens, L.N.M. and Sweers, H.E. (1963). In "Studies on Microalgae and Photosynthetic Bacteria" (Japanese Society of Plant Physiologists, ed.), pp. 353-372, University of Tokyo Press, Tokyo.
4. Haveman, J. and Mathis, P. (1976). *Biochim. Biophys. Acta* **440**, 346-355.

5. Joliot, A. and Joliot, P. (1964). *C. r. Acad. Sci. Paris* **258** D, 4622-4625.
6. Jursinic, P. and Govindjee (1977). *Biochim. Biophys. Acta* **461**, 253-267.
7. Malkin, S. (1966). *Biochim. Biophys. Acta* **126**, 433-442.
8. Mathis, P. and Haveman, J. (1977). *Biochim. Biophys. Acta* **461**, 167-181.
9. Pulles, M.P.J., Van Gorkom, H.J. and Willemsen, J.G. (1976). *Biochim. Biophys. Acta* **449**, 536-540.
10. Renger, G. and Wolff, C. (1976). *Biochim. Biophys. Acta* **423**, 610-614.
11. Van Gorkom, H.J. (1976). Thesis, University of Leiden.
12. Vredenberg, W.J. and Duysens, L.N.M. (1963). *Nature* **197**, 355-357.

REDUCTION OF PHEOPHYTIN IN THE PRIMARY LIGHT REACTION OF PHOTOSYSTEM II

V.V. KLIMOV, A.V. KLEVANIK, V.A. SHUVALOV and A.A. KRASNOVSKY*

Institute of Photosynthesis, USSR Academy of Sciences, Pushchino, and *A.N. Bakh Institute of Biochemistry, USSR Academy of Sciences, Moscow (USSR)

Introduction

The photochemical reduction of pheophytin and bacteriopheophytin has been shown *in vitro* [2,10,12]. In reaction centres of photosynthetic bacteria bacteriopheophytin a [7,13,16,17,19,22] and bacteriopheophytin b [8,18] act as an intermediary electron carrier between a bacteriochlorophyll dimer and the "primary" electron acceptor, a complex of ubiquinone and Fe. When the ubiquinone is in the reduced form, the photoaccumulation of reduced bacteriopheophytin can be observed [7,8,16-19,22]. In various species of green plants from 1.5 to 2.3 molecules of pheophytin has been found per 100 molecules of chlorophyll [9]. In photosystem II of green plants the photoreduction of the primary electron acceptor, Q (plastoquinone), is accompanied by a blue shift of absorption bands at 545 and 685 nm [20,21], which can belong to a bound or aggregated form of pheophytin in reaction centres of photosystem II [20]. The photoreduction of pheophytin may be observed in photosystem II in preparations from pea chloroplasts at 20°C [6]. In this work a reversible reduction of pheophytin in the primary light reaction of photosystem II in pea subchloroplast particles at redox potentials ($E_{1/2}$) from -50 to -550 mV (when Q is in the reduced form) is demonstrated. This photoreaction is observed at -170°C as well as at

20°C and is accompanied by a 2-3-fold decrease in the chlorophyll fluorescence yield.

Material and Methods

The "heavy" chloroplast fragments, enriched in photosystem II, were isolated using a treatment of pea chloroplasts with digitonin (0.4%) and Triton X-100 (0.1%) followed by fractional centrifugation [14]. The fraction precipitated for 45 min at 20000 g and designated as "DT-20 fragments" [14] was used. The chlorophyll-protein complexes of photosystems I and II and the "accessory" complex were prepared using the chromatography on DEAE-cellulose of pea chloroplasts treated with 3% Triton X-100 [15]. The absorbance changes (ΔA) and the changes in the chlorophyll fluorescence yield (ΔF), induced by continuous actinic light, were measured with the phosphoroscopic technique described earlier [6,7]. The measurements were made in a 10 mm cuvette, in which the $E_{1/2}$ value of the medium was registered under anaerobic conditions [7,16].

Results

In the DT-20 fragments at the medium with $E_{1/2}$ of +400 mV the actinic light induces a reversible increase, related to the photoreduction of Q [1], in the chlorophyll fluorescence yield (Fig. 1). Under these conditions the light-induced ΔA observed (Fig. 1) are similar to the positive ΔF in their kinetic according to experiments with higher concentrations of fragments. The spectrum of these ΔA is characterized by negative bands at 550 and 690 nm and by two positive bands at 542 and 680 nm (Fig. 2 A). It corresponds to the blue shift of absorption bands at 545 and 685 nm, which accompanies the photoreduction of Q [20, 21].

At $E_{1/2}$ below -50 mV the chemical reduction of Q [1,5] is accompanied by the increase in the chlorophyll fluorescence yield to its maximum level (Fig. 1). The actinic

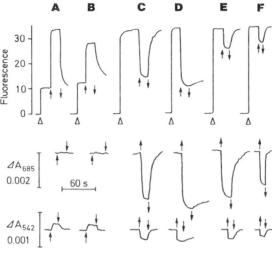

Fig. 1. Kinetics of the light-induced fluorescence changes (ΔF) and of the absorbance changes (ΔA) at 685 and 542 nm of pea photosystem II preparations. (A-d), DT-20 fragments suspended in 20 mM TRIS buffer, pH 8.5 (chlorophyll concentration 20 μg/ml), at 20°C. (A), Without additions. (B), In the presence of 5 μM ferricyanide and 1 μM DCMU at $E_{1/2}$ = +400 10 mV. (C,D), In the presence of 1 μM indigodisulphonate, 0.5 μM methylviologen and 0.2-2.0 mg/ml dithionite: (C), at $E_{1/2}$ = -200±30 mV, (D), at $E_{1/2}$ = -400±20 mV. (E), The chlorophyll-protein complex of photosystem II suspended in the same buffer (chlorophyll concentration 7 μg/ml) in the presence of 1 μM methylviologen and 2.0 mg/ml dithionite at $E_{1\,2}$ = -490±10 mV at 20°C. (F), The film, obtained by drying the suspension of DT-20 fragments in the presence of 0.5 mg/ml dithionite and 1 μM DCMU under anaerobic conditions (absorbance of the film at 680 nm was 1.6), the measurements were carried out at -170°C. Δ, monochromatic light (480 nm, a half bandwidth of 10 nm, intensity of 7 erg·cm^{-2}·s^{-1}), exciting the chlorophyll fluorescence (λ>660 nm), on. ↑, actinic light (λ>620 nm, 1.9·10^5 erg·cm^{-2}·s^{-1}), on; ↓, actinic light off.

light induces a reversible decrease in the fluorescence almost to the level of "dark" fluorescence registered at $E_{1/2}$ of +400 mV (Fig. 1). (The effect of the irreversible light-induced decrease in the chlorophyll fluorescence yield in both the pea chloroplasts and digitonin fragments of chloroplasts under reductive conditions has been described in detail earlier [5]). Under these conditions no ΔA, related to the photoreduction of Q, are observed but new reversible light-induced ΔA, which are similar in their kinetics to the negative ΔF, are detected (Fig. 1).

The spectrum of these ΔA at an $E_{1/2}$ of -200 mV is characterized by bleaching of the absorption bands at 545 and 685 nm as well as at 408, 422-428 and 518 nm and by the development of a band at 676 nm and of broad bands at 450, 658 and with $\lambda > 695$ nm (Fig. 2 B). The dark decay of the light-induced ΔF and ΔA at 685 and 542 nm is slowed down when $E_{1/2}$ is lowered from -200 to -400 mV (Fig. 1). This indicates the reductive nature of the photoprocess. At an $E_{1/2}$ below -450 mV these spectral changes in DT-20 fragments were largely irreversible in the dark. However, in the chlorophyll-protein complex of photosystem II the light-induced ΔA, accompanied by the decrease in the fluorescence yield, were reversible in the dark up to $E_{1/2}$ = -550 mV. The spectrum of these ΔA at an $E_{1/2}$ of -490 mV (Fig. 2 C) is similar to the difference absorption spectrum of DT-20 fragments at an $E_{1/2}$ of -200 mV (Fig. 2 B). The value of the light-induced ΔA (per mg chlorophyll) in the complex of photosystem II is three times higher than that in DT-20 fragments (Figs. 1 and 2). In the "accessory" complex and in the complex of photosystem I the photoreaction described was not observed. This can show its relation to photosystem II.

In DT-20 fragments the initial rate of the light-induced decrease in the chlorophyll fluorescence yield at $E_{1/2}$ = -450 mV $(\frac{\Delta Fi}{\Delta t})_{-450}$ and that of the light-induced increase in the fluorescence yield, following the Q photoreduction, at $E_{1/2}$ = +300 mV $(\frac{\Delta Fi}{\Delta t})_{+300}$ were just the same when the intensity of actinic light used at $E_{1/2}$ = -450 mV (I_{-450}) was 500 times higher than that used at $E_{1/2}$ = +300 mV (I_{+300}). When the rate of direct photoreaction is higher than that of dark reactions (that was observed in our experiments), $\frac{\Delta Fi}{\Delta t}$ is directly proportional to $\varphi \cdot I \cdot \Delta F_{max}$. Since the values of the maximum ΔF (ΔF_{max}) at $E_{1/2}$ = -450 mV and at $E_{1/2}$ = +300 mV are nearly the same (Fig. 1), the ratio of quantum yields of the

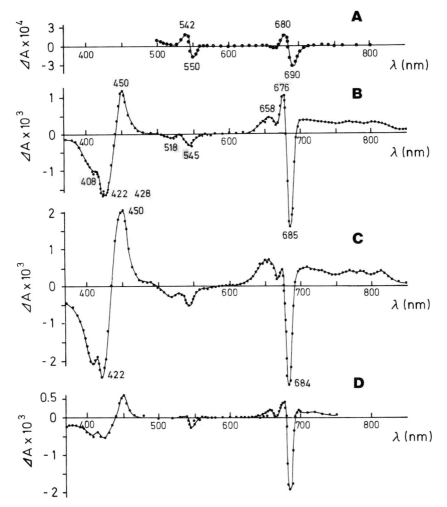

Fig. 2. Difference absorption spectra ("light minus dark") of pea photosystem II preparations. (A, B), Suspension of DT-20 fragments (chlorophyll concentration 20 µg/ml) at $E_{1/2}$ = +400 mV (A) and at $E_{1/2}$ = -200 mV (B). (C), Suspension of the chlorophyll-protein complex of photosystem II (chlorophyll concentration 7 µg/ml) at $E_{1/2}$ = -490 mV at 2°C. The additions as in Fig. 1. (D), The film, obtained by drying the suspension of DT-20 fragments in the presence of 0.5 mg/ml dithionite and 1 µM DCMU under anaerobic conditions (absorbance of the film at 680 nm was 1.6), the measurements were carried out at -17°C. To avoid ΔA related to the P_{700} photooxidation the spectrum of Fig. 2A was measured in the presence of weak far-red background light (λ>710 nm) [20,21] which oxidized P_{700} and did not induce ΔA related to the Q photoreduction.

photoreactions at -450 mV and +300 mV $(\frac{\varphi-450}{\varphi+300})$, when $(\frac{\Delta Fi}{\Delta t})_{-450} = (\frac{\Delta Fi}{\Delta t})_{+300}$, is inversely proportional to the ratio $\frac{I_{-450}}{I_{-300}}$, i.e. $\frac{\varphi_{-450}}{\varphi_{+300}} \approx \frac{1}{500}$. Thus, if $\varphi_{+300} \approx 1$ [21] then $\varphi_{-450} \approx 0.002$.

The light-induced A and F of DT-20 fragments at an $E_{1/2}$ below -50 mV were observed also in the presence of 3-(3,4-dichlorophenyl)-1,1-dimethylurea (DCMU). In 80% glycerol they were recorded at -80°C as well as at 20°C. In the film obtained by drying the suspension of DT-20 fragments in the presence of dithionite and DCMU this photoreaction is also observed at -170°C (Figs 1 F and 2 D) as well as at 20°C. The difference absorption spectrum ("light minus dark") of this film at -170°C resembles that of the suspension of DT-20 fragments at 20°C (Figs 2 B and 2 D), but the ΔA values of the film in the region of 400-500 nm are relatively smaller (probably, due to more light scattering).

Discussion

The light-induced bleaching of the absorption bands near 420 and 680 nm and the appearance of bands near 450, 660 and with λ>695 nm at low redox potentials (Fig. 2) show that the photoreduction of either chlorophyll or pheophytin occurs in photosystem II preparations. The bleaching of two small bands at 518 and 545 nm (Fig. 2) strongly indicates the photoreduction of pheophytin [4]. The development of the broad band near 450 nm (Fig. 2) is characteristic of the radical anion of pheophytin *a* [2]. In ether solution the absorption bands of pheophytin *a* are at 408, 505, 534 and 667 nm [4]. The shift of these bands *in vivo* to 422, 518, 545 and 685 nm, respectively, can be interpreted as an environmental shift. However, the development of the narrow band at 676 nm can show that the pheophytin photoreduction results in both the blue shift of the chlorophyll absorption band at 680 nm and the bleaching of the pheophytin band at 670 nm. Such an interpretation is similar to that suggested for the photoreduction of bacteriopheophytin in *Chromatium minutissimum* [7,16,17] and *Chromatium vinosum* [19,22].

Pheophytin is probably reduced in the primary light reaction of photosystem II rather than in a photoreduction of antenna pigments, since this completely reversible photoprocess occurs at -170°C and is accompanied by a 2-3-fold decrease in the chlorophyll fluorescence yield of photosystem II when only approx. 0.1% of all the pigment molecules bleach. However, one cannot exclude completely the possibility that pheophytin, which can appear during the isolation of the chloroplast particles and which is not related directly to reaction centres of photosystem II, is also reduced under illumination at 20°C. This question is under study now.

The pheophytin photoreduction in photosystem II is similar to the earlier described [7,16,17,19,22] photoaccumulation of the radical anion of bacteriopheophytin (Bph$^-$) in *Chromatium*. (This photoaccumulation is a result of a fast (~1 µs) reaction of ferrocytochrome with a biradical of the reaction centre, [P$_{890}^+$·Bph$^-$] [16, 19, 22]). Really, both the photoreactions occur only when the "primary" electron acceptor (ubiquinone or plastoquinone) is in the reduced form. Both photoreactions are accompanied by the decrease in the chlorophyll fluorescence yield and are observed at low temperatures but have a low quantum yield. From this comparison, one can assume that the photoreduction of pheophytin (Ph) in photosystem II is also a result of electron transfer from a secondary electron donor of photosystem II to the biradical [P$_{680}^+$·Ph$^-$] which is formed in the primary photoact of photosystem II preceding the reduction of Q. (Here P$_{680}$ is the primary electron donor of photosystem II [3,11,20,21]). Then by analogy with the photoreaction in *Chromatium* [7,16], the fluorescence increase under the reduction of Q can be interpreted as the appearance of short-lived luminescence which is a result of the P_{680} excitation in the charge recombination in the biradical $[P_{680}^+ \cdot Ph^-]$. The decrease in the luminescence, when pheophytin is photoreduced, can be due to the photoaccumulation of the inactive state of the reaction centre, $[P_{680} \; Ph^-]$.

We wish to note that the formation of pheophytin during the procedure of photosystem II preparations should be studied to reveal the possible pheophytin photoreduction outside the reaction centres.

Acknowledgements

We are very grateful to N.I. Shutilova and Dr I.N. Krakhmaleva for their help during this study.

References

1. Duysens, L.N.M. and Sweers, H.E. (1963). *In* "Studies on Microalgae and Photosynthetic Bacteria" (S. Miyachi, ed.), pp. 353-372, University of Tokyo Press, Tokyo.
2. Evstigneev, V.B. and Gavroliva, V.A. (1954). *Dokl. Acad. Nauk SSSR* **96**, 1201-1204.
3. Gläser, M., Wolf, C., Buchwald, H.-E. and Witt, H.T. (1974). *FEBS Lett.* **42**, 81-85.
4. Goedheer, J.C. (1966). *In* "The Chlorophylls" (L.P. Vernon and G.R. Seely, eds), pp. 143-184, Academic Press, New York and London.
5. Karapetyan, N.V. and Klimov, V.V. (1973). *Physiol. rastenij* **20**, 545-553.
6. Klevanik, A.V., Klimov, V.V., Shuvalov, V.A. and Krasnovsky, A.A. (1977). *Dokl. Acad. Nauk SSSR*, in press.
7. Klimov, V.V., Shuvalov, V.A., Krakhmaleva, I.N., Karapetyan, N.V. and Krasnovsky, A.A. (1976). *Biochimija* **41**, 1435-1441.
8. Klimov, V.V., Shuvalov, V.A., Krakhmaleva, I.N., Klevanik, A.V. and Krasnovsky, A.A. (1977). *Biochimija* **42**, 519-530.
9. Krasnovsky, A.A. and Shaposhnikova, M.G. (1970). *Physiol. rastenij* **17**, 436-439.
10. Krasnovsky, A.A. and Voinovskaya, K.K. (1951). *Dokl. Acad. Nauk SSSR* **81**, 879-882.
11. Lozier, R.H. and Butler, W.L. (1974). *Biochim. Biophys. Acta* **333**, 465-480.
12. Pakshina, E.V. and Krasnovsky, A.A. (1974). *Biofizica* **19**, 238-242.
13. Rockley, M.G., Windzor, M.W., Cogdell, R.J. and Parson, W.W. (1975). *Proc. Natl. Acad. Sci. Washington* **72**, 2251-2255.

14. Shutilova, N.I., Klimov, V.V., Shuvalov, V.A. and Kutjurin, V.M. (1975). *Biofizica* **20**, 844-847.
15. Shutilova, N.I. and Kutjurin, V.M. (1975). *Biofizica* **20**, 246-249.
16. Shuvalov, V.A. and Klimov, V.V. (1976). *Biochim. Biophys. Acta* **440**, 587-599.
17. Shuvalov, V.A., Klimov, V.V., Krakhmaleva, I.N., Moscalenko, A.A. and Krasnovsky, A.A. (1976). *Dokl. Acad. Nauk* **227**, 984-987.
18. Shuvalov, V.A., Krakhmaleva, I.N. and Klimov, V.V. (1976). *Biochim. Biophys. Acta* **449**, 597-601.
19. Tiede, D.M., Prince, R.C. and Dutton, P.L. (1976). *Biochim. Biophys. Acta* **449**, 447-467.
20. Van Gorkom, H.J. (1974). *Biochim. Biophys. Acta* **347**, 439-442.
21. Van Gorkom, H.J., Tamminga, J.J. and Haveman, J. (1974). *Biochim. Biophys. Acta* **347**, 417-438.
22. Van Grondelle, R., Romijn, I.C. and Holmes, N.G. (1976). *FEBS Lett.* **72**, 187-191.

LIGHT-DEPENDENT BINDING OF p-NITROTHIOPHENOL TO PHOTOSYSTEM II: OSCILLATORY REACTIVITY UNDER ILLUMINATION WITH FLASHES

Y. KOBAYASHI, Y. INOUE and K. SHIBATA

Laboratory of Plant Physiology,
The Institute of Physical and Chemical Research,
Wako-shi, Saitama (JAPAN)

Abstract

The treatment of chloroplasts with p-nitrothiophenol (NphSH) in the light specifically inhibited the photosystem II activity, whereas the same treatment in the dark did not affect the activity at all. The photosystem I activity was not inhibited with this reagent. The inhibition was accompanied by a change of fluorescence at liquid nitrogen temperature. The relative height of the 695-nm (F695) to the 685-nm (F685) increased on the treatment. Incorporation of $Nph^{35}SH$ to chloroplast proteins and the change of F695/F685 ratio were highly accelerated by blocking the electron transport on the oxidation site of photosystem II, whereas these were suppressed by blocking the transport on the reduction site or by addition of electron donors to photosystem II. When dark-adapted chloroplasts were illuminated with sequential flashes in the presence of NphSH, the fluorescence ratio oscillated against flash number with maxima at 2nd and 6th flashes with a cycle of four flashes. The oscillation was not observed in the presence of carbonylcyanide-*m*-chlorophenylhydrazone, but was observed with NH_4Cl. Pretreatment of chloroplasts with 0.8 M TRIS-HCl changed the quadruple oscillation to a binary oscillation with maxima at 2nd, 4th, etc. flashes. It was deduced that photosystem

II particles undergo conformational changes periodically, being coupled with the S-state changes in the oxygen evolving system.

Introduction

Oscillatory oxygen evolution from dark-adapted chloroplasts illuminated with flashes provided invaluable information about the process of water oxidation. It was postulated from such information [6,7,11] that the formation of molecular oxygen from two molecules of water requires accumulation of four positive charges by four successive photoreactions. Changes of fluorescence and delayed luminescence [3,19] reflected, to some extent, the four activation steps with reverse deactivation steps. However, the chemical events taking place in the water-splitting system are still unknown. Application of p-nitrothiophenol (NphSH) to chloroplasts and their subunits in previous studies [8,9] demonstrated that this reagent is a useful tool to study the reactions involved in the oxidation site of photosystem II. It was found that the photosystem II activities such as oxygen evolution, 2,6-dichlorophenol indophenol (DCPIP) photoreduction and C550 photoreduction are abolished by the treatment of spinach chloroplasts with NphSH in the light but not in the dark. The photosystem I activity measured as the oxygen uptake mediated by the electron transport from reduced DCPIP to methylviologen was not affected by the same treatment both in light and in darkness. During this process of inhibition, the relative height, F695/F685, of the 695-nm fluorescence band to the 685-nm band at liquid nitrogen temperature increased progressively. The coexistence of carbonylcyanide-m-chlorophenylhydrazone (CCCP) with NphSH in the light accelerated the fluorescence change while the co-existence of 3-(3,4-dichlorophenyl)-1,1-dimethylurea (DCMU) or addition of an electron donor such as diphenylcarbazide suppressed the change with NphSH alone. This suggested that oxidation of the water side of photosystem II is essential for NphSH to cause the inhibition or the fluorescence change. It was presumed that, on the oxidation, photosystem II proteins undergo conformational changes to be modified with NphSH.

The present study was undertaken for two purposes: i) to see the labelling of chloroplast proteins with Nph^{35}SH in various conditions and ii) to see the action of NphSH under illumination with flashes. It was hoped that we might correlate the action to the S-state changes in the oxygen-evolving system.

Material and Methods

Chloroplasts were isolated from spinach leaves and suspended in the isolation medium (50 mM TRIS-HCl, pH 7.4, 0.4 M sucrose and 10 mM NaCl). Chloroplasts in the suspension (0.1 ml) were added to 5 ml of a 100 µM NphSH solution in the isolation medium, and the mixture was incubated at 20°C under illumination with continuous light or flashes as described previously [8] or in darkness as control. Continuous illumination was made with red light ($\geq$670 nm, 140 µW/cm^2) from a 300-Watt slide projector through a Toshiba VR-67 filter and a 10 cm water layer.

Xenon flashes from a Sugawara stroboscope (MSL-1A) with a half width duration of 2 µs were used for intermittent illumination. The maximal energy of a single flash on the sample was 2.0×10^2 erg/cm^2/flash. The dark interval between successive flashes was controlled with Omron twin timers. An aliquot of the reaction mixture was cooled to liquid nitrogen temperature for the fluorescence measurement with a Shimadzu recording fluorometer model RF-502. Fluorescence spectra were not corrected for the variation with wavelength of photomultiplier sensitivity and monochromator efficiency. It was confirmed in a separate experiment that NphSH in the reaction mixture does not interfere with the fluorescence measurement.

Nph^{35}SH was prepared from p-chloronitrophenol and hot sodium disulfide by the method of Price and Stacy [13] with slight modification. To a solution of 250 mg of p-chloronitrophenol in 25 ml of boiling ethanol was added portionwise a mixture of 300 mg sodium disulfide nonahydrate in 5 ml water and 41 mg (18.1 mCi) of sulfur in 15 ml ethanol over a period of 10 min. An alcoholic solution (25 ml) of 4 g of sodium hydroxide was then added dropwise to the mixture in a hot bath over a period of one hour. The mixture was cooled and, then, ethanol and water were removed by evaporation in a Kjeldahl flask. The solid product was dissolved in ethanol, purified two times by the procedure of Prince and Stacy [13], and dried in a vacuum desiccator. The yield calculated from the radioactivity was about 10%. The Nph^{35}SH thus prepared showed the same inhibitory effect on chloroplasts as that observed with authentic NphSH.

The chloroplasts treated with Nph^{35}SH in light or in darkness were collected by centrifugation, and sonicated for 30 s after addition of 80% acetone. This was repeated twice to remove free and adsorbed Nph^{35}SH not chemically bound. The protein fraction sedimented by centrifugation after the acetone treatment was suspended in water (1 ml) and sonicated for 1 min. Incorporation of Nph^{35}SH to this protein fraction was determined by counting the radioactivity with a gas-flow counter.

Experimental Results

Binding of Nph^{35}SH to Proteins and Fluorescence Change

Typical examples of the fluorescence change on the treatment of chloroplasts with NphSH are shown in Fig. 1, where the peak at the longest wavelength of 735 nm is taken as unity for the three curves to show the relative heights of other two peaks. It is evident from curves A and B that the middle peak at 695 nm of intact chloroplasts is intensified and shifted to longer wavelengths by the treatment with 100 μM NphSH in red light for 10 min and the shortest wavelength peak at 685 nm was lowered by the treatment. Such a fluorescence change took place more dramatically (curve C) on the chloroplasts pretreated with 0.8 M TRIS-HCl (pH 8.2). Throughout the experiments in the present study, the ratio, F695-700/F685, of the middle peak height to the 685-nm height was determined as a measure of these fluorescence changes.

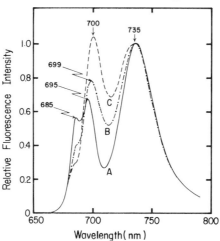

Fig. 1. Changes of fluorescence spectrum at 77 K by the treatment with 100 μM NphSH in red light for 10 min; curve A for intact chloroplasts before the NphSH treatment, curve B for chloroplasts after the treatment and curve C for chloroplasts pretreated with 0.8 M TRIS (pH 8.2) prior to the NphSH treatment. The figures with arrows on these curves show the maximum wavelength in nm, and the height of the 735-nm peak was taken as unity on these curves. The chlorophyll concentration in the reaction mixture was 12 μg/ml.

Table 1 summarizes the fluorescence ratios and the degrees of incorporation of Nph^{35}SH determined in various conditions. Chloroplasts were treated with 100 μM Nph^{35}SH for 5 min in red light (140 μW/cm^2) in the presence of inhibitor and/or electron donor. Treatments were made in the same conditions with cold NphSH for fluorescence measurements. An appreciable amount of Nph^{35}SH was bound to chloroplast proteins on the 5 min treatment of intact chloroplasts in the dark, but the binding was doubled by the same treatment in the light. The light-induced

TABLE 1

Incorporation of Nph^{35}SH to chloroplast proteins and the fluorescence ratio (F695-700/F685) obtained for chloroplasts treated with 100 μM NphSH in red light for 5 min. in various conditions

Sample chloroplasts	Additive	NpH^{35}SH binding (nmoles/mg Chl)		F695-700/F685	
		Dark	Light	Dark	Light
Intact	None	8	17	1.16	1.50
	DCMU	8	10	1.17	1.21
	CCCP	8	35	1.17	3.40
	CCCP + DPC	——	15	1.18	1.52
	Ascorbate	——	——	1.18	1.40
	DCPIP + Ascorbate	——	——	1.18	1.19
TRIS-treated	None	11	43	1.17	2.95
	DCMU	——	10	1.16	1.54
	DPC	——	24	1.18	1.67
Heat-treated	None	11	16	1.19	1.70

The reaction mixture contained 20 μM DCMU, 100 μM CCCP, 100 μM diphenylcarbazide (DPC), 1 mM ascorbate and 100 μM DCPIP. The chlorophyll concentration was 56 μg/ml in the incorporation experiment and 17 μg/ml in the fluorescence measurement.

increment of binding was suppressed significantly by the addition of DCMU, which is an inhibitor on the reduction site of photosystem II. Contrary to this effect of DCMU, the presence of carbonyl-cyanide-*m*-chlorophenylhydrazone (CCCP) during the treatment promoted the binding 4-fold, and this promotion was suppressed by the co-existence of diphenylcarbazide (DPC). Pre-treatment of chloroplasts with 0.8 M TRIS-HCl (pH 8.2) prior to the Nph^{35}SH treatment accelerated the binding in the light strikingly (about 4-fold) but the binding in the dark only slightly. The addition of DPC or DCMU greatly suppressed the acceleration. Heat pretreatment of chloroplasts showed a similar effect but the effect was less pronounced.

These data of Nph^{35}SH incorporation are well reflected as the changes of F695-700/F685 ratio. The ratio of intact chloroplasts, which was 1.17 ± 0.02, was not affected by the treatment with NphSH in the dark but was increased to 1.50 by the treatment in the light. The addition of DCMU or DCPIP plus ascorbate suppressed the increase, while the addition of CCCP further enhanced the ratio. This enhancement with CCCP was suppressed by the presence of DPC. The ratio obtained for TRIS-treated chloroplasts in the light was as high as 2.95, and this high ratio was decreased appreciably by the presence of DPC or DCMU. A high ratio was also obtained for heat-treated chloroplasts.

These data shown in Table 1 indicate that the binding of NphSH as well as the fluorescence ratio is largely dependent on the redox state of the water side of photosystem II. The binding increases when electron carriers on the water side are oxidized by illumination. An active site on the water side may be exposed to be modified with NphSH when positive charges are formed by illumination.

Fluorescence Changes by NphSH under Illumination with Flashes

As a preliminary experiment, chloroplasts were treated with 100 µM NphSH under illumination with flashes at uniform intervals. Three hundred flashes were given in each experiment, and the interval (denoted as t_d in the figure) was varied between 100 ms and 120 s to determine the life time of the activated state(s). Curve A in Fig. 2 shows the result obtained for intact chloroplasts, which indicates that the NphSH-reactive state induced by flashes becomes inactive after 60 s in darkness. Curve B is the result obtained in the control experiment in complete darkness. The half life time was estimated from curve A to be 7.5-8.0 s. The pretreatment with TRIS or the presence of CCCP on the NphSH treatment appreciably shorten the life time, as seen from curves C and D, respectively. These curves show saturation at $t_d \approx 10$ s, and the half life time was approximately 0.6-0.8 s in these cases.

Chloroplasts, which had been kept in darkness for 30 min after a few minutes of exposure to weak light, were treated with 100 µM NphSH under repeated illumination

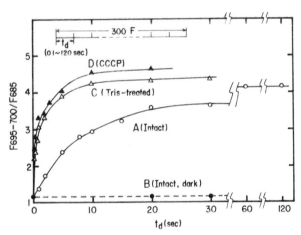

Fig. 2. Effect of the dark interval (t_d) between flashes on the fluorescence change of chloroplasts (12 µg Chl/ml) treated with 100 µM NphSH under illumination with 300 flashes at uniform intervals; curves A and B for intact chloroplasts illuminated with flashes and kept in darkness, respectively, curve C for TRIS-pretreated chloroplasts before the NphSH treatment with flashes, and curve D for intact chloroplasts treated with NphSH in the presence of 100 µM CCCP.

with a group (cluster) of flashes at a saturating intensity. As shown on the margin of Fig. 3, a fixed number (N) of sequential flashes spaced 300 ms apart was given as a flash cluster, and such a cluster was repeated 25 times at longer intervals of 60 s, which is the life time of the NphSH-reactive state (Fig. 2). The cluster repetition was made to obtain a detectable change of fluorescence. Immediately after the flash illumination, the reaction mixture was cooled to liquid nitrogen temperature for fluorescence measurement. The F695-700/F685 ratio thus obtained was plotted against N, the flash number in a cluster. As seen from Fig. 3, the fluorescence ratio changed in an oscillatory manner against N with a cycle of $N = 4$. The maxima were found at 2nd and 6th flashes, and the ratio became progressively higher on every cycle.

Oxygen evolution starting from dark-adapted chloroplasts measured with flashes by Joliot [7] showed its 1st maximum at 3rd flash, and the oxygen evolution on 1st flash was practically zero. In the present case of the NphSH effect, the 1st flash caused a considerable change of fluorescence, and 1st maximum was found at 2nd flash.

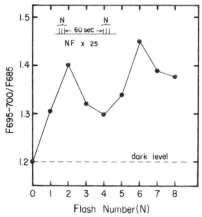

Fig. 3. Effect of the flash number (N) on the fluorescence change of intact chloroplasts (12 μg Chl/ml) treated with 100 μM NphSH under illumination with flash clusters. The illumination with a flash cluster made up of N flashes spaced 300 ms apart was repeated 25 times with uniform dark intervals of 60 s to obtain an appreciable fluorescence change.

Fig. 4 shows the results obtained in the presence of 25 μM CCCP or 20 mM NH_4Cl and for TRIS pretreated chloroplasts. The quadruple oscillation pattern was smoothed out to be a uniformly rising curve by the presence of CCCP, while the quadruple oscillation was obtained, though less pronounced, in the presence of NH_4Cl. The pattern obtained for TRIS-pretreated chloroplasts was different from the above two. The fluorescence ratio changed in an interesting manner regarded as a binary oscillation. The ratio increased on 1st and 2nd flashes, but 3rd flash decreased the ratio slightly. Coming after these flashes, the ratio changed with maxima at 4th, 6th and 8th flashes and with minima at 5th and 7th flashes. It is interesting to note that the chloroplasts isolated from summer spinach leaves did not show the above oscillation clearly. The distorted pattern of summer spinach chloroplasts was changed to the normal pattern of quadruple oscillation by addition of 2 mM dithiothreitol to the chloroplast suspension. This suggests requirement of SH groups to hold active structures of some proteins in chloroplasts.

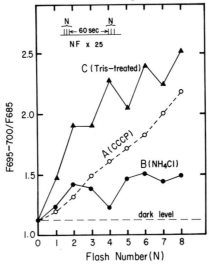

Fig. 4. Effect of flash number (N) on the fluorescence change in the flash-cluster experiments for intact chloroplasts treated with 100 μM NphSH in the presence of 25 μM CCCP (curve A) or 20 mM NH_4Cl (curve B) and for TRIS-pretreated (0.8 M, pH 8.8) chloroplasts (curve C). Other experimental conditions were the same as in Fig. 3.

Discussion

The oscillation with a cycle of four flashes and with its maxima at 2nd and 6th flashes has been reported for delayed light emission observed after flashes and for the luminescence induced by pH jump or temperature jump from flashed chloroplasts [1,4,19]. We have recently observed a similar type of oscillation for the thermoluminescence from flashed chloroplasts [5]. The delayed light emission and luminescence in oscillation are considered as due to recombination of negative charges on the reduction side of photosystem II and positive charges on its oxidation side [4,13], so that this oscillation reflects the changes in charge accumulation around the reaction center of photosystem II. The oscillation found in the present experiment is concerned with the changes in the reactivity between NphSH and photosystem II proteins.

Two different mechanisms may be proposed for the light-induced incorporation of NphSH to photosystem II, being accompanied by the fluorescence change.

1) Light induces conformational changes of photosystem II particles, being coupled with the changes of S species in water-splitting system. This opens up the site for NphSH to attach, and the modification of photosystem II structure by this NphSH attach changes the fluorescence spectrum.

2) The second mechanism for the NphSH incorporation is the modification of photosystem II proteins by oxidized NphSH, which might be produced by positive charges in the water-splitting system. This mechanism may be less probable a) because NphSH is incorporated to some extent in darkness into chloroplast proteins (Table 1), and b) because NphSH is incapable of donating electrons to photosystem II.

The first mechanism is inconsistent with the observation by Satoh [16,17], who treated chloroplasts with a mixture of a chaotropic reagent (urea or guanidine) and 1,10-phenanthroline and found the shift of the 695-nm fluorescence band to 700 nm, the same phenomenon found on the NphSH treatment. He attributed this to alternation of the apoprotein for the fluorophore. The 695-nm band is the emission from the chlorophylls associated with the energy trap in photosystem II [2,10,12]. If the apoprotein of the reaction center chlorophyll associated with the water-splitting system, the positive charge on the S species in the water-splitting system will influence the apoprotein structure. As shown in Table 1, the fluorescence change as well as $Nph^{35}SH$ binding was suppressed in the presence of diphenylcarbazide or ascorbate. The more effective reductant, reduced DCPIP with ascorbate, caused a greater suppression. It is deduced from these results that photosystem II particles undergo conformational changes, being coupled with the changes in redox state in the water-splitting system.

According to Kok et al.[11], the formation of oxygen requires accumulation of four positive charges on the active site of S by four sequential photoreactions. They assumed that S_0 with no positive charge and S_1 with a single positive charge are stable in the dark and that S_1 is oxidized stepwise by flashes to more charged species, S_2, S_3 and S_4. The final S_4 state returns back to S_0, evolving oxygen. If we assume that various S species reflect the oscillatory reactivity with NphSH, S_0 and S_1 are note reactive with NphSH because NphSH was not effective at all in the dark for the fluorescence change. The fluorescence change caused by the first flash was about half that caused by the second flash (Fig. 3). This suggests that S_2 and S_3 are active in causing the fluorescence change.

The ratio of fluorescence change increased uniformly with increasing the flash number in the presence of CCCP, while the quadruple oscillation was observed with NH_4Cl. These different responses to CCCP and NH_4Cl, both being an uncoupler, suggest that the smoothing action of CCCP does not result from uncoupling. Renger [14] found that CCCP at a low concentration accelerates the deactivation of the water-splitting enzyme system. He assumed that the ADRY agent such as CCCP makes a shunt between the plastoquinone pool and positive charges in the water-splitting system [15]. It was also suggested that CCCP or the TRIS treatment increases the redox potential of cytochrome b_{559}, thus inducing a cyclic electron flow in photosystem II (18). It was found in the present study that the life time of the NphSH reactive state is greatly shortened by CCCP or the TRIS treatment. This may be explained also as the ADRY effect of CCCP or the TRIS treatment. It seems, however, difficult to explain, in terms of the ADRY effect, the acceleration of the fluorescence change and the Nph^{35}SH incorporation by CCCP or the TRIS treatment. The binary oscillation is also difficult to explain. We may have to assume a smaller cyclic shunt on the S cycle or two latent sites to be activated for a single S site.

Acknowledgements

The present study was supported by a research grant on "Photosynthetic oxygen evolution" given by the Ministry of Education and by a grant for "Life sciences" at The Institute of Physical and Chemical Research (Rikagaku Kenkyusho).

References

1. Barbieri, G., Delosme, R. and Joliot, P. (1970). *Photochem. Photobiol.* **12**, 197-206.
2. Cederstrand, C.N. and Govindjee (1966). *Biochim. Biophys. Acta* **120**, 177-180.
3. Etienne, A.L. (1974). *Biochim. Biophys. Acta* **333**, 320-330.
4. Hardt, H. and Malkin, S. (1973). *Photochem. Photobiol.* **17**, 433-440.
5. Inoue, Y. and Shibata, K. (1977). To be published in the Proceedings of the 4th International Congress on Photosynthesis, Reading.
6. Joliot, P. (1965). *Biochim. Biophys. Acta* **102**, 116-134.
7. Joliot, P. (1968). *Photochem. Photobiol.* **8**, 451-463.
8. Kobayashi, Y., Inoue, Y. and Shibata, K. (1976). *Biochim. Biophys. Acta* **423**, 80-90.
9. Kobayashi, Y., Inoue, Y. and Shibata, K. (1976). *Biochim. Biophys. Acta* **440**, 600-608.
10. Kok, B. (1963). *In* "Photosynthetic Mechanism of Green Plants" (B. Kok and A.T. Jagendorf, eds), pp. 45-55. Natl. Acad. Sci. and Natl. Res. Council, Washington.
11. Kok, B., Forbush, B. and McGloin, M. (1970). *Photochem. Photobiol.* **11**, 457-475.
12. Murata, N. (1968). *Biochim. Biophys. Acta* **162**, 106-121.
13. Price, C.C. and Stacy, G.W. (1946). *J. Am. Chem. Soc.* **68**, 498-500.
14. Renger, G. (1969). *Naturwissenschaften* **56**, 370.
15. Renger, G. (1973). *Bioenergetics* **4**, 491-505.
16. Satoh, K. (1972). *Plant Cell Physiol.* **13**, 23-34.
17. Satoh, K. (1974). *Biochim. Biophys. Acta* **333**, 107-126.
18. Wada, K. and Arnon, D.I. (1971). *Proc. Natl. Acad. Sci. Washington* **68**, 3064-3068.

19. Zankel, K.L. (1971). *Biochim. Biophys. Acta* **245**, 373-385.

NEW ASPECTS ON THE FUNCTION OF NAPHTHOQUINONES IN PHOTOSYNTHESIS

H.K. LICHTENTHALER and K. PFISTER

Botanisches Institut (Pflanzenphysiologie), Universität Karlsruhe, 7500 Karlsruhe (BRD)

Abstract

The effect of 3-halogenated naphthoquinones (2-chloro-3-methyl-1.4-naphthoquinone; 2-bromo-3-methyl-1.4-naphthoquinone; 2-bromo-3-isopropyl-1.4-naphthoquinone) on the photosynthetic light reactions and *in vivo* chlorophyll fluorescence was investigated in the attempt to get new information on the possible functional site of the endogenous naphthoquinone phylloquinone (vitamin K_1). All 3 naphthoquinones were found to be good inhibitors of photosynthetic electron flow in isolated chloroplasts (pI_{50}-values between 4.8-5.7), leaves and *Scenedesmus* cultures; the inhibition site lying, similar to that of DCMU, between Q and PQ-9. It is assumed that they block at the functional site of the endogenous naphthoquinone K_1. Besides this the applied naphthoquinones are very strong quenchers of pigment system II fluorescence and thus inhibit photosynthetic oxygen evolution by a double mechanism.

The results are discussed in view of new findings on the endogenous phylloquinone. The working hypothesis that phylloquinone K_1 may be a functional constituent of the linear electron transport chain and may also support cyclic electron flow is discussed.

Introduction

Green plants always contain several lipophilic quinones. Ubiquinone plays an important role as redox carrier in the mitochondrial biomembrane. In contrast to mitochondria chloroplasts possess three prenylquinones as regular constituents of the photochemically active thylakoids. These are the naphthoquinone derivative phylloquinone (vitamin K_1) and the benzoquinones plastoquinone-9, α-tocoquinone and its reduced cyclic form α-tocopherol [19]. Plastoquinone-9 functions as terminal electron acceptor of PS II and as proton translocator [1,34,37]. The function of phylloquinone and α-tocoquinone are not yet known. As quinones they are potential endogenous electron carriers of the photosynthetic membrane, their participation in the photosynthetic electron transport can thus be expected.

In the case of naphthoquinone vitamine K_1, which is in the centre of the present communication, several observations indicate a possible function in the photosynthetic electron transport. Some of the most important data about phylloquinone are summarized in the first part of this paper. By studying the effects of K_1 analogues (halogenated naphthoquinones) on photosynthetic electron flow and chlorophyll fluorescence, which is described in the second part of this paper, we made a new approach to get further information on the possible function of vitamin K_1 in photosynthesis.

Material and Methods

The extraction and determination of prenylquinones and pigments was carried out as described before [25]. Phylloquinone was separated by one- or two-dimensional thin-layer chromatography on silica plates [21], identified and quantitatively determined by its absorption spectrum before and after reduction with potassium borohydride in a 0.1 M Tricine-buffered ethanol solution at pH 7. The difference spectrum is shown in Fig. 1. In addition to this K_1 was isolated and estimated by the sensitive method of high performance liquid chromatography [27].

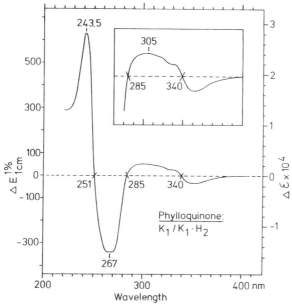

Fig. 1. Difference spectrum of phylloquinone $K_1/K_1 \cdot H_2$ in 0.1 M Tricine-buffered ethanol, pH 7.0. The absorbance decrease at 268 nm $\Delta E\ _{1\ cm}^{1\%} = 336$ ($\varepsilon=15100$), the absorbance increase at 243.5 nm $\Delta E\ _{1\ cm}^{1\%} = 621$ ($\varepsilon=28000$) and at 305 nm $\Delta E\ _{1\ cm}^{1\%} = 53.7$ ($\varepsilon=2420$).

Chloroplasts were isolated from 5 day old *Raphanus* cotyledons by differential centrifugation. Electron transport was assayed in a medium described by Strotmann and Gosseln [32] and measured as change in oxygen concentration with a Clark-type electrode. The green alga *Scenedesmus obliquus* D_3 was cultured in a nutrient solution after Kessler *et al.* [14] at 22°C and about 3000 lux light intensity.

Fluorescence was excited with a blue HeCd - Laser (Fa. Liconix, Mod. 401) and detected at 685±10 nm with a RCA 7265 Multiplier. The fluorescence signal was stored on a Telequipment DM 64 storage oscilloscope and photographed from the screen.

The halogenated naphthoquinones were synthesized by Professor Musso in the Institute of Organic Chemistry, University of Karlsruhe.

The Endogenous Naphthoquinone Phylloquinone (vitamin K_1)

Green plant tissues are particularly rich in phylloquinone. This had first been described using the biological assay method for vitamin K_1 [8] and was later confirmed by applying chromatographic and chemical determination methods [15,17,23]. Phylloquinone, like chlorophylls, contains the phytyl side chain. This and its presence in green plant tissues suggested that the phylloquinone K_1 may be associated with the chlorophylls within the chloroplasts forming a functional unit [9,15,36]. Using the biological vitamin assay dam et al. [10] had reported that isolated chloroplasts possess vitamin K activity. The actual presence of phylloquinone in chloroplasts was, however, demonstrated much later by two independent research groups [13,16]. Within the chloroplasts vitamin K_1 is quantitatively associated with the photochemically active thylakoids [23]. Its concentration amounts to 0.5 to 2.5 moles per 100 moles of chlorophyll, depending on the plant species and on the growth conditions [17,19,21]. Chlorophyll-free plant tissues do not contain vitamin K_1 or only in trace amounts [18]. Etiolated tissues have some K_1 and also a second lipophilic naphthoquinone 'K' [22], the probable biosynthetic K_1 precursor, with a more unsaturated side chain [33]. Light induces parallel to chlorophyll and thylakoid formation an enhanced synthesis of phylloquinone [20], which is initiated by active phytochrome [21,26]. The level of the precursor naphthoquinone 'K' decreases with increasing illumination time (Fig. 2) parallel to the accumulation of K_1. Fully greened chloroplasts do no more contain this second naphthoquinone 'K'.

It is of particular interest that during the early stages of greening of etiolated plant tissues the level

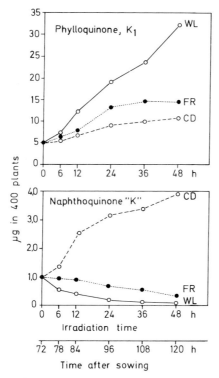

Fig. 2. Promotion of phylloquinone K_1 synthesis in etiolated *Raphanus* seedlings by continuous far-red (FR) and white light (WL) as compared to darkness (CD). The level of the second naphthoquinone 'K', which accumulates in darkness, is decreased by light.

of phylloquinone is very high as compared to that of chlorophyll a. This is the case before and after onset of photosynthetic oxygen evolution, at a stage where the bulk antenna chlorophyll is still missing, and can be seen in continuous white light and during delayed greening in intermittent (light-dark-cycles, flash light program; Table 1). In fully developed functional chloroplasts with a high capacity for photosynthetic light quanta conversion (*e.g.* 50 h continuous white light) the K_1 level on a chlorophyll basis is much lower.

After the greening phase of leaves the level of K_1 remains fairly constant throughout the vegetation period, no matter whether leaf area, one single leaf or chloro-

TABLE 1

Chlorophyll a and phylloquinone K content, Hill activity of isolated chloroplasts with DCPIP and variable fluorescence of the intact leaf from 6 d old etiolated barley leaves which greened under different light treatments (after [12]). Light-dark cycles: 2 min white light + 58 min darkness. Flash light program: 10^{-3} white flash + 15 min darkness. The variable fluorescence (v.F.), which is an indicator of the photosynthetic capacity of the leaf, was measured according to Lichtenthaler et al. [21].

	µg/100 plants			µM O_2	
	Chl.a	K_1	Chl.a/K_1	mg Chl. x h	v.F.
a) continuous white light					
1 h	68.9	4.5	15.3	0	0
2 h	109.1	6.0	18.2	0	0
5 h	489.6	8.0	61.2	137	1.52
10 h	1221	10.0	122.1	51	1.62
50 h	4371	15.0	291.4	39	2.55
b) light-dark cycles (LDC)					
5 LDC	64	7.2	15.3	0	0
10 LDC	126	8.1	15.6	12	0.34
50 LDC	328	12.5	26.3	24	0.88
c) flash light program					
40 flashes	74.3	5.9	12.6	0	0
100 flashes	128.5	8.6	14.9	0	0
200 flashes	245.0	11.0	22.3	0	0

phyll is taken as reference basis (Fig. 3). This is in contrast to the chloroplast benzoquinones (plastoquinone-9, α-tocopherol, α-tocoquinone) of which the reduced forms (plastohydroquinone-9, α-tocopherol) are continuously accumulated even after the greening phase (Fig. 3). These excess amounts of prenylbenzoquinones are then stored extrathylakoidal in the osmiophilic plastoglobuli of the chloroplast stroma (Fig. 4). During natural breakdown of thylakoids, e.g. at the end of the vegetation period (Fig. 3) or during an artificially induced thylakoid degradation, K_1 is destroyed together with the chlorophylls [24].

The strong correlation between chlorophyll and K_1 concerning formation and concentration points to a basic role of phylloquinone in the photosynthetic light react-

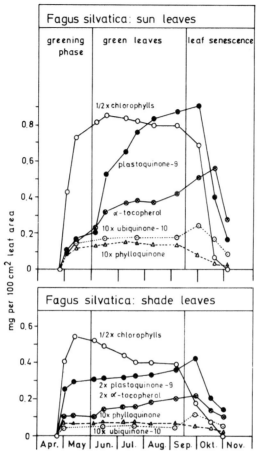

Fig. 3. Kinetic of prenylquinone and pigment accumulation in *Fagus* leaves showing the maintenance of a fairly constant level of phylloquinone during the green vegetation period. The level of plastoquinone-9 (+ plastohydroquinone-9) and of α-tocopherol, in turn, increases continuously.

tions. As a naphthoquinone K_1 is a potential photosynthetic electron carrier. Its redox potential lies in the range of ± 0 Volt (*e.g.* - 25 mV, [36]). In contrast to plastoquinone-9 which exhibits a large pool size in thylakoids, the level of K_1 is lower and similar to that of other photosynthetic electron carriers such as P 700 or cytochromes [19]. Since synthetic naphthoquinones like menadione (K_3) or K_5 catalyze a cyclic electron flow in isolated chloroplasts [3,4], it was assumed that K_1 may

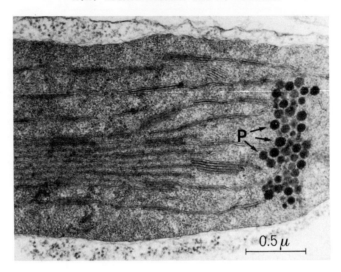

Fig. 4. Section of a chloroplast from 8 d old barley seedlings showing grana and stroma thylakoids and a large group of plastoglobuli (P). (5% Glutardialdehyd, 1% OsO_4; 20000 x).

be the endogenous cofactor of an *in vivo* cyclic electron flow. This hypothesis was further supported by the finding that K_1 is enriched in PS I - particles [19]. However, a smaller portion of the total vitamin K_1 was found to be associated with the pigment system II particle fraction.

A *Scenedesmus* mutant, which lacks PS I activity, has only a low level of K_1 as compared to the wild type, while another mutant with PS II deficiency has a normal K_1 level (Table 2). This again indicates that within the thylakoids the major part of the K_1 may be associated with PS I.

Recent work with *Scenedesmus* cultures, grown under different culture conditions, again points to a possible role of K_1 in cyclic electron flow. Thus under photoheterotrophic growth conditions (- CO_2, + glucose), which does not allow photosynthesis with linear electron transport and water splitting, but should favour cyclic electron flow, the level of K_1 on a chlorophyll basis is increased, while that of plastoquinone-9 decreases. This is found in both synchronized and non-synchronized *Scenedesmus*

TABLE 2

Lipoquinone content on a chlorophyll basis of Scenedesmus obliquus D_3 *cultures and of 2* Scenedesmus *mutants (No. 8 and 11 of Dr. N.I. Bishop), which are kept under mixotrophic growth conditions in presence of glucose. The PS I deficient mutant exhibits a low level of phylloquinone K_1 and the PS II deficient mutant a low level of plastoquinone-9.*

	Wild-type	PS I deficient (No.8)	PS II deficient (No.11)
chlorophyll a	100	100	100
chlorophyll b	19.3	30.4	18.5
a/b	5.2	3.3	5.4
phylloquinone	0.026	0.008	0.020
plastoquinone-9	2.7	7.6	1.1
α-tocopherol +α-tocoquinone	1.2	4.9	1.5

cultures (Fig. 5).

Though the analytical data available hint to a possible function of K_1 in cyclic electron flow, they do not exclude the possibility for a participation in linear electron transport near PS II.

There is one further recent observation, which gives some information on the endogenous function of phylloquinone. When *Chlorella* suspensions are allowed to photosynthetically fix $^{14}CO_2$ under steady state conditions one finds a very high labelling degree for α- and β-carotene, but also for phylloquinone, while the labelling of the remaining chloroplast prenyl-lipids is much lower (Table 3). This indicates that phylloquinone, like carotenes, has a very low half-life time in the photosynthetic membrane. A fast turnover of carotenes may be understandable since, in particular in algae, they have a light protection function for chlorophylls and other components of the thylakoid membrane. Thus the well-established energy

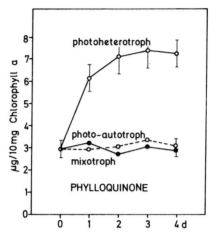

Fig. 5. Increase of the phylloquinone K level on a chlorophyll basis when autotrophic *Scenedesmus* cultures are transferred to photoheterotrophic growth conditions ($-CO_2$, $+0.5\%$ glucose).

TABLE 3

^{14}C-*labelling degree (in %) of prenyllipids in* Chlorella pyrenoidosa *after 1 and 2 hours of* $^{14}CO_2$ *photosynthesis*

The experiment was performed under steady state conditions with a constant pigment and lipoquinone content of the algae suspension (after [12]).

	1 h	2 h
Phylloquinone	12.2	18.0
α-Carotene	13.1	18.8
β-Carotene	13.1	18.5
Plastoquinone-9	0.5	1.6
α-Tocopherol	0.1	0.4
Lutein	0.4	1.4
Chlorophyll *b*	0.5	1.3

transfer from excited chlorophyll molecules to carotenoids at high light intensities [28,37] may result in a photooxidative degradation of the carotenes involved. Whether the phylloquinone participates in similar energy dissipation reactions or possibly acts as quencher of chlorophyll fluorescence is not known. In any case the labelling data indicate that phylloquinone has a function with a high turnover rate.

Inhibition of the Photosynthetic Light Reactions by Naphthoquinones

In the attempt to study the function of an endogenous substance it is a common method in biology to block the function of this endogenous component by applying structural analogues. Thus the group of Trebst developed and described a substituted and halogenated benzoquinone, DBMIB, as an inhibitor against the endogenous benzoquinone plastoquinone-9 [7]. Another benzoquinone derivative (2,3-dimethyl-5-dihydroxy-6-phytol-benzoquinone) was also found to be an antagonist of plastoquinone function [5].

In analogy to this we tried to block the function of the endogenous naphthoquinone vitamin K_1 by substituted halogenated naphthoquinones (Fig. 6). If so, the halogenated naphthoquinones should not inhibit the electron transport at the same position as DBMIB. Therefore the aim of our investigations was to localize and characterize the inhibition site of the halogenated naphthoquinone in the photosynthetic electron transport chain. In addition to this, we studied the quenching properties of the compounds used in order to get information about the extent and mechanism of the quenching of chlorophyll fluorescence.

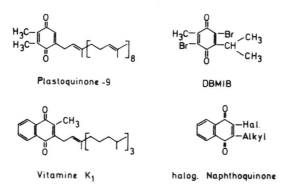

Fig. 6. Structures of plastoquinone-9, DBMIB, vitamin K_1 and of the halogenated naphthoquinone.

Photosynthetic Electron Transport

The influence of some naphthoquinone derivates on the photosynthetic electron flow was studied by testing the activity of several electron donor and acceptor systems with isolated *Raphanus* chloroplasts in presence of added naphthoquinones. All applied naphthoquinones exhibit an inhibitory action on photosynthetic oxygen evolution. The inhibition of the methylviologen reduction (MV reduction, measured as O_2 uptake) is expressed as I_{50} or pI_{50}-value, which indicates the concentration of the substrate needed for 50% inhibition of this Hill reaction (Fig. 7).

The results show a remarkably high inhibitory efficiency of the halogenated naphthoquinones, whereas the efficiency of the non-halogenated ones is essentially poorer.

Substance	pI_{50}	I_{50}
1 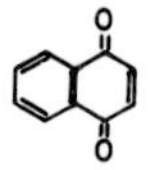	2,7	$2000 \cdot 10^{-6}$ M
2 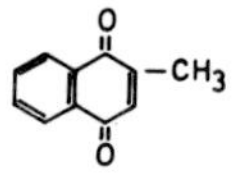	3,3	$500 \cdot 10^{-6}$ M
3 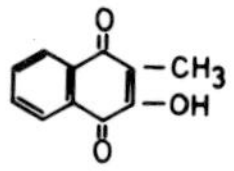	3,8	$160 \cdot 10^{-6}$ M
4 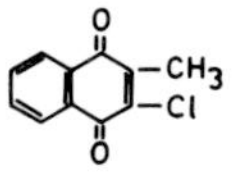	4,8	$16 \cdot 10^{-6}$ M
5 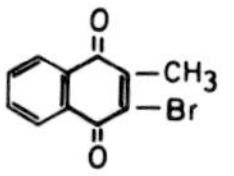	5,1	$7,9 \cdot 10^{-6}$ M
6 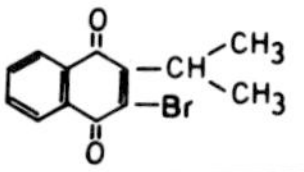	5,7	$2 \cdot 10^{-6}$ M
7 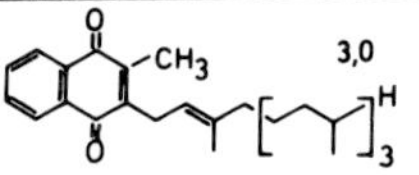	3,0	$1000 \cdot 10^{-6}$ M

Fig. 7. Inhibition of methylviologen reduction by naphthoquinones. Material: Isolated *Raphanus* chloroplasts. 1 : 1.4-naphthoquinone; 2 : Menadione, vitamin K_3; 3 : 2-methyl-3-hydroxy-1.4-naphthoquinone, phthiocol; 4 : 2-chloro-3-methyl-1.4-naphthoquinone; 5 : 2-bromo-3-methyl-1.4-naphthoquinone; 6 : 2-bromo-3-isopropyl-1.4-naphthoquinone, BIN; 7 : phylloquinone, vitamin K_1, 2-methyl-3-phytyl-1.4-naphthoquinone.

The most effective inhibitor is the 2-bromo-3-isopropyl-1.4-naphthoquinone (BIN), which has a pI_{50}-value of 5.7. The chlorophyll concentration independent inhibitor constant K_i for this compound is 5.8 for *Raphanus* and 5.5 for spinach chloroplasts. Together with DBMIB (pI_{50}=6.5 for ferricyanide reduction; calculated after [7] or DBMIB-analogues [29] the halogenated naphthoquinones are the best-known quinoid inhibitors of photosynthetic electron flow. We were surprised to see that the endogenous naphthoquinone K_1 also blocks the electron transport, although at much higher concentrations. The effect of 1.4 naphthoquinone, menadione, phthiocol and vitamin K_1 seems to be an unspecific action because of the high unphysiological concentrations.

Our results, using MV as electron acceptor, clearly indicate an inhibition effect on the photosynthetic electron transport chain between the O_2 evolving system and the acceptor site of PS I. A closer determination of the site of action in the electron transport chain was obtained by comparison of three electron donor and acceptor systems: MV reduction, representing the activity of both pigment systems, BQ reduction, mainly system II dependent, and of the system DCPIP/Asc→ MV, which in presence of

TABLE 4

Inhibition of partial reactions of the photosynthetic electron transport by naphthoquinones. Values in µM O_2 mg Chl x h

Substance	Conc.	H₂O→BQ	H₂O→MV	DCPIP/Asc→MV
control	-	92	80	96
2-chloro-3-methyl-1.4-naphthoquinone	10^{-5}M	52	47	92
2-bromo-3-methyl-1.4-naphthoquinone	10^{-5}M	36	32	82
2-bromo-3-isopropyl-1.4-naphthoquinone	10^{-5}M	14	12	86
menadione (K_3)	10^{-3}M	33	22	90
vitamin K_1	10^{-3}M	49	43	91

DCMU is a test exclusively of PS I activity (Table 4).

The results show a similar degree of inhibition on all PS II dependent reactions, whereas the activity of PS I remains nearly unaffected. The most effective substances again are the halogenated naphthoquinones. At higher concentrations a complete inhibition of PS II dependent reactions can be obtained.

One important question is, whether the naphthoquinones block the electron transport chain at the same or at a different position as DBMIB does. Photosynthetic BQ reduction is possible to about 60% in presence of 4×10^{-6} M DBMIB. From the fact that the O_2 evolution with BQ as electron acceptor in the presence of DBMIB can be completely abolished by addition of higher concentrations of halogenated naphthoquinones, one can conclude that the inhibition site of these naphthoquinones is prior to the inhibition site of DBMIB (Table 5).

A possible inactivation of the water splitting system by the halogenated naphthoquinones was excluded with the following experiments. Heat-inactivated chloroplasts (5 min, 45°C) only exhibit a very small rate of electron transport to MV. This rate can be increased by donation

TABLE 5

Inhibition of benzoquinone reduction by DBMIB and by halogenated naphthoquinones

Inhibitors	µM O_2/mg Chl x h	Reaction rate (% of control)
control, no additions	75	100
+DBMIB (4×10^{-6} M) without naphthoquinone	46	59
+DBMIB (4×10^{-6} M) +2-bromo-3-methyl-1.4-naphthoquinone (10^{-4} M)	<2	<3
+DBMIB (4×10^{-6} M) +2-bromo-3-isopropyl-1.4-naphthoquinone (10^{-4} M)	<2	<3

TABLE 6

Inhibition of photosynthetic electron transport in the system: donor→PSII→PSI→MV

Chloroplasts heat inactivated (5 min, 45°C). Inhibitor concentration 8×10^{-4} M (a) and 8×10^{-5} M (b,c). Values in μM O_2/mg Chl x h.

Inhibitor	intact chloroplasts	heat-inactivated chloroplasts		
	H$_2$O→MV	H$_2$O→MV	PhD/Asc→MV	HQ/Asc→MV
control	80	4	67	61
2-chloro-3-methyl-1.4-naphthoquinone	6	-	13*	9*
2-bromo-3-methyl-1.4-naphthoquinone	4	-	8*	6*
2-bromo-3-isopropyl-1.4-naphthoquinone	~0	-	11*	5*

*remaining reaction rates mainly DCMU-insensitive

of electrons at the oxidizing site of PS II by suitable donors (PhD or HQ). The electron transport in these systems (PhD/Asc →MV and HQ/Asc→ MV) can, however, be inhibited to a similar extent as in the system H$_2$O→MV, indicating the location of the inhibition site after the water splitting system (Table 6).

In our electron transport measurements we found no indication for cyclic reactions around PS I. To test the absence of cyclic electron flow in a different way, we studied the formation of the light-induced pH gradient in presence of halogenated naphthoquinones. There is no stimulation of this gradient, which could be expected in case of cyclic, proton-translocating reactions. In contrast to this, we found an inhibition of pH gradient formation which parallel the inhibition of the electron transport [31].

In summarizing the results on the inhibition of the electron transport reactions by halogenated naphthoquinones, it was shown, that their site of action is on the acceptor side of PS II, presumably at a similar place, where DCMU blocks the electron transfer between the primary electron acceptor Q and the PQ pool. Further proof for this conclusion was obtained by kinetic fluorescence measurements.

Fluorescence measurements

Typical changes in chlorophyll fluorescence intensity after onset of illumination are well known as fluorescence induction curves or as Kautsky effect. (For a recent review see [6]). These induction curves reflect in a detailed manner oxidation and reduction processes connected with the function of the two photosynthetic pigment systems. An example for the modification of a normal fluorescence induction curve after additions of DCMU or DBMIB is given in Fig. 8 and 9. Since the halogenated naphthoquinones block photosynthetic electron flow at the reducing site of PS II this should be seen as a fast fluorescence rise in the induction curve. The interpretation of fluorescence induction curves in presence of the halogenated naphthoquinones is, however, impeded by their strong quenching of chlorophyll fluorescence. Quenching occurs *in vitro* (pure chlorophyll in organic solvents) and also with a much higher efficiency *in vivo* with algae and isolated chloroplasts. As one example, a concentration of 1.5×10^{-6} M of the 2-bromo-3-methyl-1.4-naphthoquinone, which blocks the electron transport to about 20%,

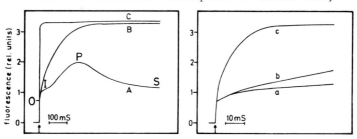

Figs. 8 and 9. Fluorescence induction in presence of 5×10^{-5} M DCMU (C, c) and 10^{-5} M DBMIB (B, b). Control: A,a. Letters at curve A mark the origin level of fluorescence (O), initial fluorescence (I), peak (P) and steady state (S).

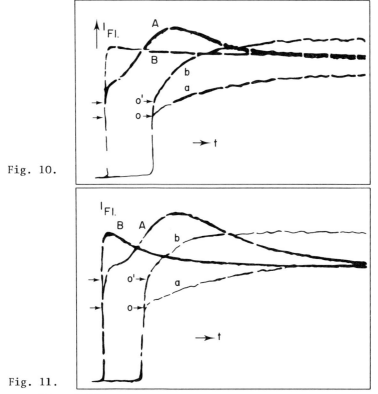

Fig. 10.

Fig. 11.

Figs 10 and 11. Fluorescence induction in presence of 2-bromo-3-methyl-1.4-naphthoquinone. Material: *Scenedesmus obliquus* D_3. A,a: Control; **Fig. 10** B,b: +5 x 10^{-6} M inhibitor; **Fig. 11** B,b: +8 x 10^{-5} M inhibitor. Capitals: time resolution: 100 ms/division, small letters: 10 ms/division.

quenches chloroplast fluorescence already by 50%. The effect of fluorescence increase in the induction curves produced by inhibition of electron transport is therefore always superimposed by a fluorescence decrease due to the quenching activities of the corresponding naphthoquinone.

From all halogenated naphthoquinones the rise in fluorescence, due to the inhibition of photosynthetic electron flow, can be seen most clearly with the 2-bromo-3-methyl-1.4-naphthoquinone (Figs 10 and 11). The OI - fluorescence rise is much faster in presence of this naphthoquinone than in the control. In addition, the O-mark is shifted to a somewhat higher level (Fig. 11, curve b). This type of affection of the induction curve points clearly to a DCMU-like inhibition of the electron transport. An inhibition of a DBMIB-type would not produce such a fast fluorescence rise in the first milliseconds after onset of illumination (compare Fig. 9, curve b).

In case of BIN quenching of the variable part of fluorescence becomes more predominant than the first fluorescence increase, due to electron transport inhibition. The latter can, however, be seen in the first 100 ms of the induction curve. At higher BIN concentrations (8 x 10^{-5}M) the variable part is mostly quenched and the ground fluorescence is affected, too, as seen by a decreased 0-level (Fig. 12, curve d). Quenching of the ground fluorescence must therefore be considered as unspecific reaction occurring only under high quinone concentrations or after longer incubation times (e.g. more than 15 min).

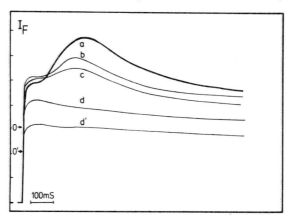

Fig. 12. Fluorescence induction in presence of 2-bromo-3-isopropyl-1.4-naphthoquinone (BIN). Material:*Scenedesmus obliquus* D . a: control; b + 4 x 10^{-6} M BIN; c: + 8 x 10^{-6} M. BIN; d + 8 · 10^{-5} M BIN; d': same as d, but after 30 min. Incubation time 3 min in darkness (curves b - d).

The results of the electron transport and fluorescence studies after addition of the halogenated naphthoquinones indicate a twofold action of these naphthoquinones (Fig. 13). At first they inhibit the electron transport chain

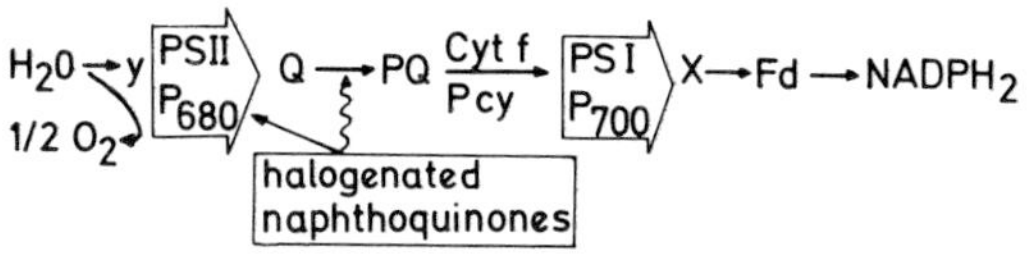

$H_2O \rightarrow y \begin{vmatrix} PSII \\ P_{680} \end{vmatrix} Q \rightarrow PQ \dfrac{Cyt\ f}{Pcy} \begin{vmatrix} PSI \\ P_{700} \end{vmatrix} X \rightarrow Fd \rightarrow NADPH_2$
$1/2\ O_2$
halogenated naphthoquinones

Fig. 13. Mode of action of the halogenated naphthoquinones on the photosynthetic electron transport chain.

before the plastoquinone pool, presumably at a similar place as DCMU does, which is seen as fast fluorescence rise. This is in agreement with the results derived from electron transport measurements. Neither in the fluorescence induction curves nor in the electron transport measurements did we find indications for a DBMIB-like inhibition of the halogenated naphthoquinones. The second property of halogenated naphthoquinones is their strong quenching of the chlorophyll fluorescence. From the fact, that low naphthoquinone concentrations only quench the variable part of the fluorescence and that this occurs with high efficiency, we assume an action of the halogenated naphthoquinones at or near the reaction centres of PS II. Their efficiency as inhibitors of the electron transport is increased by dissipating excitation energy, which is lost for conversion into chemical energy. They apparently form artificial traps, which has also been proposed for the action of some quinones [2].

Conclusion

The halogenated naphthoquinones have been shown to be effective inhibitors of the photosynthetic electron transport at the acceptor side of PS II between Q and PQ. Their inhibition site is thus clearly different from that of DBMIB. In analogy to the fact that a halogenated benzoquinone (DBMIB) blocks the plastoquinone function, we conclude that the halogenated naphthoquinones specifically block the function of the endogenous naphthoquinone phylloquinone. One basic requirement for this hypothesis, that halogenated naphthoquinones should block at a different site as DBMIB, is fulfilled. The studies with halogenated naphthoquinones thus point to a function of the phylloquinone K_1 in the linear electron transport chain (Fig. 14).

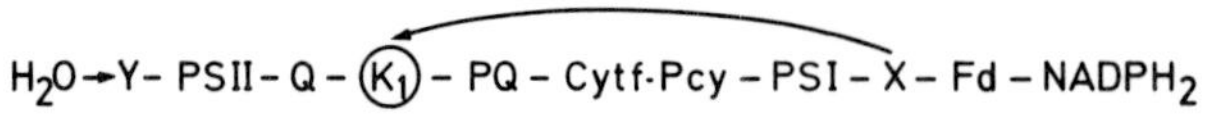

$$H_2O \to Y - PSII - Q - \overset{\frown}{(K_1)} - PQ - Cytf \cdot Pcy - PSI - X - Fd - NADPH_2$$

Fig. 14. Hypothetical localization of vitamin K_1 in the photosynthetic electron transport chain.

On the other hand the halogenated naphthoquinones proved to be strong quenchers of chlorophyll fluorescence in solution and *in vivo*. Since vitamin K_1 has similar quenching properties in solution and also when applied to intact chloroplasts (though at higher concentration) it can be deduced that the endogenous phylloquinone not only functions as an electron carrier but, in addition to this, acts as a quencher of PS II fluorescence. If so, K_1 by its quenching properties would exhibit a protective function for PS II. The high phylloquinone content on a chlorophyll basis at a stage where the pigment system II activity is already present (Table 1) - complete photosynthesis including the water splitting system are not yet functional - would be consistent with such a view. The high turnover rate of K_1, which can be deduced from its high labelling degree (Table 3) could also be explained by such a quencher function in which K_1 is degraded and gets continuously resupplied by new formation.

If phylloquinone is a component of the linear electron transport between Q and PQ it should be possible to see its reduction and reoxidation by *in vivo* spectroscopy. The difference spectrum of $K_1/K_1 \cdot H_2$ (Fig. 1) shows absorbance changes in a wavelength range where in fact changes, which are attributed to the formation of plastosemiquinone, are observed [35,38]. Thus reduction of phylloquinone may have not been observed so far because of its interference with the plastoquinone signal. In solution plastoquinone becomes reduced faster than phylloquinone, this may also occur in the photosynthetic membrane. One thus should look in future experiments for a slower component in the *in vivo* UV absorption changes.

Some data favour the view that K_1 may also participate in cyclic electron flow around PS I, e.g. synthetic naphthoquinones stimulate cyclic electron flow, enrichment of K_1 in PS I particles, low K_1 level in a PS I deficient *Scenedesmus* mutant, increased K_1 level in *Scenedesmus* cultures under photoheterotrophic growth condition which is thought to favour cyclic electron flow. If so, this would mean that electrons coming from the primary acceptor of PS I enter the linear electron transport chain at the position of K_1, between Q and PQ. The model (Fig. 14) is consistent with a function of K_1 in the linear as well as in the cyclic electron flow.

The increased K_1 level under photoheterotrophic conditions (Fig. 5) could mean that the thylakoids formed in this case contain more K_1 sites, where cycling electrons can re-enter the electron transport chain. Considering the quenching capacity of K_1, the photoheterotrophically increased K_1 pool could also indicate that more quencher molecules are formed to dissipate excitation energy absorbed by pigment system II, which is not required because this particular growth condition favours cyclic electron flow. In any case the present concept opens a new interesting view on the function of K_1 in the control of the photosynthetic light reactions.

Acknowledgements

This work was sponsored by a grant from the Deutsche Forschungsgemeinschaft. We wish to express our thanks to Professor Dr H. Musso, Institute of Organic Chemistry, University of Karlsruhe, for synthesizing the halogenated naphthoquinones and to Mrs M. Lehmann and Mrs U. Prenzel for excellent technical assistance. We also thank Dr K.

Grumbach, Dr H. Hanigk and Mr D. Meier for the disposal of unpublished data. We are especially indebted to Mrs U. Widdecke for valuable advice during the preparation of the manuscript.

References

1. Amesz, J. (1973). *Biochim. Biophys. Acta* **301**, 35-51.
2. Amesz, J. and Fork, D.C. (1967). *Biochim. Biophys. Acta* **143**, 97-107.
3. Arnon, D.I. (1961). *Federation Proc.* **20**, 1012-1014.
4. Arnon, D.I., Whatley, F. and Allen, M. (1955). *Biochim. Biophys. Acta* **16**, 1026-1027.
5. Arntzen, C.J., Neumann, J. and Dilley, R.A. (1971). *Bioenergetics* **2**, 73-83.
6. Bensasson, R. and Land, E.J. (1973). *Biochim. Biophys. Acta* **325**, 175-181.
7. Böhme, H., Reimer, S. and Trebst, A. (1971). *Z. Naturforschg.* **26b**, 341-352.
8. Dam, H. and Glavind, J. (1938). *Biochem. J.* **32**, 485-487.
9. Dam, H., Glavind, J. and Nielsen, N. (1940). *Hoppe-Seyler's Z. physiol. Chem.* **265**, 80-87.
10. Dam, H., Hjorth, E. and Kruse, I. (1948). *Physiol. Plantarum* **1**, 379-381.
11. Govindjee and Papageorgiou, G. (1971). *In* "Photophysiology". (A.C. Giese, ed). pp. 1-44, Academic Press, New York.
12. Grumbach, K.H. (1976). Thesis, University of Karlsruhe.
13. Kegel, L.P. and Crane, F.L. (1962). *Nature*, **194**, 1282.
14. Kessler, E., Langner, W., Ludwig, I. and Wiechmann, H. (1963). *In* "Studies on Microalgae and Photosynthetic Bacteria" (Japanese Society of Plant Physiologists, ed.), pp. 7-20, University of Tokyo Press, Tokyo.
15. Lichtenthaler, H.K. (1962). *Planta* **57**, 731-753.
16. Lichtenthaler, H.K. (1963). *Plant Physiol.* **38**, Suppl. XI.
17. Lichtenthaler, H.K. (1968). *Planta* **81**, 140-152.
18. Lichtenthaler, H.K. (1968). *Z.f. Pflanzenphysiol.* **59**, 195-210.
19. Lichtenthaler, H.K. (1969). *In* "Progress in Photosynthesis Research" (H. Metzner, ed.). Vol. I, pp. 304-314, International Union of Biological Sciences, Tübingen.
20. Lichtenthaler, H.K. (1969). *Biochim. Biophys. Acta* **184**, 164-172.
21. Lichtenthaler, H.K. (1977). *In* "Lipid and Lipid Polymers in Higher Plants" (M. Tevini and H.K. Lichtenthaler, eds), pp.231-258, Springer-Verlag, Berlin.
22. Lichtenthaler, H.K. and Becker, K. (1975). *Z. f. Pflanzenphysiol.* 296-302.
23. Lichtenthaler, H.K. and Calvin, M. (1964). *Biochim. Biophys. Acta* **79**, 30-40.
24. Lichtenthaler, H.K. and Grumbach, K.H. (1974). *Z. Naturforschg.* **29c**, 532-540.
25. Lichtenthaler, H.K., Karunen, P. and Grumbach, K.H. (1977). *Physiol. Plantarum* **40**, 105-110.

26. Lichtenthaler, H.K. and Kleudgen, H.K. (1975). *Z. Naturforschg.* **30c**, 64-66.
27. Lichtenthaler, H.K. and Prenzel, U. (1977). *J. Chromatography* **135**, 493-498.
28. Mathis, P. (1969). *Photochem. Photobiol.* **9**, 55-63.
29. Öttmeier, W., Reimer, S. and Trebst, A. (1977). *In* Abstracts of the IV. International Congress on Photosynthesis, p. 279, Reading.
30. Papageorgiou, G. (1975). *In* "Bioenergetics of Photosynthesis" (Govindjee, ed.)., pp. 319-371. Academic Press, New York.
31. Pfister, K. (1977). Thesis, University of Karlsruhe.
32. Strotmann, H. and Gösseln, C. (1972). *Z. Naturforschg.* 27b, 445-455.
33. Threlfall, D.R. and Whistance, G.R. (1977). *Phytochemistry*, in press.
34. Trebst, A. (1974). *Ann. Rev. Plant Physiol.* **25**, 423-458.
35. Van Gorkom, H.J. (1974). *Biochim. Biophys. Acta* **347**, 439-442.
36. Wessels, J.S.C. (1954). *Rec. des Trav. chim. des Pays-Bas* **73**, 529-536.
37. Witt, H.T. (1971). *Quarterly Rev. Biophys.* 4, 365-477.
38. Witt, H.T. (1973). *FEBS Lett.* **38**, 116-118.

OXYGEN-DEPENDENT PHOTOOXIDATIONS IN PHOTOSYSTEM II OF ISOLATED CHLOROPLASTS

P.H. HOMANN

Department of Biological Science and Institute of Molecular Biophysics, Florida State University, Tallahassee, Florida (USA)

Introduction

The light-dependent charge separations in photosystem II of chloroplasts result in the reduction of electron carriers in the electron transport chain between the photosystems, and in the formation of oxidized intermediates on the water-splitting site of photosystem II. According to our present knowledge (for review see ref. [10]), the oxidized intermediates have to be stored by each PS II reaction center until four of them have been accumulated to allow the formation of one oxygen molecule from water. The exact nature of the stored oxidants, and their oxidation potentials, are not known. Even though they must be assumed to be of high oxidative power, they are surprisingly stable and react only slowly with components of the thylakoids. However, certain chemicals destabilize the stored oxidized equivalents. Among these, two major types can be distinguished. The first serves as direct substrate for the oxidants, while for the second neither a permanent oxidation of the chemical, nor a net gain of electrons in the electron transport chain beyond the reaction center complex of PS II has been observed. Agents of the latter type which divert the oxidizing equivalents in a yet unknown fashion have been called ADRY ("acceleration of the deactivating reactions of the water-splitting system Y") reagents by Renger [14]. Their structural prerequisite is an acidic OH or NH group attached to a hydrophobic, usually aromatic skeleton which makes such compounds proton carriers across energy transducing membranes, and thus uncouplers of electron transport from phosphorylation [16].

One class of ADRY reagents are the carbonylcyanide-phenyl-hydrazones. Yamashita *et al.* [24] have shown that illuminated chloroplasts in the presence of these compounds oxidatively destroy carotenoid components of their membranes. For some other ADRY reagent Vater [22] found that it facilitates the oxidation by PS II of reduced electron carriers between the photosystems.

In the present investigation it is shown that agents which are capable of acting as ADRY reagents do not necessarily possess uncoupling activities. A common feature appears to be that they stimulate the light-dependent oxygen uptake by isolated chloroplasts. In contrast to the Mehler-type [12] oxygen utilization which is supported by electrons donated directly to PS II by the added chemicals, the stimulated oxygen consumption in the presence of ADRY reagents is more or less independent of photosystem I activity. It is also sensitive to very low concentrations of PMS. A close correlation exists between the lowering of the chloroplast fluorescence, which has been noted previously in the presence of certain ADRY reagents [6,15,24], and the light-dependent oxygen consumption.

The experimental data suggest that certain ADRY reagents divert oxidized equivalents from the water-splitting site of photosystem II into oxygen requiring reactions involving chloroplast constituents and to some extent even components of the suspension medium.

Material and Methods

Usually, chloroplasts were prepared from leaves of field grown pokeweed (*Phytolacca americana* L.] in a medium at pH 7.8 containing 400 mM sucrose, 25 mM Tricine, 6 mM

Mg(OH)$_2$ and 5 mM NaCl. Qualitatively similar results were obtained with chloroplasts isolated from pea leaves [*Pisum sativum* L., var. Progress 9). Details of the procedure have been described earlier [16].

The inactivation of photosystem I was carried out using the cyanide treatment of Ouitrakul and Izawa [13] as modified by Yokum and Guikema [25]. Any PS I activity which reappeared during the subsequent washing was suppressed by an inclusion of 1.5 mM KCN in the reaction medium [3]. The NH$_2$OH treatment to inactivate oxygen evolution was done according to the procedure of Cheniae and Martin [2].

Measurements of light intensities, oxygen uptake, electron transport, and chloroplast fluorescence were performed as previously [7]. Flash groups were obtained with an EG & G Multiflash Model 533 as used earlier by Schmid and Gaffron [17], and chlorophyll was determined according to MacKinney [11].

Unless otherwise stated, all experiments were performed at room temperature (ca 26°C) and with green light (λ_{max} = 560 nm, bandwidth at half-maximal intensity 100 nm). The reaction medium usually was identical to the isolation medium. PCB$^-$ as its tetramethylammonium salt was kindly supplied by Dr M.F. Hawthorne, ANT 2s by Dr G. Renger (Technical University, Berlin) and by Dr K.H. Büchel (Bayer A.G., Leverkusen, F.R.G.), and X464, a nigericin analog. by Dr R.L. Harned of the Commercial Solvents Corp., Terre Haute, Indiana. All other chemicals were obtained commercially.

Experimental Results

The fluorescence yield of chloroplasts is markedly lowered by the presence of the ADRY reagents CCCP and ANT 2s [6,12,16]. In the course of studies with membrane-active chemicals I noted that PCB$^-$ had a similar fluorescence lowering capacity. Very low concentrations of PMS reversed this effect almost entirely (Fig. 1). Because of

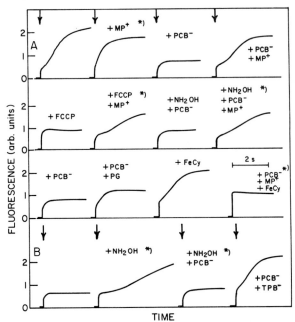

Fig. 1. Effects of various agents, and combinations of agents, on the fluorescence induction of isolated chloroplasts. Pokeweed chloroplasts (11 µg Chl/ml) in normal medium; 2 mW/cm^2 green light. Additions where indicated: MP$^+$: 1 µM; PCB$^-$: 0.1 µM; FCCP : 15 µM; n-propylgallate (PG) : 600 µM; NH$_2$OH : 2 mM; ferricyanide (FeCy) : 0.5 µM; TPB$^-$: 200 µM. Traces designated by *) were recorded with the same suspension subsequent to the preceding trace.
A : normal chloroplasts; B : NH$_2$OH treated chloroplasts.

the tendency of borane-derived anions to bind organic and inorganic cations, the action of PMS could have been due to a complex formation between PCB$^-$ and MP$^+$. However, also the FCCP induced fluorescence lowering was significantly inhibited by PMS (Fig. 1) and similarly, but to a varying degree, that induced by ANT 2s. This indicated some common mechanism in the action of these chemicals on chloroplasts.

Since the fluorescence lowering was observed only in the absence of the inhibitor DCMU (unless the concentrations of the agents were very high), it was concluded that the added chemicals facilitated an electron transport which kept the primary acceptor Q of PS II oxidized. Under my experimental conditions, oxygen was the only abundant electron acceptor. Its participation in the reaction was confirmed by the absence of any significant fluorescence lowering under nitrogen (Fig. 2) and an increased oxygen consumption by illuminated chloroplasts when PCB^-, FCCP, or ANT 2s were added (Table 1).

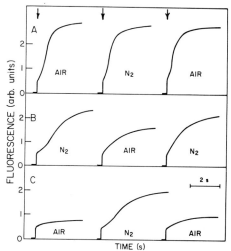

Fig. 2. Effect of anaerobiosis on the fluorescence quenching by FCCP and PCB^-. Pokeweed chloroplasts (10 µg Chl/ml) in standard medium; 1.5 mW/cm² green light.
A: control; B: +12 µM FCCP; C: +0.1 µM PCB^-.

TABLE 1

Effect of an inactivation of photosystem I on light-dependent oxygen consumption in chloroplast suspensions

Addition Chloroplast preparation	10 mM Me-NH$_2$ 0.1 mM Me-viol	10 mM NH$_2$OH	0.25 µM PCB$^-$	13 µM FCCP	0.5 µM ANT 2s	None
Control	16.0	8.5	7.0	5.0	5.0	1.5
CN$^-$-treated	<1	1.5	6.0	4.5	1.5	0.5
Inhibition by CN$^-$ treatment	>95%	80%	15%	10%	70%	70%

Oxygen values in µmoles O_2 consumed/mg Chl x h. Pokeweed chloroplasts, 30 µg Chl/ml; green light 1 mW/cm².

The reducing site of photosystem I was considered the most likely site of electron acceptance by oxygen. Surprisingly, however, an inhibition of photosystem I activity by a cyanide treatment did not much affect the oxygen consumption in the presence of FCCP or PCB$^-$, but it significantly lowered that in the presence of ANT 2s. Methylviologen is known to be reduced in photosystem I and, consequently, the associated oxygen consumption was completely abolished by a cyanide treatment. Electrons donated to PS II of illuminated chloroplasts by NH_2OH apparently also reached molecular oxygen largely through PS I. The same was found with the electron donor TPB$^-$ (not shown). DCMU inhibited the oxygen consumption in all cases.

If oxygen utilization and fluorescence lowering were directly related, they should, under all experimental conditions, respond similarly to a cyanide treatment. From Table 2 it can be seen that the correlation was qualitatively as expected, but the sensitivity of the fluorescence effect to a cyanide treatment was lower in the presence of ANT 2s and higher in the presence of FCCP. On the other hand, Fig. 3 documents a close correlation between the dependence of the fluorescence lowering, and the rate of oxygen consumption, on the concentrations of FCCP and ANT 2s. Figs. 1 and 3 and Table 2 show, furthermore, that both

TABLE 2

Effect of an inactivation of photosystem I on the fluorescence lowering effect of various chemicals

Addition Chloroplast preparation	10 mM Me-NH$_2$ 0.1 mM Me-viol	0.10 µM PCB$^-$	0.10 µM PCB$^-$ 1.5 µM PMS	0.25 µM ANT 2s	5 µM FCCP
Control	63	52	80	73	55
CN$^-$ treated	98	49	90	82	63
Inhibition by CN$^-$ treatment	95%	0%	50%	35%	40%

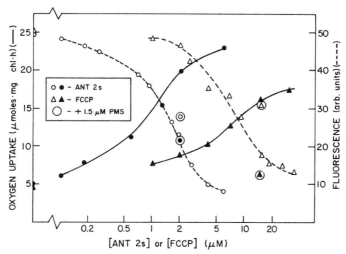

Fig. 3. Dependence of oxygen consumption by chloroplasts, and of the fluorescence quenching, on the concentration of ANT 2s and FCCP, and effect of PMS. Pokeweed chloroplasts (17 μg Chl/ml) in a medium containing 350 mM sucrose, 25 mM Na-phosphate buffer pH 7.6, 1 mM NaN_3 and 10 mM CH_3NH_3Cl; 2 mW/cm^2 green light.

processes were equally sensitive to additions of small amounts of PMS. For still unexplained reasons, the response to an addition of PMS was quite variable (compare Figs 3 and 4). Its prevention by low concentrations of ferricyanide (Fig. 1) suggested an involvement of products formed by a photoreduction of the dye.

According to the experimental results, PCB$^-$, FCCP, and ANT 2s stimulated oxygen requiring photoreactions in chloroplasts of which at least some were confined to photosystem II. Yamashita et al.[24] had previously noted a CCCP induced oxygen uptake by illuminated chloroplasts which they correlated with a destruction of carotenoids. Heath and Packer [4] studied the peroxidation of thylakoid lipids during an illumination of suspensions of isolated chloroplasts and linked it to actions of photosystem II. In the course of an extension of these studies using chloroplasts in the absence and in the presence of various inhibitors, including CCCP, Takayama and Nishimura [20,21] reached a similar conclusion. That my observations might be related to peroxidation reactions in chloroplasts was indicated by the partial reversal of the fluorescence lowering when the antioxidant n-propylgallate had been added (Fig. 1). Additional evidence came from the stoichiometry of the oxygen consumption and its dependence on the suspension medium.

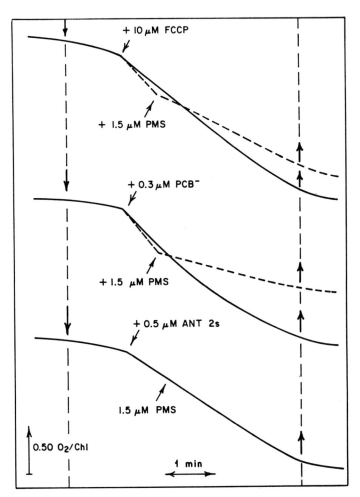

Fig. 4. Oxygen uptake by illuminated chloroplasts in the presence of FCCP, PCB⁻, or ANT 2s, and the inhibitory effect of PMS. Pokeweed chloroplasts (30 µg Chl/ml) in a medium containing 300 mM sucrose, 25 mM Na-phosphate buffer pH 7.6, 1 mM NaN$_3$, and 10 mM CH$_3$NH$_3$Cl (except the PCB⁻ experiment). 2.5 mW/cm^2 green light. Note that in this experiment PMS did not inhibit the ANT 2s stimulated oxygen uptake.

The data of Fig. 4 reveal that 1 PCB⁻ induced the uptake of several oxygen molecules, and that more oxygen was taken up in suspensions containing sucrose as osmoticum. The same observations were made with other oxidation stimulating agents, including TPB⁻. H_2O_2 was formed in all cases (see also Fig. 5).

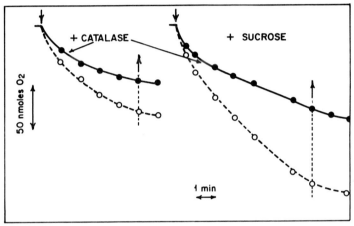

Fig. 5. Dependence of the PCB⁻ stimulated oxygen uptake by chloroplasts on the composition of the suspension medium, and the presence of catalase. Pokeweed chloroplasts (20 μg Chl/ml) in 50 mM Na-phosphate buffer pH 7.6 plus or minus 300 mM sucrose and 0.1 mg catalase/ ml. 1 mM NaN$_3$ had been added when no catalase was present. 30 mW/cm² red light.

Both ANT 2s and FCCP are well-established representatives of ADRY reagents [14,15]. It was of interest to see, therefore, whether PCB⁻ belonged to the same class of compounds. An ADRY reagent is expected to lower the average yield of oxygen per light flash in a series of flashes when the individual flashes are separated by a relatively long dark time. The reason is that a lengthened dark interval increases the chance for stored oxidants to be reduced in side reactions [14]. Table 3 provides data from flash experiments which strongly suggest that PCB⁻ indeed acts like an ADRY reagent. In accordance with this conclusion, PCB⁻ also prevented the cyclic reoxidation of Q⁻ in DCMU poisoned chloroplasts by removing the oxidized

TABLE 3

Oxygen yields in flash groups with differently spaced flashes

Number of flashes per group	Flash spacing (ms)	Number of groups	Relative amounts of oxygen evolved		
			no addition	+0.75 μM FCCP	+0.60 μM PCB$^-$
90	16	2	1.0 ± 0.1	0.9 ± 0.1	0.9 ± 0.1
15	100	12	0.9 ± 0.1	0.4 ± 0.0	0.6 ± 0.0
% inhibition in slower flash sequence			10	55	33

Reaction medium: 300 mM sucrose, 25 mM Na-phosphate buffer, pH 7.6, 5 mM $MgCl_2$, 15 mM methylamine, 15 mM K-ferricyanide, 15 mM K-ferrocyanide, 2.5 mM Tricine-MgO, pH 7.6. 250 μg chlorophyll in 4 ml. Spacing of flash groups approximately 300 ms. Oxygen evolution = 1 for 1 O_2 per 3500 chlorophylls. Data are averages from 4 determinations.

In order to eliminate the inwardly directed proton release due to the oxidation of plastoquinone by photosystem I we added the plastoquinone antagonist dibromothymoquinone (DBMIB) [30,31]. DBMIB does not inhibit the

TABLE 4

Reoxidation of Q^- in DCMU-poisoned chloroplasts measured as percent restoration of area over the fluorescence induction curve

Duration of first illumination	No addition	20 μM PMS	0.1 μM PCB$^-$	0.1 μM PCB$^-$ 20 μM PMS	1 μM FCCP	1 μM FCCP 20 μM PMS
5 s	60	40	<10	25	<10	25
30 s	40	10				

Reaction medium same as isolation medium, plus 8 μM DCMU; illumination with 1.0 mW/cm^2 green light.

period [16]. The table also shows that PMS inhibited the back reaction, especially after an extended light period, yet allowed the oxidation of some Q^- even in the presence of PCB^- or CCCP. Some inhibition of the reoxidation of Q^- was observed after a prolonged light period also in the absence of any retarding agent. The degree of these effects varied significantly between different chloroplast preparations.

The partial reversal by PMS of the inhibition of the reoxidation of Q^- in the presence of CCCP had been observed independently by Schmidt [18]. More data are depicted in Fig. 6 from an experiment in which the oxidizing action of PMS on Q^- was ascertained in the presence of CCCP, an ADRY reagent, and NH_2OH, an electron donor to photosystem II.

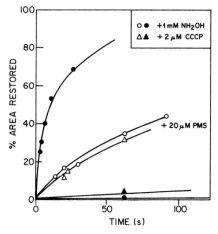

Fig. 6. Stimulation of the cyclic back reaction in photosystem II after inhibition with CCCP or NH_2OH. Tobacco chloroplasts (10 μg Chl/ml) in normal medium +8 μM DCMU; control illuminated with 1 mW/cm^2, others with 0.5 mW/cm^2 green light. Time scale gives time in darkness after the first illumination; relative size of restored area was measured in a second illumination of the same intensity.

The action of PMS is difficult to explain, Slovacek and Bannister [19] have observed that PMS increased the fluorescence yield of DCMU poisoned chloroplasts in weak light when some ascorbate was present. I saw the same phenomenon even in the absence of ascorbate during a prolonged illumination with weak red light. These findings supported the notion [8,18] that PMS photoreduced and that reduced PMS acts as electron donor to PS II. They also agreed with the observed inhibition of the back reaction in the presence of PMS when the illumination period was lengthened (Table 4). If the fluorescence yield of DCMU poisoned chloroplasts under the prevailing experimental conditions was solely determined by the oxidation state of Q^-, one has to postulate that the reoxidation of Q^- by PMS became increasingly inhibited, perhaps by the formation of a stable charge transfer complex between Q^- and the PMS derived cation MP^+.

It was established above that PCB^- has ADRY activity since it induced a loss of oxidizing equivalents from the water-splitting site of PS II. The data of Table 5 show that PCB^-, like other ADRY reagents, did not restore noncyclic electron flow in chloroplasts with an impaired photosystem II. In fact, PCB^- inhibited noncyclic electron flow from NH_2OH to DCPIP (now shown) like that from H_2O to DCPIP (Table 5). Its successful competition with NH_2OH for the oxidants in PS II can also be deduced from the ability of PCB^- to lower the fluorescence of normal chloroplasts, and of chloroplasts with an inactivated oxygen evolving system, even after an addition of NH_2OH as electron donor. This is shown in Fig. 1 which also reveals that, in contrast, electron donation to the oxidants by TPB^- won over the oxidant inactivation induced by PCB^-. This was confirmed by measurements of electron transport. From other data of Table 5 it is obvious that PCB^- is an inhibitor of electron transport even at very low concentrations and not, like other ADRY reagents [14], an uncoupler.

Discussion

The experiments reported here establish a clear relation between the ability of certain chemicals to deactivate the

TABLE 5

Effects of PCB^- on the Hill reaction

	Initial rate of electron transport ($\mu eq \cdot mg^{-1}\ Chl \cdot h^{-1}$)			
	No uncoupler added		Uncoupler added	
	no addition	plus addition	no addition	plus addition
Control chloroplasts + 0.4 µM PCB^-	250	210	930	650
NH_2OH treated chloroplasts				
+ 2.5 µM PCB^-	45	45*		
+15 µM PCB^-	45	20*		
+20 µM TPB^-	45	105	72	310

Control chloroplasts were assayed with ferricyanide as electron acceptor, the NH_2OH treated chloroplasts with DCPIP. Reaction mixtures same as used for isolation of the chloroplasts plus 0.6 mM ferricyanide or 0.035 mM DCPIP, and 15 mM CH_3NH_3Cl as uncoupler with the NH_2OH treated chloroplasts, or 1 µg X464 with the control chloroplasts where indicated. Pokeweed chloroplasts, 3-7 µg Chl/ml.

*In either case the reaction stopped after less than 1/2 min.

oxidant of the water-splitting enzyme in chloroplasts, their capacity to lower the chloroplast fluorescence, and their stimulating activity on the light-dependent oxygen consumption by chloroplast suspensions. The data prove that oxygen requiring photooxidations in isolated chloroplasts can occur in the absence of any photosystem I activity and thus support a similar claim made by Bekina *et al.*[1]. The study confirms that the reactivity of stored oxidants in PS II can be directed towards surrounding molecules other than those normally involved in the photosynthetic mechanism of water oxidation.

    The mechanism of the photooxidation reactions in chloroplasts remains unclear, however. Takayama and Nishimura [20,21] have presented evidence for the participation of singlet oxygen and superoxide radicals in these reactions, but even with the autoxidizable benzylviologen as electron acceptor for PS I and source for peroxyradicals, the inhibition by superoxide dismutase was only partial. In my experiments, I have not seen any measurable effect of 100 units [5] superoxide dismutase on the PCB⁻ stimulated oxygen uptake. Either the superoxide anion was no oxidation propagating intermediate, or the enzyme did not reach the site of the oxidation reactions.

It is not difficult to formulate peroxidation reactions when peroxyradical formation *via* the reductant of PS I is invoked [20]. When one restricts any considerations to PS II, the task is much more difficult. Furthermore, it has to be explained why a loss of oxidizing equivalents to reactions other than an oxidation of water or added electron donors should lead to a reduction of oxygen, and suppress noncyclic electron flow.

The almost stoichiometric formation of 1 hydrogen peroxide per oxygen taken up (Fig. 4) suggests that the products of oxygen reduction are peroxyradicals which can dismutate to oxygen and hydrogen peroxide, or that easily hydrolizable organic hydroperoxides are formed. Unless the unlikely possibility is considered that peroxyradicals can be generated by a reaction between the photooxidant in a perturbed PS II, molecular oxygen, and water, one has to postulate that the added ADRY reagent permits the oxidant to produce radicals by reacting directly, or *via* the ADRY reagent, with surrounding molecules such as carotenoids. These then could be responsible for the formation of peroxyradicals which might propagate oxidative reactions (perhaps *via* singlet oxygen as suggested by Takayama and Nishimura [20]) or dismutate to yield oxygen and hydrogen peroxide. The mechanism of the initial formation of radicals by stored oxidants in photosystem II may be similar to that postulated for the manganese-requiring oxidation of diketogulonate by illuminated chloroplasts [5]. Some of these ideas are presented in the scheme of Fig. 7.

The fluorescence lowering in the presence of ADRY reagents was shown to be directly related to the oxygen

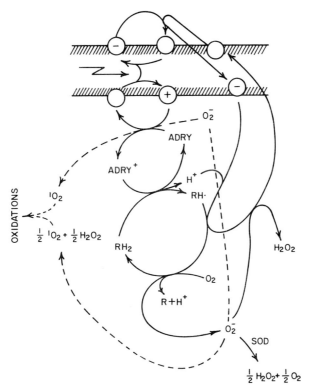

Fig. 7. Possible mechanism of photosystem II dependent photooxidations in chloroplasts. In the scheme are indicated pathways for radical production, for a cyclic electron flow in PS II, and for the formation of hydrogen peroxide and singlet oxygen (see also refs [4,20,21]).

utilizing reaction. Two possible explanations shall be considered briefly. The first would postulate that some intermediate in the oxidation reactions has fluorescence quenching properties. The somewhat different response to a cyanide treatment of the fluorescence lowering, and the rate of oxygen consumption, may support such a contention.

Alternatively, a cyclic electron flow between the reducing site, and intermediates of the photooxidation reaction, may keep the primary acceptor Q of PS II oxidized and, consequently, the fluorescence low. Since the added ADRY reagents inhibit cyclic electron flow in the presence of DCMU, any such process would have to involve electron acceptors on the reducing site of PS I, or between Q and PS I. This is essentially what Renger *et al.*[16] and Vater [22] have postulated, but these authors assumed a more direct reduction of the photooxidant in PS II.

PMS inhibited oxygen uptake as well as fluorescence lowering induced by the ADRY reagents even though PMS itself can accept electrons from the photosynthetic electron transport chain and transfer them to oxygen [9,19]. The reversal of the fluorescence lowering by PMS was sensitive to low concentrations of ferricyanide which themselves were unable to produce a significant oxidation of Q^-. This indicated an involvement of reduced PMS as intermediate. It is possible that PMS cycles electrons to intermediates of the oxidation reaction, thereby cutting short the reaction chain, and eliminating the quenching substance. If, however, the fluorescence lowering is the result of a rapid cyclic reoxidation of reduced electron carriers, reduced PMS must be assumed to retard such reoxidations conceivably by complexing with Q or a secondary acceptor as suggested before to explain the action of PMS on the back reaction in DCMU poisoned chloroplasts. This would slow the formation of oxidants in photosystem II and, consequently, the oxidative reactions. In support of such a hypothesis, it should be mentioned that PMS is an unusually poor Hill oxidant in spite of its high activity as cofactor of cyclic photophosphorylation in PS I, and inhibits noncyclic electron transport to certain other oxidants (unpublished observations; see also refs. [9,19]).

At present it is impossible to decide whether the hypothesis of an oxidant-dependent fluorescence quenching is correct, or that of a cyclic electron flow. Because it can explain the experimental data with the least number of necessary assumptions, the quenching hypothesis is to be favoured. It postulates an ADRY dependent additional path of excitation dissipation which could possibly account for the observation that a diversion of stored oxidizing

equivalents from an impaired water-splitting system by
ADRY reagents does not restore any significant electron
flow to added Hill oxidants. However, an increased cyclic
electron flow would have the same effect. In either case,
one would predict a measurable electron flow under anaerobic
conditions when the fluorescence remains high, but
this has never been observed. The problem may be that
electron transport studies are usually made in rather
strong light and with relatively high concentrations of
the active chemical. Under such conditions the rate of
destructive oxidations may preclude a normal functioning
of the photosynthetic electron transport system on the
thylakoid membrane.

The agents used in this study are likely to exert their
effects by changing membrane properties, or interacting
with electron carriers in photosystem II. The resulting
photooxidative processes clearly are not identical for
the three tested chemicals, but may serve as models for
events which may occur as the result of other experimental,
or natural disturbances of photosystem II.

Acknowledgements

This investigation was supported by Grant No. PCM 76-16975 from the National Science Foundation and by U.S. Energy Research and Development Administration Contract No. E-(40-1)-2690 with the Institute of Molecular Biophysics, Florida State University.

References

1. Bekina, R.M., Lebedeva, A.F. and Rubin, B.A. (1976). *Biochimija* **41**, 815-821.
2. Cheniae, G.M. and Martin, I.F. (1971). *Plant Physiol.* **47**, 568-575.
3. Cohen, W.S. and McCarty, R.E. (1976). *Biochem.Biophys.Res.Comm.* **73**, 679-686.
4. Heath, R.L. and Packer, L. (1968). *Arch. Biochem. Biophys.* **125**, 850-857.
5. Homann, P.H. (1965). *Biochemistry* **4**, 1902-1911.
6. Homann, P.H. (1971). *Biochim. Biophys. Acta* **245**, 129-143.
7. Homann, P.H. (1972). *Biochim. Biophys. Acta* **256**, 336-344.

8. Homann, P.H. (1975). Abstracts of the 3rd Annual Meeting of the American Society of Photobiology, p. 58.
9. Homann, P.H. (1977). *Biochim. Biophys. Acta* **460**, 1-16.
10. Joliot, P. and Kok, B. (1975). *In* "Bioenergetics of Photosynthesis" (Govindjee, ed.), pp. 388-412. Academic Press, New York.
11. MacKinney, H.G. (1941). *J. Biol. Chem.* **140**, 315-322.
12. Mehler, A.H. (1951). *Arch. Biochem. Biophys.* **33**, 65-77.
13. Ouitrakul, R. and Izawa, S. (1973). *Biochim. Biophys. Acta* **305**, 105-118.
14. Renger, G. (1972). *Europ. J. Biochem.* **27**, 259-269.
15. Renger, G. (1973). *Biochim. Biophys. Acta* **292**, 796-807.
16. Renger, G., Bouges-Bocquet, B. and Büchel, K.H. (1973). *Bioenergetics* **4**, 491-505.
17. Schmid, G.H. and Gaffron, H. (1968). *J. Gen. Physiol.* **52**, 212-239.
18. Schmidt, B. (1976). *Biochim. Biophys. Acta* **449**, 516-524.
19. Slovacek, R.E. and Bannister, T.T. (1976). *Biochim. Biophys. Acta* **430**, 165-181.
20. Takayama, U. and Nishimura, M. (1975). *Plant Cell Physiol.* **16**, 737-745.
21. Takayama, U. and Nishimura, M. (1976). *Plant Cell Physiol.* **17**, 111-118.
22. Vater, J. (1973). *Biochim. Biophys. Acta* **325**, 149-156.
23. Winterbourn, C.C., Hawkins, R.E., Brian, M. and Carrell, R.W. (1975). *J. Lab. Clin. Med.* **85**, 337-342.
24. Yamashita, K., Konishi, K., Itoh, M. and Shibata, K. (1969). *Biochim. Biophys. Acta* **172**, 511-524.
25. Yokum, C.F. and Gnikema, J.A. (1977). *Plant Physiol.* **59**, 33-37.

PROTON RELEASE DURING PHOTOSYNTHETIC WATER OXIDATION: KINETICS UNDER FLASHING LIGHT

W. JUNGE and W. AUSLÄNDER

Max-Volmer-Institut für Physikalische Chemie und Molekularbiologie, Technische Universität Berlin, Berlin, Germany.

Abstract

The release of protons into the internal phase of thylakoids is detected by spectrophotometric techniques. Under flash excitation of chloroplasts proton release inside follows a complex kinetic pattern. Half of the extent occurs with a half-rise-time of approximately 20 ms. It is attributable to the oxidation of plastohydroquinone and will not be furthermore discussed. The other half, with rise times ranging from 100 μs to 2 ms, is related to the oxidation of water by photosystem II. Resolution of the extent and the kinetics of proton release due to water oxidation under excitation with a series of short flashes (with the water oxidizing enzyme system in its dark-adapted state) produces a pattern which deviates from the well known pattern of oxygen evolution. While the oxygen evolution follows a stoichiometry 0:0:0:4 when related to the transitions between the four oxidation states of the enzyme system $(S_0-S_1-S_2-S_3-(S_4--S_0)-S_1-)$ proton release can be approximated by a stoichiometric pattern 0:1:1:2. At least three different kinetic phases of proton release can be distinguished. The slowest one (half-rise-time 1-2 ms) is correlated with the transition S_3-S_0. The fastest phase (half-rise-time less than 100 μs) seems to be attributable to the transition S_2-S_3 and a third phase (300-600 μs) possibly belongs to the transition S_1-S_2. These data clearly show that the reaction step which leads to the liberation of di-oxygen is not the only transition of the water oxidizing enzyme system which liberates protons. This is compatible with a suggested model, (see [28]) which assumes that the enzyme system interacts with water already at its earlier oxidation states. On the other hand, further kinetic studies are required for an unequivocal attribution of the proton release to the well known transitions of the water oxidizing enzyme system.

Introduction

For several years measurements of the oxygen evolution and of luminescence phenomena (for reviews, see [17,23]) were the only means of access for students of the water oxidizing enzyme system. Based on such studies it was suggested that the enzyme system cycles through five oxidation states ($S_0,\ldots,S_4$), with four transitions requiring the input of one quantum into photosystem II, before oxygen is liberated [22]. However, the chemical identity of the only formally defined S states has hitherto remained unknown. More recently the role of manganese was probed by water-proton MR [33] and one intermediate between the S states and the photochemical center of PS II became visible by EPR signal II_{vf} [6]. In this communication we are discussing the release of protons inside thylakoids during the transitions of the water oxidizing enzyme system from one oxidation state to the other. This approach may eventually help to define which of the S states involve water and its oxidation products (see [28]).

We have shown previously by spectrophotometric techniques that water oxidation causes proton release into the internal phase of thylakoids. First indirect evidence for this was that the protons generated during electron transport from water to the non-proton-binding system II acceptor ferricyanide (high concentration) appeared in the outer phase at a velocity which correlated with the membrane's proton conductivity [3,20]. The same conclusion was drawn by Fowler and Kok [13] based on work with a rapid glass electrode. Later we introduced the pH-indicating dye neutralred [4] to obtain an extremely sensitive and kinetically highly resolving indication of flash-induced proton release into the internal phase.

We studied proton release inside by this technique under excitation of chloroplasts with a group of short flashes, with the water oxidizing enzyme system in its dark equilibrium before firing of the flash group [21]. We observed that the pattern of proton release deviates from the well known pattern of oxygen evolution. While the former could be approximated by a stoichiometric sequence 0:1:1:2 the latter followed the sequence 0:0:0:4, as usual (related to the transitions, S_0-S_1, ..., S_3-S_4--S_0). In this paper we report on the time resolution of the proton release during the subsequent flashes. It is shown that two kinetic components are compatible with the expected velocity of the transitions between states S_1-S_2 and S_3-S_4, while one component (half-rise-time 100 μs) which correlates with the S_2-S_3 transition is faster than the expectation for the electron transport as based on EPR data [6].

Since experimental details and some of the results are published elsewhere [21], we will discuss here only those items which may become relevant in a later comparison with related results from other authors. Fowler [10,11,13] has published conflicting data on proton release in a group of short flashes which ranged from a reasonable fit to the pattern of oxygen release [10] to a pattern with some proton liberation during all transitions between S states [11]. Saphon and Crofts [29] obtained a pattern of oxygen release (1:0:1:2) which deviates from our observations as well. The reasons for this discrepancy are unknown at present.

Methods

Detection of inwardly liberated protons by neutralred:

Neutralred is a pH-indicating dye (pK=6.8) which distributes over the external, the internal and the membrane phase of thylakoids. Under flash excitation of chloroplasts in a sensitive spectrophotometer (for instrumentation, see [18]) absorption changes of neutralred are observed which may be due to pH transients but also to other reactions of this dye in response to the light flash. We have previously shown, [4] that the pH_{in}-indicating absorption changes of neutralred can be extracted under the following conditions:
1.) The external phase is strongly buffered by a non-permeating buffer as, for instance, bovine serum albumin.
2.) The absorption changes are recorded twice, first in the absence and then in the presence of a permeating buffer, for instance, imidazole. 3.) Subtraction of the latter signals from the former yields the pH_{in}-indicating absorption changes of neutralred, while eliminating the response of the dye to any other event.

This technique has the following valuable properties:
1.) The pH_{in}-indicating absorption changes of neutralred are very probably due to pH transients in the internal osmotic volume of thylakoids and not to pH changes within the thylakoid membrane. We were led to this conclusion by the observation that highly hydrophilic buffers like, for instance, phosphate can substitute for imidazole in buffering away the pH_{in}-indicating absorption changes, (these results are published elsewhere, see [5]). Phosphate will not likely accumulate within the thylakoid membrane.
2.) The response of neutralred to the small pH-changes which are induced on flash excitation of chloroplasts can be calibrated. As described elsewhere [5] there is good evidence that the inside located neutralred molecules conserve their pH-value (6.8). As the buffering capacity of

the internal phase is approximately constant from pH 8 to
pH 6.5 and as the amount of neutralred inside is not
changed markedly due to the inwardly directed diffusion of
this dye (in response to the light-induced acidification),
the response of neutralred is quantitative in this narrow
pH range. The inwardly directed diffusion of the weak acid
can be neglected only because the amount of neutralred inside (in the dark) is approximately tenfold higher than
expected for equal distribution between the two aqueous
phases. This accumulation was observed in our laboratory
as well as by Pick and Avron [24]. As described in greater
detail elsewhere [5] we calibrated the pH change induced
by excitation of both photosystems with a short flash of
light to 0.15 units (at pH around 7). Below an internal pH
of 6.5 units, neutralred can be used only as a qualitative
indicator (see [24]).

3.) The response of neutralred is very fast. We resolved
rise-times of the internal acidification down to 100 μs
(see below).

The scheme for the interrelationship between photosynthetic electron transport and proton uptake and release
which resulted from our flash spectrophotometric studies
with pH-indicating dyes is represented in Fig. 1 (for a
review, see [19]). The stoichiometric features of this
scheme and the site attribution of protolytic reactions
were independently established by other techniques (for
a review, see [14]). This communication is focussed on
the site of rapid proton release (half-rise-times 100 μs
to 2 ms) attributable to the water oxidizing enzyme system.

Blocking proton release between the two photosystems by DBMIB

In order to eliminate the inwardly directed proton
release due to the oxidation of plastoquinone by photosystem I we added the plastoquinone antagonist dibromothymoquinone (DBMIB) [30,31]. DBMIB does not inhibit the

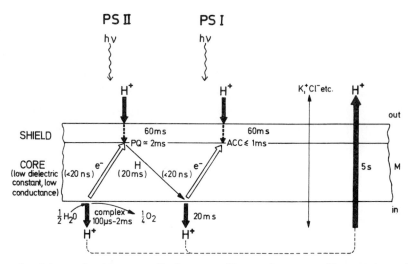

Fig. 1. Schematic diagram of electron flow and proton flow across the thylakoid membrane under flash excitation of chloroplasts based on work published elsewhere (see [3,4,20]) and on data on the electric potential generation which were reviewed recently [19,32]. Two sites of protolytic reactions are linked to each photosystem as indicated by fat arrows, each with a stoichiometry of 1 H^+/e^- under excitation with single turnover flashes. While proton release inside is rather fast, proton uptake from the outer phase is delayed due to a diffusion barrier for protons shielding the redox reactions from the outer phase [4]. The times indicated in the figure denote half-rise-times of the respective events (for the protons, see [3,4,20] - for the electron transport and for the electric field, see [32]).

slow dark reduction of chlorophyll a_I by some unknown donor, however, it inhibits its rapid re-reduction *via* plastoquinone and it inhibits proton release coupled therewith, as shown previously [2,4]. Fig. 2 again illustrates how DBMIB abolishes the slow phase of proton release inside while not affecting the rapid phase which we attribute to the water oxidizing enzyme system.

Acceleration of the relaxation between the S states by addition of ANT 2s:

To improve the signal-to-noise ratio in the spectrophoto-

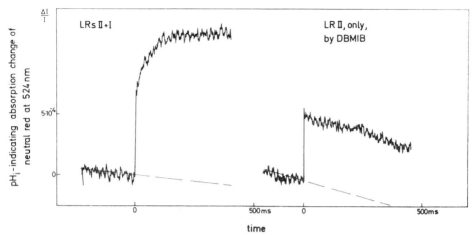

Fig. 2. Time course of the pH_{in}-indicating absorption changes of neutralred at 524 nm under flash excitation of chloroplasts at t=0. Left: with both photosystems active in proton release inside, right: with only photosystem II active in proton release, due to the presence of DBMIB. The experimental conditions were given elsewhere (see [4]).

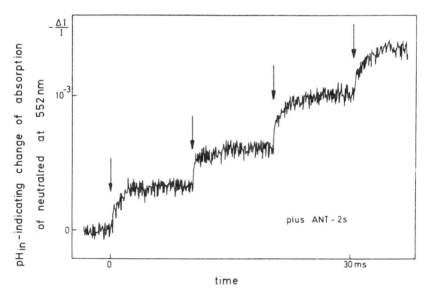

Fig. 3. Time course of the pH_{in}-indicating absorption changes of neutralred at 552 nm under excitation of chloroplasts with groups of four short flashes (duration: 15 µs, spacing: 10 ms, repetition rate: 0.1 Hz). DBMIB (3 µM) and ANT 2s (0.6 µM) were present in the reaction medium. The reaction conditions were given elsewhere [21].

metric experiments at high time resolution we preferred repetitive excitation of chloroplasts with flash groups and averaging over the repetitive signals. To make sure that the S states relaxed to their dark equilibrium distribution during the interval between successive flash groups (5-10 s), 2(3,4,5,-trichloro)-anilino-3,5-dinitrothiophene (ANT 2s) was added to the chloroplast suspension. ANT 2s is an ADRY-agent as defined by Renger [25]. As our experiments were carried out in the presence of bovine serum albumin the ANT 2s concentrations had to be higher than usual to overcome the adsorption of this agent to BSA. That ANT 2s was effective under the given conditions is evident from the oscillatory pattern of the oxygen liberation (see [21]).

Results

Fig. 3 shows the pH_{in}-indicating absorption changes of

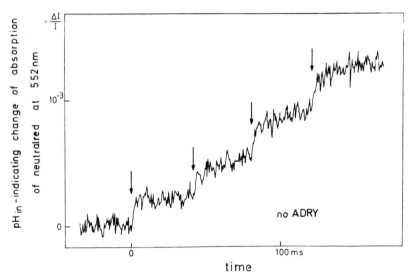

Fig. 4. Time course of the pH_{in}-indicating absorption changes of neutralred at 552 nm under excitation of dark adapted chloroplasts by a group of four short flashes (duration: 15 μs, spacing 40 ms, no repetition). DBMIB (3 μM) was present in the reaction medium, but no ADRY agent (Ant2 s). The trace represents the average over 10 signals, which were obtained with chloroplasts dark adapted as described in the text. Further chemical conditions as in [21].

neutralred under excitation of chloroplasts with a group of four short flashes spaced 10 ms apart from each other. The experiment was carried out under repetitive excitation (repetition frequency of flash groups: 0.1 Hz) in the presence of ANT 2s to accelerate the relaxation of the S states. It is evident that protons are released after each flash. In a parallel experiment, Dr Renger measured the amount of oxygen liberated per flash [21]. The result is shown in Fig. 5 by full points connected by a solid line. It reveals the well known pattern with low oxygen yield in

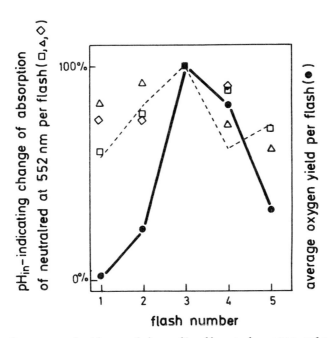

Fig. 5. Oxygen production and inwardly directed proton release as function of the flash number normalized at the third flash. Full circles: oxygen [21], open symbols: protons. The broken line characterizes the expectation for the proton release under the specific assumptions given in the text.

flashes number 1 and 2 and a high yield in the third flash. This demonstrates clearly that ANT 2s was effective as an ADRY-agent under our experimental conditions. As another check for possible artefacts due to the repetitive measuring technique, we repeated the experiment shown in Fig. 4, but now with dark-adapted chloroplasts, which were exposed only to one flash group. The reaction mixture was pipetted in the dark, then inserted into an absorption cell mounted in the spectrophotometer while still kept in the dark. Then the measuring light was turned on with an energy of only 20 $\mu J/cm^2$ (at 552 nm) during an interval of 200 ms, during which a group of four short flashes was fired. The energy of each flash was 2 mJ/cm^2, sufficient to saturate photosystem II. As half saturation occurred at an energy input of 150 $\mu J/cm^2$ (at wavelength matched to the peak absorption of chloroplasts), the input of 20 $\mu J/cm^2$ at a wavelength, where the action spectrum is low, could not have altered the dark equilibrium between the S states. The result of this experiment is shown in Fig. 4. It is obvious that the pattern of proton release of dark-adapted chloroplasts is very similar to the one observed with repetitive technique in the presence of an ADRY-agent.

These results are summarized in Fig. 5. Open symbols describe the extent of proton release inside in dependence of the flash number, closed circles give the extent of oxygen production and the dotted line shows the theoretically expected proton release under the following assumptions: 1.) the dark equilibrium between the S states ($S_0:S_1:S_2:S_3$) is 0.25:0.75:0:0, 2.) the probability of double hits and of misses, respectively, is 10% (see [1,9]) and 3.) the stoichiometric ratio of proton release during the transitions S_0-S_1, S_1-S_2, S_2-S_3, S_3-S_4--S_0 is 0:1:1:2. In view of the relatively large scatter in the experimental points on proton release it is certainly premature to accept the stoichiometric pattern of proton release (0:1:1:2) as quantitative, however, it is evidently a fair approximation.

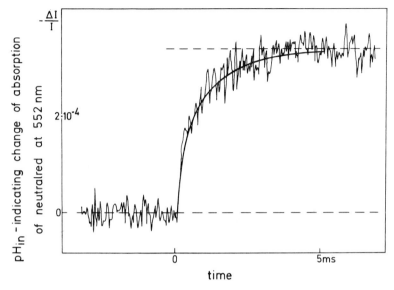

Fig. 6. Time course of the pH_{in}-indicating absorption changes of neutralred at 552 nm under excitation of chloroplasts with a short flash at t=0 (duration: 15 µs, repetition rate: 0.2 Hz). DBMIB (3 µM) was present in the reaction medium to eliminate proton release by photosystem I, but no ADRY agent was present. Therefore the S states were equally populated at the chosen repetition rate.

We asked whether proton release during the subsequent oxidation reactions of the oxygen evolving enzyme system occurs at different relaxation rates. From earlier studies [1] we knew that proton release by the water oxidizing enzyme system under repetitive excitation with no ADRY agent present (and hence with all S states equally populated) is kinetically multiphasic. This is shown in Fig. 6. The solid line represents a theoretical fit to the experimental curve based on the superposition of three exponentials with the following half-rise-times (and relative weights): 100 µs (1), 300 µs (1) and 1 ms (2). This led us to resolve the rise of the internal acidification for each flash of a group of four with the enzyme system in its dark equilibrium before the first flash (in the presence of an ADRY agent). The result is shown in Fig. 7. At least three distinct kinetic phases were

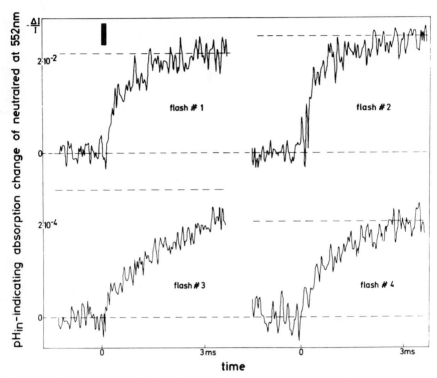

Fig. 7. Time course of the pH_{in}-indicating absorption changes of neutralred at 552 nm under excitation of chloroplasts with groups of four short flashes (duration: 15 μs, spacing 10 μs, repetition rate of groups: 0.2 s). DBMIB (3 μM) was present in the reaction medium and ANT 2s (0.6 μM) was present to enhance the relaxation rate of the S states to their dark equilibrium. The experimental conditions were as in [21].

observed in similar experiments: one with a half-rise-time between 300 and 600 μs (depending on the chloroplast preparation) dominated the first flash, one rising at ≤ 100 μs dominated the second flash and a third component at between 1 and 2 ms dominated the third flash.

Discussion

The results are schematically summarized in Fig. 8. With S_1 being the most stable of the S states in the dark, we find proton release inside thylakoids during the first

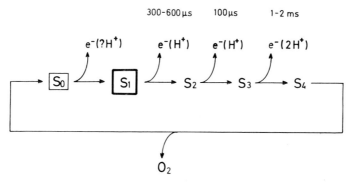

Fig. 8. Tentative scheme for the stoichiometry and the kinetic properties of inwardly directed proton release after flash excitation of chloroplasts in the presence of DBMIB to deactivate proton release by photosystem I. (For details, see text.)

three transitions of the water oxidizing enzyme system. We cannot exclude that protons are released during the transition S_0-S_1. However, our data are much better to be fitted by assuming a sequence of 0:1:1:2 for proton release than by assuming 1:1:1:1 or even 1:0:1:2. It is noteworthy, that the stoichiometric ratio is only relative. We have not yet absolutely calibrated the number of protons released per number of electrons transferred by photosystem II.

The kinetics of proton release during the first four flashes only partly fulfils the expectation based on data from the literature. Several authors have determined the relaxation time of the rate-limiting step in the reaction path from water through the reaction center of the photosystem II into plastoquinone for the subsequent oxidation steps of the water oxidizing enzyme system [9,25,29]. More relevant perhaps are the EPR data by Blankenship *et al.* [6], since these authors claim that "signal II$_{vf}$" is indicative of a component which is rapidly oxidized by photosystem II and then re-reduced while oxidizing the S states. On dark-adapted chloroplasts they observed the following half-relaxation times as function of the flash number:

flash 1 ($\leq$100 μs), flash 2 (400 μs), flash 3 (1 ms) and flash 4 ($\leq$100 μs). For flash numbers 1, 3 and 4 our data on proton release are compatible with the interpretation that proton release from the water oxidizing enzyme system has to follow, but not to precede, the respective electron transfer (as indicated by EPR signal II_{vf}). However, proton release which follows the second flash is faster than expected (100 μs against 400 μs!). If one accepts the proposed interpretation for signal II_{vf}, this implies that proton release during the transition S_2-S_3 occurs from a site preceding the site of signal II_{vf} in the sequence of the electron transport chain. In the light of this complexity we consider it as premature to judge the stoichiometric pattern of proton release that we have observed [21] in comparison with the different patterns which will appear in forthcoming papers by Saphon and Crofts [29] and by Fowler [12]. Further studies are required to quantitatively account for and to unequivocally attribute the protons which are released inside by the water oxidizing enzyme system. However, it is evident that the pattern of proton release markedly deviates from the one of oxygen evolution. Hence the reaction step liberating oxygen is not the only one producing protons. The mechanism of proton release is unknown. One possible mechanism, which is compatible with our data, was proposed by Renger [26-28]. It visualizes water and its oxidation products "crypto-hydroxyl" and "crypto-peroxyde" as acceptors of oxidizing equivalents in states S_1, S_2 and S_3.

Acknowledgements

We are very grateful for the collaboration of Dr Gernot Renger. Ilse Columbus and Allison McGreer have participated in some experiments. The work was financially supported by the EC (Solar Energy program) and by the DFG.

References

1. Ausländer, W. (1977). Thesis, Technical University Berlin.
2. Ausländer, W., Heathcote, P. and Junge, W. (1974). *FEBS Lett.* **47**, 229-235.
3. Ausländer, W. and Junge, W. (1974). *Biochim. Biophys. Acta* **357**, 285-298.
4. Ausländer, W. and Junge, W. (1975). *FEBS Lett.* **59**, 310-315.
5. Ausländer, W. and Junge, W. (1978). *In* "Mechanisms of Proton and Calcium Pumps" (G.F. Azzone *et al.*, eds), p. 31-44, Elsevier, Amsterdam.
6. Babcock, G.T., Blankenship, R.E. and Sauer, K. (1976). *FEBS Lett.* **61**, 286-289.
7. Bouges-Bocquet, B. (1973). *Biochim. Biophys. Acta* **292**, 772-785.
8. Diner, B. (1974). *Biochim. Biophys. Acta* 368, 371-385.
9. Forbush, B., Kok, B. and McGloin, M. (1971). *Photochem. Photobiol.* **14**, 307-321.
10. Fowler, C. (1973). *Biophys. Soc. Abstr.* of the 17th Ann. Meeting, Ohio, p. 64a.
11. Fowler, C. (1977). *Biochim. Biophys. Acta* **459**, 351-361.
12. Fowler, C. (1977). In press.
13. Fowler, C. and Kok, B. (1974). *Biochim. Biophys. Acta* **357**, 299-309.
14. Hauska, G. and Trebst, A. (1977). *Current Topics Bioenergetics* **6**, 151-220.
15. Joliot, P., Barbieri, G. and Chabaud, R. (1969). *Photochem. Photobiol.* **10**, 309-329.
16. Joliot, P., Joliot, A., Bouges, B. and Barbieri, G. (1971). *Photochem. Photobiol.* **14**, 287-305.
17. Joliot, P. and Kok, B. (1975). *In* "Bioenergetics of Photosynthesis" (Govindjee, ed.), 387-412. Academic Press, New York.
18. Junge, W. (1976). *In* "Biochemistry of Plant Pigments" (T.W. Goodwin, ed.), vol. II, 233-333. Academic Press, New York.
19. Junge, W. (1977). *Ann. Rev. Plant Physiol.* **28**, 503-536.
20. Junge, W. and Ausländer, W. (1974). *Biochim. Biophys. Acta* **333**, 59-70.
21. Junge, W., Renger, G. and Ausländer, W. (1977). *FEBS Lett.* **79**, 155-159.
22. Kok, B., Forbush, B. and McGloin, M. (1970). *Photochem. Photobiol.* **11**, 457-475.
23. Lavorel, J. (1975). *In* "Bioenergetics of Photosynthesis", (Govindjee, ed.), 224-318. Academic Press, New York.
24. Pick, U. and Avron, M. (1976). *FEBS Lett.* **65**, 48-53.
25. Renger, G. (1972). *Europ. J. Biochem.* **27**, 259-269.
26. Renger, G. (1977). *Topics in Current Chem.* **69**, 39-90.
27. Renger, G. (1977). *FEBS Lett.* **81**, 223-228.
28. Renger, G. (1977). This volume.
29. Saphon, S. and Crofts, A.R. (1977). *Z. Naturforschg.* **32c**, 617-626.
30. Trebst, A., Harth, E. and Draber, W. (1970). *Z. Naturforschg.* 25b, 1157-1159.
31. Trebst, A. and Reimer, S. (1973). *Biochim. Biophys. Acta* **305**, 129-139.
32. Witt, H.T. (1971). *Quart. Rev. Biophys.* **4**, 365-477.
33. Wydrzynski, T., Zumbulyadis, N., Schmidt, P.G. and Govindjee

(1975). *Biochim. Biophys. Acta* **408**, 349-357.
34. Zankel, K.L. (1973). *Biochim. Biophys. Acta* **325**, 138-148.

THEORETICAL STUDIES ABOUT THE FUNCTIONAL AND STRUCTURAL ORGANIZATION OF THE PHOTOSYNTHETIC OXYGEN EVOLUTION

G. RENGER

Max-Volmer-Institut für Physikalische Chemie und Molekularbiologie der Technischen Universität, Berlin 12, (F.R.G.)

Introduction

Water cleavage into molecular oxygen and protons by photosensitized electron abstraction plays a central role in the conversion of sunlight into useful chemical energy in all higher photoautotrophic organisms above the evolutionary level of blue-green algae. Irrespective of the mode of mechanistic realization, water oxidation to molecular oxygen requires in any case the generation as well as the co-operation of four redox equivalents of sufficient oxidizing power. Both basic functions were found to be performed by different operational units: a) The formation of the positive charges occurs at special chlorophyll a complexes, referred to as Chl a_{II} [12,50] *via* the primary photochemical events within the reaction center of PS II. b) The indispensible charge co-operation takes place in a special storage device, which was found to be charged up sequentially by photo-oxidized Chl a_{II}^{+} *via* univalent electron transfer reactions, until after the accumulation of four redox equivalents molecular oxygen can be evolved (Kok mechanism, see [24]). Accordingly, the reaction scheme of this operational unit, referred to as the water splitting enzyme system Y, can be formally described by eq. (1):

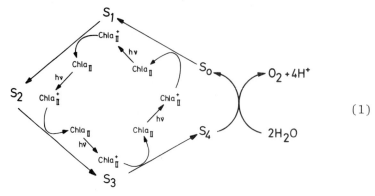

$$(1)$$

where S_i symbolizes the charge accumulation state and index i gives the number of positive charges stored in system Y.

In order to reconcile eq. (1) with the experimentally detected pattern of oxygen evolution in dark-adapted algae and chloroplasts [21,24], one has additionally to assume that S_1 is very stable in the dark (the lifetimes of S_2 and S_3 are of the order of seconds) and that the electron transfer from Y to Chl a_{II}^+ is incomplete, with an average probability of misses of 0.1-0.2 under normal conditions [21,24]. The formulation by eq. (1) suggests a four-step univalent electron transfer mechanism for the photosynthetic water cleavage, with the S states being the intermediate stages of the overall process. Hence, if one takes the common features [15] of the free intermediates of the four-step univalent water oxidation (OH-radical, H_2O, superoxide radical) as a guiding line for theoretical considerations, then at least 7 fundamental questions have to be answered for an understanding of photosynthetic water oxidation:

i) Which molecular mechanism realizes the "taming" of the highly reactive free intermediates in order to prevent the immediate decay of the higher S_i states and to exclude the photodynamic destruction of the sensitive biological material?

ii) How is the high stability of S_1 achieved?

iii) At which step occurs the formation of the dioxygen bond?
iv) What is the chemical nature of the S_i states?
v) What is the functional and structural arrangement of the storage places for the oxidizing redox equivalents within the water splitting enzyme system Y?
vi) At which step does the proton release occur?
vii) What is the underlying mechanism for the occurrence of misses?

Despite recent progress of photosynthesis research the molecular mechanism of photolytical water cleavage still remains an unsolved problem. Therefore, in the present paper a hypothetical model is proposed which might answer some of the above-mentioned questions.

The Model

In agreement with earlier suggestions [30,31] the water splitting enzyme system Y is basically assumed to be a mangano-protein, which forms a special microscopic "reaction vessel" with manganeses as the central ions surrounded by a specific ligand shell building up the functional storage places. On the basis of this supposition the following mechanistic properties are postulated (see [36]):

(i) The intermediates of the water oxidation are complexed in a special way as ligands of the central manganese ions. As the interaction with d-orbitals of transition metal ions (*e.g.* copper, manganese, iron) was shown to play a pivotal role in the complexation of hydrogen peroxide, superoxide and molecular oxygen [19], the intermediary species are assumed to be stabilized as inner sphere complexes (for other types of possible complexation, see [37]). Furthermore, as described in [32], a charge delocalization between the central manganese and the intermediates of water oxidation is assumed to occur leading to a "smearing out" of the stored positive charges.

(ii) The preformation of the dioxygen bond occurs at the level of hydrogen peroxide, which is binuclearly complexed by two manganeses located sufficiently close together.

(iii) Besides the functional manganese groups there exists in system Y a further 1-electron donor component, denoted by M, whose oxidized form M^+ is stable in the dark. The oxidation of M^+ is not coupled with a deprotonation (see [36]).

(iv) Molecular oxygen is primarily formed as a binuclearly complexed species *via* the oxidation by M^+ of a complexed superoxide.

(v) Oxygen release from system Y occurs *via* an exergonic exchange reaction with water leading to the replacement of binuclearly complexed dioxygen by two mononuclearly complexed water molecules. The ΔG-value for this reaction is estimated to be about 10-20 kJ/mol [36].

(vi) The redox properties of the complexed species are so drastically changed in comparison to the corresponding free intermediates (as shown in Fig. 1), that only a formal analogy exists between both types of oxidation states. Accordingly, the free enthalpy content of the complexed "hydroxyl radical" (denoted as "cryptohydroxyl radical" see footnote (a)) related to water complexed at the functional manganese groups is reduced by 100-200 kJ/mol in comparison to that of the free OH radical related to bulk water. Similarly, the free enthalpy of the binuclearly complexed intermediates "cryptoperoxide" and "cryptosuperoxide" are lower by 70-80 kJ/mol in comparison to the corresponding free substances. Thus, complexation changes completely the energetics of the four-step univalent water oxidation.

(vii) Corresponding to the energetic modification, the kinetic behaviour of the complexed species is also drastically changed, so that the lifetime of the "cryptohydroxyl radical" and of "cryptohydroperoxide" becomes rather long (order of a few seconds), whereas "crypto-superoxide" is rapidly oxidized by M^+ (see postulate (iv)).

(viii) The pH-values of the complexed intermediates of water oxidation are shifted not drastically enough as to lead to a protonization of the "cryptohydroxyl radical" and of "cryptohydroperoxide" (no estimation is required for the protonization state of "cryptosuperoxide").

The Reaction Scheme

The above-mentioned postulates give rise to the following molecular reaction mechanism:

$$M \xrightarrow{+} M^+ \tag{2a}$$

$$(H_2O)^* \xrightarrow{+} (OH)^* + H^+ \tag{2b}$$

$$(OH)^* + (H_2O)^* \xrightarrow{+} (H_2O_2)^{**} + H^+ \tag{2c}$$

$$M^+ + (H_2O_2)^{**} \xrightarrow{+} \left[(H\cdots O_2)^{**}\cdots M^+ + H^+\right] \longrightarrow (O_2)^{**} + 2H^+ + M \tag{2d}$$

$$2 H_2O + (O_2)^{**} \longrightarrow O_2 + 2(H_2O)^* \tag{2e}$$

In this scheme + symbolizes an oxidizing redox equivalent produced by photo-oxidation of Chl a_{II} and transferred to the water splitting enzyme system Y; one or two asterisks indicate mono- or binuclear complexation at one or two manganeses, respectively. The "cryptohydroxyl radical", $(OH)^*$, has the structure

$$\left[\widehat{\bigcirc Mn}^{(n+\delta)+} \; \widehat{OH}^{\delta-}\right]^{n+}$$

where $\widehat{\bigcirc Mn}^{n+}$ is a manganese center in its reduced form (characterized by the oxidation number n) surrounded by a special ligand shell, except for the water ligand; δ is a charge distribution parameter (a). Similarly, $(H_2O_2)^{**}$ has the structure

$$\left[\underset{H}{\overset{(n+\delta')+}{Mn}} \cdots \overset{\delta'-}{\underset{H}{O}} \overset{H}{\underset{\delta'-}{-O}} \cdots \overset{(n+\delta')+}{\underset{Mn}{\ldots}} \right]^{2n+}$$

where δ' is now the charge distribution parameter for the "cryptohydroperoxide" symmetrically complexed by two manganese centers (probably $\delta' \ll \delta$). The complexed "cryptosuperoxide" is assumed to form an unstable transition state together with M^+, which instantaneously leads to the formation of binuclearly complexed oxygen:

$$\left[\overset{(n+\delta'')+}{Mn} \cdots \underset{\delta''-}{O} \overset{\delta''-}{-O} \cdots \overset{(n+\delta'')+}{\underset{Mn}{\ldots}} \right]^{2n+}$$

A comparison of the present molecular reaction mechanism with Kok's scheme shows, that the states S_o and S_1 represent the redox state of the component M ($S_o \triangleq M$, $S_1 \triangleq M^+$), whereas S_2 corresponds to a "cryptohydroxyl radical" and S_3 to a "cryptohydroperoxide". For S_4 two different states appear, the transition state (superoxide$\cdots M^+$) and the complexed oxygen.

In this context it must be emphasized that S_1 in the present scheme does not symbolize the first oxidation state of water, which would be expected to be unstable in the dark, but S_1 rather reflects the oxidized state of a univalent redox component M with a moderate redox potential. M^+ enters as oxidant into the overall process only at the last oxidation step, which requires less free energy than the other electron abstraction reactions (Fig. 1).

The midpoint redox potential of M/M^+ has been estimated to fall into the range of + 200 to + 400 mV [37,36]. This value would make the high potential form of cytochrome b_{559} with an $E_{m,7}$ of about 350 mV [8] a likely candidate for the substantiation of the component M.

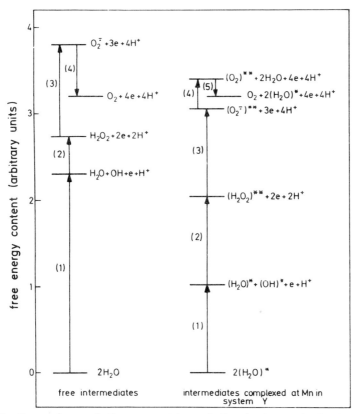

Fig. 1. Comparison of the energetics of the 4-step univalent water oxidation *via* free radicals and *via* the catalytic pathway in photosynthesis, respectively. The states H_2O and $(H_2O)^*$ are arbitrarily chosen to define the zero point of the energy scale. It must be clearly emphasized that the absolute energies of both states are not of the same value. The complexed intermediates (symbolized by stars) differ in their chemical nature from the corresponding free intermediates (for details see text and footnote (a)).

Furthermore, cytochrome b_{559} was shown to act as an efficient PS II electron donor at cryogenic temperatures [49]. On the other hand, there exists a number of experimental data, which suggests cytochrome b_{559} not to be directly involved into the electron transfer reactions leading to water oxidation [2,7,28]. However, as according to the present model M^+ participates at the last step of water oxidation in a very rapid reaction, the contribution of cytochrome b_{559} might have escaped the detection. Therefore, further experiments are required to clarify the function of cytochrome b_{559} and to substantiate the hypothetical substance M.

The Proton Release Pattern of System Y

As is evident from the overall reaction, summarized by eq. (1), water oxidation concomitantly leads to a release of four protons per dioxygen molecule. With respect to the coupling mechanism two different modes of proton release patterns have to be distinguished:

A) The *intrinsic* proton release pattern, which is exclusively determined by the molecular mechanism of water oxidation and the chemical nature of the storage places. This H^+ release pattern is unambiguously detectable only by indicators located within the water splitting enzyme system Y.

With respect to the intrinsic proton release pattern two different coupling mechanisms are possible: a) The exclusive coupling to O_2 formation, which occurs if the charging of the storage places in system Y by univalent electron transfer to Chl a_{II}^+ is not accompanied by a deprotonation, unless after the accumulation of four charges a simultaneous co-operation of four oxidizing redox equivalents with two water molecules takes place. b) The non-exclusive coupling to O_2 formation, which occurs if the charging of a functional group at charge accumulation states of system Y below S_3 is accompanied by proton release, *i.e.* if the functional groups are acids in their oxidized form with a pK value well below the internal pH.

Accordingly, the reaction mechanism proposed in this paper involves an intrinsic proton release of the non-exclusive coupling type to O_2 formation.

B) The *extrinsic* proton release pattern, which can be experimentally determined by suitable detectors acting outside of system Y.

With respect to the extrinsic proton release pattern also two different types have to be considered: a) The synchronous proton release pattern. In this case the oscillatory pattern for proton release across the boundaries of the water splitting enzyme system Y has to coincide with that for the oxygen evolution observed by excitation of dark-adapted algae or chloroplasts with a train of single turnover flashes [21,24]. b) The asynchronous proton release pattern. In this case the oscillatory patterns do not coincide for the release of protons and oxygen, respectively. The extrinsic proton release pattern is not necessarily congruent with the intrinsic pattern. A correspondence can only occur, if the protons dissociated from the oxidized functional groups are exchangeable with the bulk water phase outside of system Y.

According to the present model a non-exclusive coupling of proton release to oxygen formation is postulated (see eq. 2a-2d). Therefore, the extrinsic proton release pattern should be asynchronous with a stoichiometry (c) of $[0,1,1,2]H^+$ for the charging sequence $[S_0 \rightarrow S_1, S_1 \rightarrow S_2, S_2 \rightarrow S_3, S_3 \rightarrow (S_4) \rightarrow S_0]$ of the water splitting enzyme system Y, provided that an equilibration with the water phase outside of Y is possible. This mechanism is schematically depicted on the top of Fig. 2. However, the functional groups could be shielded up by a hydrophobic barrier surrounding the water splitting enzyme system Y. If one assumes that for the sake of simplicity system Y attains two conformational states (symbolized by Y and Y^o, see Fig. 2, bottom) and only one state, which is also required for the oxygen - water ligand-exchange (see eq. (2e)), allows the free protonization/deprotonization, then a synchronous extrinsic proton release pattern could arise even for an intrinsic non-exclusive coupling of proton release to oxygen evolution, as is shown in Fig. 2, bottom. In this way, the protons

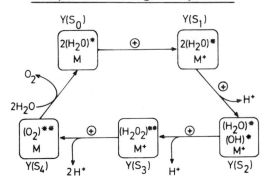

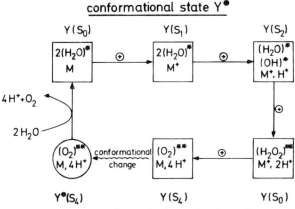

Fig. 2. Schematic representation of the relations between intrinsic and extrinsic proton release due to photosynthetic water oxidation in system Y (for details and explanation see text).

released by the functional manganese groups would be bound to internal protonizable groups (*e.g.* aminoacid residues) which are accessible to equilibration with the outer bulk water phase only in the conformational state Y^o. The variation of the degree of exchangeable groups at different conformational states of an enzyme has been already shown for the ATPase [42].

Another important effect, which could severely mask the intrinsic proton release pattern, has to be taken into account: If not all of the protons generated due to water oxidation are extruded into the inner thylakoid and if this effect depends on the charge accumulation state of system Y, then an apparent stoichiometry arises, which does not correspond to the intrinsic pattern, even if a complete equilibration with the bulk water phases is assured.

These simple considerations show that the determination of the extrinsic proton release pattern does not allow to draw unequivocal conclusions about the intrinsic mechanism of water oxidation, unless the above-mentioned factors are clarified. Experimental data about the proton release pattern became available only recently. In their first study using very sensitive glass electrodes for detection of pH-changes KOK and FOWLER found a synchronous pattern for the release of oxygen and protons, respectively [14]. However, in a later communication, this was reported probably not to be the case [13]. By a completely different technique, applying neutral red as indicator and 2(3,4,5-trichloro)-anilino-3,5-dinitrothiophene (ANT 2 s) as ADRY-reagent [33,34] for the fast relaxation of the charge accumulation states S_2 and S_3 [38], under repetitive flash group excitation conditions results were obtained [22] which resemble that of ref. [13]. Therefore, the extrinsic proton release pattern seems to be asynchronous. Accordingly, the intrinsic mechanism has to involve a non-exclusive coupling between proton release and oxygen evolution. In Fig. 3 the experimental data of ref. [13,21,22,24] are compared with the theoretical curves obtained by evaluation of the present model according to the simple Kok procedure (d), with the assumption of [M] = 0.35 ([M$^+$] = 0.65) and an average probability of misses of 0.1 (see ref. [36]). The experimental data are in fair agreement with the predictions of the model, except for the proton release induced by the first flash, where a large scatter exists. Despite these uncertainties the present data favour the conclusion,

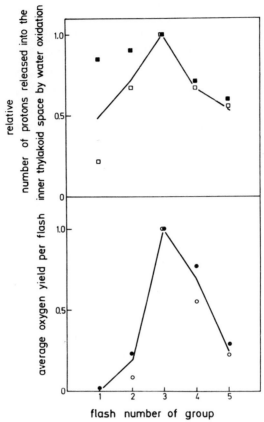

Fig. 3. Oscillation pattern of oxygen evolution and proton release due to photosynthetic water oxidation induced by a train of single turnover flashes. The curves give the theoretical values obtained for the present model (see eq. 2a-2d) with the additional assumption of a dark oxidation degree of M to be 0.65 and average probabilities of misses and double hits of 0.1 and 0.05, respectively. The experimental data are redrawn from ref. [21,24] (open circles), ref. [13] (open squares) and ref. [22] (closed symbols). The activity of the third flash was used for the normalization.

that the intrinsic proton release is non-exclusively coupled with oxygen formation and that the intrinsic pattern roughly coincides with the extrinsic proton release pattern. If the latter conclusion is correct then one can additionally infer, that the protonation state of the functional groups is in equilibrium with the bulk water phase outside of the water-splitting enzyme system Y. However, in order to become able to draw unambiguous conclusions about the mechanism of proton release due to water oxidation further information about the kinetics of deprotonation is required.

The Functional and Structural Organization of the Photosynthetic Water Oxidation

In the preceding sections a general molecular mechanism for photosynthetic water oxidation *via* inner sphere complexes at functional manganese groups has been discussed. However, for an understanding of the overall process a more detailed knowledge is required about the functional and structural organization scheme of all reactive groups on the donor side of system II, including Chl a_{II}. Generally, two main questions have to be answered: a) How is the functional connection realized between Chl a_{II} and the water-splitting enzyme system Y? b) What is the structural arrangement of all reactive groups on the donor side of system II? The mode of functional connection between Chl a_{II} and system Y determines the efficiency as well as the kinetics of the charging processes, described in eq. (2a-d) as the transfer of holes. It has been found that the reduction of photooxidized Chl a_{II}^+ occurs in the μs range with an at least triphasic kinetics characterized by half-life times of ≤1 μs, 35 μs and 200 μs, respectively [12,16,50]. The relative amplitudes of these reactions were shown to be dependent on the charge accumulation state S_i of system Y [16] as well as on the inner thylakoid proton concentration $[H^+]_{in}$ [40]. The latter effect is easily understandable by a localization of the donor site of system II near the inner side of the thylakoid membrane [14,35,46,50], whereas the former effect is in correspondence with the findings of varying kinetics for the univalent charging steps [$S_0 \rightarrow S_1$, $S_1 \rightarrow S_2$, $S_2 \rightarrow S_3$, $S_3 \rightarrow (S_4) \rightarrow S_0$] of system Y [1,3]. Furthermore, it has been discussed that Chl a_{II}^+ does not directly oxidize the functional groups of system Y, but the electron transfer occurs *via* a connector molecule D acting as univalent redox carrier, whose function as primary electron donor for Chl a_{II}^+ depends on the charge accumulation state S_i of system Y. As the reduction of photooxidized Chl a_{II} is fast in comparison to the reoxidation of the primary electron acceptor of system II [45,48], the stabilization of the primary photoproducts occurs primarily *via* the former process. Therefore, for the sake of efficiency, the functional connection between Chl a_{II} and system Y should be as perfect as possible. However, in order to interpret the damping of the oscillation of the oxygen yield in dark-adapted algae and chloroplasts, it is now widely accepted that according to Kok the charging process of system Y has an average efficiency of only 80-90%. The imperfections, referred to as misses, are explained by two different modes of functional connections: i) If a permanent, structurally fixed, coupling exists between Chl a_{II} and system Y, then the proposed misses might be due to photochemical losses or could be caused by a statistical blockage of the reaction centers with an average probability of 0.1-0.2. A more refined analysis claims the S_2 state being nearly exclusively responsible for the misses due to a rapid equilibration with an inactive state S_2' [9]. ii) If system Y is mobile with only a temporary contact with Chl a_{II} then the misses are readily understandable by an imperfect connection (depending on the association constant) and/or by a stoichiometric ratio different from 1:1 [25].

Both interpretations seem to be unsatisfactory, because the first one involves the question about the physiological role of the dissipative losses of redox equivalents, whereas the second one needs further strong restrictions of the contact conditions, because the lifetime of Chl a_{II} is of the order of 100-200 μs, if the functional connection to system Y is interrupted (*e.g.* by disconnection or degradation of Y at low pH or by TRIS-treatment [11,20, 41,47]). In a previous theoretical study [30] it was shown that a damping pattern for oxygen evolution can be rationalized without the assumption of losses, if one assumes that a specific structural array of the functional groups in system Y regulates the probability for the electron transition to Chl a_{II}^+. It can be shown that a further elaboration of this idea might explain the oscillatory pattern of oxygen evolution without the "need" of "dissipative" misses, but rather by "conservative" (see ref. [25]) misses. Two types of "conservative" misses can arise: a) If one supposes, that the photooxidized Chl a_{II}^+ component of two reaction centers co-operate for electron transfer with two systems Y, each containing only one functional group (see Fig. 4, left), then one obtains a probability of "conservative" misses, which has to be counterbalanced by the same amount of double hits. Obviously, this type of "conservative" misses does not account for the experimental data. b) If, however, each system Y contains different functional groups acting independently of each other with respect to the charging processes as well as of oxygen evolution, then another type of "conservative" misses arises, which does not require the concomitant occurrence of double hits. Accordingly, this second type of "conservative" misses is appropriate to explain the experimental data. In the present model it is assumed that two different types of components contribute to water oxidation in system Y: a redox couple M/M^+ and a manganese containing functional group, which binds the intermediary stages of oxidized water. As the higher oxidized forms are assumed to be binuclearly complexed, each functional group has to contain two manganese centers. On the other hand, 4-6 manganeses were found to be bound per system Y [5]. Therefore, it is reasonable to assume, that in system Y there exist 2 functional groups each containing a pair of

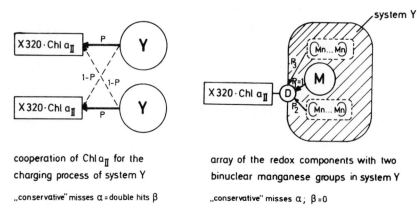

Fig. 4. Schematic representation of the mode of structural and functional organization of the donor side of PS II. Left side: Co-operation of two reaction centers of PS II (symbolized by X 320·Chl a_{II}) for charging of system Y with an electron transfer probability of P or 1 - P. Right side: Structure of system Y with two functional groups for water oxidation. The transfer probability P_1 describes the oxidation of M by Chl a_{II}, whereas P_2 and P_3 give the probabilities for the oxidation of the functional groups (Mn····Mn) with $P_2 + P_3 = 1$ (charge conservation) and $P_3 < P_2$ (determined by geometry).

manganese centers. If one additionally assumes an asymmetric array of two functional groups (symbolized by (Mn·····Mn)), one M-component and one molecule D connected with Chl a_{II}, as is shown in Fig. 4, right, then different electron transfer probabilities could arise which lead to the occurrence of "conservative" misses. A suitable numerical fitting by variation of the electron transfer probabilities P_1, P_2 and P_3 leads to a good correspondence with the experimental data. The scheme becomes even more complex, if additionally "conservative" misses of the first type (co-operation of two Chl a_{II}^+ centers) are taken into account (e). Experimental data supporting the idea of a co-operation of two reaction centers of PS II have been given earlier [10,43].

Next the question arises about the valence state of the manganese centers. Until now unambiguous data do not exist. Earlier experiments indicate that a photoactivation of the manganese centers is required, which was shown to be an oxidation. This would suggest an oxidation state higher than +2 [29]. Measurements of water proton spin relaxation led to the conclusion that manganese exists as a mixture of the oxidation states +2 and +3 [51]. Furthermore, evidence was presented for a change of the valence state of manganese centers during the redox cycle leading to water oxidation [52]. According to the present model, the oxidation state should not be an integer, because of the charge delocalization effect. Accordingly the "cryptohydroxyl-radical" is a hydroxyl ion complexed to Mn(III) rather than a hydroxyl radical stabilized at Mn(II), whereas the "cryptoperoxide" is assumed to be shifted towards a complexation at centers with a more Mn(II)-"character". This would be in correspondence with latest findings of S_3 being more reduced than S_2 with respect to the oxidation state of the manganese centers [17]. Further experiments are required to clarify this effect.

The present model predicts, that only S_2 and S_3 represent stabilized intermediates of oxidized water (S_4 is a transition state), whereas S_1 is a completely different oxidation state. In this respect it is interesting to note, that S_2 is very sensitive to degradation by high pH [4] or TRIS-treatment [6]. Similarly, system Y in the states S_2 and S_3 was recently found to be attacked by *p*-nitrothiophenol [23]. Furthermore, the stability of S_2 and S_3 is markedly modified by ADRY-reagents [39] probably due to ligand-exchange reactions [32,34]. On the other hand, S_1 appears to be much less affected or even invariant to these treatments.

As a last point it remains to be noted, that the nature of the ligand shell of the manganese groups, which might be an essential site for a possible anion effect *e.g.* HCO_3^-, see ref.[27], but also ref. [44] on the function of system Y, is completely unknown.

Footnotes

(a) The charge distribution parameter δ determines the radical character: $\delta = 0$ would represent a hydroxyl radical co-ordinated to Mn^{n+}; $\delta = 1$ denotes a hydroxyl ion bound to $Mn^{(n+1)+}$. Energetic considerations suggest δ to be closer to 1.

(b) The problems arising for the definition of a pH value in a molecular system, like system Y, will not be outlined here.

(c) The stoichiometry of the proton release pattern strongly depends on the validity of postulate 3, where the oxidation of M to M^+ is assumed to be a pure electronic abstraction. If a deprotonization would occur, then the pattern is [1,1,1,1].

(d) Despite a more refined mathematical analysis [18,26] the simple procedure according to Kok seems to be appropriate for an approximate evaluation of the oscillatory pattern.

(e) It must be stressed, that the electron transfer probabilities could be influenced by chemicals, thus giving rise to different oscillatory patterns.

Acknowledgements

The author gratefully acknowledges the financial support by ERP-Sondervermögen and by Deutsche Forschungsgemeinschaft.

References

1. Babcock, G.T., Blankenship, R.E. and Sauer, K. (1976). *FEBS Lett.* **61**, 286-289.
2. Ben Hayyim, G. (1974). *Eur. J. Biochem.* **41**, 191-196.
3. Bouges-Bocquet, B. (1973). *Biochim. Biophys. Acta* **292**, 772-785.
4. Briantais, J.M., Vernotte, C., Lavergne, J. and Arntzen, C.J. (1977). *Biochim. Biophys. Acta* **461**, 61-74.
5. Cheniae, G. and Martin, I.F. (1971). *Plant Physiol.* **47**, 568-575.
6. Cheniae, G. and Martin, I.F. (1977). Proceedings of the IV. International Congress on Photosynthesis, Book of Abstracts, p. 67, Reading.
7. Cox, R.P. and Bendall, D.S. (1972). *Biochim. Biophys. Acta* **283**, 124-135.

8. Cramer, W.A., Fan, H.N. and Böhme, H. (1971). *J. Bioenerg.* **2**, 289-303.
9. Delrieu, M. (1974). *Photochem. Photobiol.* **20**, 441-454.
10. Diner, B. (1974). *Biochim. Biophys. Acta* **368**, 371-385.
11. Döring, G. (1975). *Biochim. Biophys. Acta* **376**, 274-284.
12. Döring, G., Renger, G., Vater, J. and Witt, H.T. (1969). *Z. Naturforschg.* **24**b, 1139-1143.
13. Fowler, C.F. (1977). *Biochim. Biophys. Acta* **459**, 351-361.
14. Fowler, C.F. and Kok, B. (1974). *Biochim. Biophys. Acta* **357**, 299-309.
15. George, P. (1965). In "Oxidases and Related Redox Systems" (T.E. King, H.S. Mason and M. Morrison, eds). Vol. I, pp. 3-36. Wiley, New York, London, Sidney.
16. Gläser, M., Wolff, C. and Renger, G. (1976). *Z. Naturforschg.* **31**c, 712-721.
17. Govindjee, Wydrzynski, T. and Marks, S.B. (1977). In "Membrane Bioenergentics" (L. Packer, ed.). Elsevier, Amsterdam.
18. Greenbaum, E. (1977).*Photochem. Photobiol.* **25**, 293-298.
19. Hanzlik, R.P. (1976). "Inorganic Aspects of Biological and Organic Chemistry", Chapter X, pp. 253-94, Academic Press, New York, San Francisco, London.
20. Haveman, J. and Mathis, P. (1976). *Biochim. Biophys. Acta* **440**, 346-355.
21. Joliot, P., Joliot, A., Bouges, B. and Barbieri, G. (1971). *Photochem. Photobiol.* **11**, 287-305.
22. Junge, W., Renger, G. and Ausländer, W. (1977). *FEBS Lett.* **79**, 155-159.
23. Kobayashi, Y., Inoue, Y. and Shibata, E. (1977), this volume.
24. Kok, B., Forbush, B. and McGloin, M. (1970). *Photochem. Photobiol.* **11**, 457-475.
25. Lavorel, J. (1976). *FEBS Lett.* **66**, 164-167.
26. Lavorel, J. (1976). *J. Theor. Biol.* **57**, 171-185.
27. Metzner, H. (1977), this volume.
28. Phung Nhu Hung, S. (1974). *Z. Pflanzenphysiol.* **72**, 389-394.
29. Radmer, R. and Cheniae, G.M. (1971). *Biochim. Biophys. Acta* **253**, 182-186.
30. Renger, G. (1970). *Z. Naturforschg.* **25**b, 966-970.
31. Renger, G. (1972). *Physiol. Vég.* **10**, 329-345.
32. Renger, G. (1972). *Eur. J. Biochem.* **27**, 259-269.
33. Renger, G. (1972). *Biochim. Biophys. Acta* **256**, 428-439.
34. Renger, G. (1973). *Biochim. Biophys. Acta* **314**, 390-402.
35. Renger, G. (1976). *Biochim. Biophys. Acta* **440**, 287-300.
36. Renger, G. (1977). *FEBS Lett.* **81**, 223-228.
37. Renger, G. (1977). *Topics in Curr. Chem.* **69**, 39-90.
38. Renger, G., Bouges-Bocquet, B. and Büchel, K.H. (1973). *J. Bioenerg.* **4**, 491-505.
39. Renger, G., Bouges-Bocquet, B. and Delosme, R. (1973). *Biochim. Biophys. Acta* **292**, 796-807.
40. Renger, G., Gläser, M. and Buchwald, H.E. (1977). *Biochim. Biophys. Acta* **461**, 392-402.
41. Renger, G. and Wolff, C. (1976). *Biochim. Biophys. Acta* **423**, 610-614.
42. Ryrie, I.J. and Jagendorf, A.T. (1971). *J. Biol. Chem.* **246**,

3771-3774.
43. Siggel, U., Renger, G. and Rumberg, B. (1972). *In* "Proceedings of the 2nd Congress on Photosynthesis Research" (G. Forti, M. Avron and A. Melandri, eds), Vol. I, pp. 753-762, Dr. W. Junk Publishers, The Hague.
44. Stemler, A. (1977), this volume.
45. Stiehl, H.H. and Witt, H.T. (1969). *Z. Naturforschg.* **24**b, 1588-1598.
46. Trebst, A. (1974). *Ann. Rev. Plant Physiol.* **25**, 423-458.
47. Van Gorkom, H.J., Pulles, M.P.J., Haveman, J. and Den Haan, G.A. (1976). *Biochim. Biophys. Acta* **423**, 217-226.
48. Vater, J., Renger, G., Stiehl, H.H. and Witt, H.T. (1969). *In* "Progress in Photosynthesis Research" (H. Metzner, ed.). Vol. II, pp. 1006-1008, International Union of Biological Sciences, Tübingen.
49. Vermeglio, A. and Mathis, P. (1973). *Biochim. Biophys. Acta* **292**, 763-771.
50. Wolff, C., Gläser, M. and Witt, H.T. (1975). *In* "Proceedings of the 3rd International Congress on Photosynthesis Research" (M. Avron, ed.). Vol. I, pp. 295-305, Elsevier, Amsterdam.
51. Wydrzynski, T., Zumbulyadis, N., Schmidt, P.G. and Govindjee (1976). *Biochim. Biophys. Acta* **408**, 349-354.
52. Wydrzynski, T., Zumbulyadis, N., Schmidt, P.G., Gutowsky, H.S. and Govindjee (1976). *Proc. Natl. Acad. Sci. Washington* **73**, 1196-1198.

ON THE ORIGIN OF DAMPING OF THE OXYGEN YIELD IN SEQUENCES OF FLASHES

J. LAVOREL

*Laboratoire de Photosynthèse du C.N.R.S.,
Gif-sur-Yvette (France)*

Summary

According to the linear four step model of the O_2 evolving system proposed by Kok, the progressive damping of the order 4 oscillatory pattern of O_2 yield induced by light flashes is mainly due to "misses". The latter are assumed to be photochemical; *i.e.* PS II centers may with a given probability fail to perform the primary charge separation. The photochemical deficit would amount to up to 36% in *Chlorella* (less in chloroplasts). Moreover, it is expected that the O_2 yield decreases in experimental conditions where damping due to photochemical misses increases. The first consequence seems unlikely and the second one is disproved in a number of cases. Therefore, other classes of damping mechanisms have been considered. With conservative misses, the oxidizing equivalents are produced with possibly maximal yield, but they are randomly stored in the O_2 evolving system. We are also exploring a mechanism of photochemical misses conditional to the high light density of the actinic flashes (*e.g.* involving some kind of non-linear annihilation process). Obviously, this mechanism would not be operative under low light intensity (as used for measuring the maximum quantum yield of photosynthesis). So far, the various experimental correlations found between the O_2 yield and the damping coefficient (and other PS II properties) do not seem to favour exclusively a single mechanism.

The Four States of the Oxygen Evolving System

In 1968, a most remarkable and - as it turned out - most important effect was discovered by Joliot *et al.*[9]. When submitting an active photosynthetic material (*Chlorella* cells or chloroplast) that has been dark-adapted for a few minutes to a regime of short (few microseconds), saturating, light flashes, one observes a characteristic pattern of O_2 evolution as recorded with the platinum electrode. Three points are noteworthy:

1. the pattern is oscillatory with a periodicity of 4;

2. the oscillations are damped and the pattern evolves to a steady-state after a sufficient number of flashes (10 to 20 approximately);

3. with a fully dark-adapted sample (so-called "deactivated") no oxygen is emitted at the first flash, whereas the maximum emission is observed at the third flash.

Several models have been put forward in an attempt to explain this experimental behaviour [9,11,20]. They all share in common the central idea that the 4 oxidizing equivalents which need be given to 2 H_2O molecules in order to evolve 1 O_2 molecule are produced sequentially, one per flash and stored in a special enzymatic entity, the oxygen evolving system (OES) closely associated to each PS II reaction centre[a]. Thus we seem to have given up the older implicit assumption that the oxidizing equivalents were produced by different reaction centers and had to migrate to some enzymatic site in order to oxidize water [24]. (As we shall see, abandoning altogether this assumption might be questionable). By now, the most widely accepted model is the one proposed by Kok [11], no doubt because of its magistral simplicity: the OES cyclically occupies 4 states S_0 to S_3 for each O_2 molecule evolved according to the scheme:

$$S_0 \xrightarrow{h\nu} S_1 \xrightarrow{h\nu} S_2 \xrightarrow{h\nu} S_3 \xrightarrow{h\nu} O_2 \qquad (1)$$

each step corresponding to the storage of one positive charge (or oxidizing equivalent). Thus this model obeys a simple transition rule

$$S_i \longrightarrow S_{i+1} \qquad (2)$$

the index being understood modulo 4. In order to account for the pattern of O_2 evolution (notably its damping) two other types of transitions are assumed

$$S_i \longrightarrow S_i \text{ or "misses"} \qquad (3)$$

because some centers fail to react photochemically and

$$S_i \longrightarrow S_{i+2} \text{ or "double hits"} \qquad (4)$$

because other centers may re-open after a first hit and are thus able to react twice during the actual time of the flash. It is important to realize that the foregoing transitions must be random events and as such are characterized by probabilistic transition coefficients. Finally, the initial make-up of the O_2 sequence is explained assuming that the states have unequal stabilities: actually S_2 and S_3 are unstable and decay into S_0 and S_1, which thus accumulate in the dark-adapted condition ($S_0 \simeq 30\%$, $S_1 \simeq 70\%$).

Kok's model has had a pervasive influence on all kinetic studies of PS II (*e.g.* fluorescence and luminescence) because it rationalized the functioning of the OES in terms of states and because the latter have proved to be of such easy operational value[b]. Any model however must be critically scrutinized. Is it unique? Is there a better one? To what extent are the structure and mechanism verified? Is the available experimental information sufficient to answer the above questions?

In my laboratory, we have been dealing with such

questions when we became interested by the significance of misses and their possible mechanism. In what follows, I shall attempt to answer the above questions.

The Photochemical Misses in Kok's Model

Quoting the original paper [11], it is assumed that "the flashes convert a certain percentage of the traps more than once ('double hit') while another fraction of the traps are not converted at all ('misses')". We shall not discuss here the problem of double hits because, as a first approximation, it seems to be well resolved [10]. It is immediately apparent, however, that the assumption of "misses" had no experimental support at the moment of its proposal. Actually, the only observed effect is damping and, as noted earlier [9] the elementary cause must be some kind of genuine disorder built in the mechanism. Therefore, we may at least consider *a priori* two kinds of damping mechanisms:

1. "photochemical" misses, meaning as in Kok's model that the non-reacting centers contribute to a deficit in the quantum yield of PS II and

2. non-photochemical or "conservative" misses.

The first objective should be to know which type of damping mechanism is actually operating, for instance by looking for a general test of the hypothesis of photochemical misses. There are two difficulties to this task: firstly, O_2 measurements with the platinum electrode are only relative (absolute measurements obviously would be needed both of the O_2 output and of the number of centers in the sample); secondly, at the time when we took the problem, there was no easy, straightforward means for evaluating the damping of a sequence.

We have solved this second point by proposing the method of σ analysis [14] which also permits to partly overcome the first difficulty. Following is a summary of σ analysis.

1. Any state model operating under a flash sequence (Markoff process) can be translated into a matrix recurrence relation [2]:

$$S^{(n+1)} = \tilde{K} \, S^{(n)} \qquad (5)$$

$S^{(n)}$, $S^{(n+1)}$ being state vectors at flash number n, n+1 (in Kok's model, $S^{(n)} = S_0^{(n)}, S_1^{(n)}, S_2^{(n)}, S_3^{(n)}$) and $\tilde{K}$ a transition matrix embodying the various transition coefficients of the model (in Kok's model, they are α_i (miss), β_i (single hit) γ_i (double hit) with the relations

$$\alpha_i + \beta_i + \gamma_i = 1 \qquad i = 0,1,2,3 \qquad (6) \;).$$

Note that in principle the transition coefficients are time-invariant.

2. As a consequence of eq. 5 a scalar recurrence relation may be found for any linear expression in specific components (S states) of the state vector, *e.g.* for successive O_2 yield $Y^{(n)}$ of the flash sequence. The coefficients of this relation are in the case of a 4 state model, *e.g.* Kok's model - which we shall only consider in what follows - four functions of the transition coefficients: $\sigma_1, \sigma_2, \sigma_3, \sigma_4$. We need only give the meaning of σ_1 which is the simplest and most important of them

$$\sigma_1 = \alpha_0 + \alpha_1 + \alpha_2 + \alpha_3 \qquad (7)$$

(in other words, $\sigma/4$ is the average miss coefficient). We shall take σ_1 as an objective measure - making no assumption on the distribution of misses among states - of damping. If the model is right, the σ_i's are not independent; they must obey the relation:

$$1 - \sigma_1 + \sigma_2 + \sigma_3 + \sigma_4 = 0 \qquad (8)$$

3. The σ_i's can be calculated by solving a system of linear equations built on 7 consecutive $Y^{(n)(c)}$. Note that solving for the σ_i's does not require a knowledge of the initial state distribution, $S^{(0)}$.

There are several consequences to this treatment.

1. It can be shown that the information which one may draw from any experimental $Y^{(n)}$ suite is quite limited. No matter how we proceed, we cannot know exactly the initial state distribution, the individual miss coefficient α_i and double hits coefficients γ_i. The problem is thus overloaded by the weight of the above "hidden" variables.

2. It is almost impossible to make a decision among isomorphous models, that is models distinct mechanisitcally but identical structurally (*e.g.* 4-state models).

3. One can, however, reach a firm conclusion concerning the structural aspect of the model, *i.e.* the multiplicity of states[d]. It was recently demonstrated by Thibault [28] using an extended version of σ analysis, that the relation eq. 8 was verified with a high degree of precision. This excludes models with a number of states different from 4 [20,22].[e]

Concerning the test of the hypothesis of photochemical misses, we used σ_1 in the following way. Fig. 1 shows a relation which must hold as a necessary consequence of Kok's model if the misses are photochemical (assuming negligible double hits). It is seen that Y_{ss} is inversely related to σ_1, *i.e.* if some cause acts on σ_1 in some direction (*e.g.* increase), there must result a change in Y_{ss} on the opposite direction (*e.g.* decrease). Note that 1. the opposite is not necessarily true (*e.g.* centers might be inactivated thus decreasing Y_{ss}, with no necessary change in s_1) and 2. the relation cannot be predicted precisely because we do not know the exact distribution of misses among states: the permissible region is shown as a shaded domain in Fig. 1 (thus a limited amount of direct relationship between Y_{ss} and σ_1 is allowed which might result from a change in the above distribution).

As a first application of the above test, are shown on Fig. 1 the average positions of σ_1 in chloroplasts and in *Chlorella*, taken from a survey of published O_2 sequences

[1]. Accordingly, it is deduced that the PS II quantum yield in isolated chloroplasts is higher than that in live *Chlorella* and that the latter is between 0.6 and 0.8 electron per quantum. The first conclusion seems unlikely and the second one wrong according to Sun and Sauer [27] who measured a quantum yield of 1.00 ± 0.05 for the PS II reaction. Another application has been to look for experimental conditions inducing concomitant variations of σ_1 and Y_{ss}. Table 1 includes such correlations from our own work

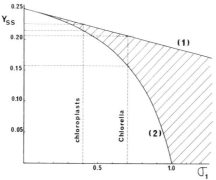

Fig. 1. Relationship between the steady-state O_2 yield Y_{ss} and the sum of miss coefficients σ_1 in Kok's model. The exact relationship depends upon the specific distribution of misses among S states; however, the representative points must always lie between curve (1) corresponding to equal miss coefficient on all states and curve (2) corresponding to a single non-zero miss coefficient. The horizontal dashed lines indicate the range of Y_{ss} corresponding to the average σ_1 found in the literature for chloroplasts and *Chlorella* (Lavorel, [15]).

as well as from published results. It is apparent that in a number of cases, the observed correlation (direct) contradicts the prediction of the model with photochemical misses (inverse). Noteworthy is the case of heat-pretreatment which permits to reach a σ_1 value even lower than found in chloroplasts (see also Fig. 2). It is interesting also to note that - with the exception of DCMU - the predicted correlation is indeed found in all cases where the yield is decreased due to a blockage of electron flow on the acceptor side of PS II. Obviously centers in the state of Q^- cannot react photochemically and should

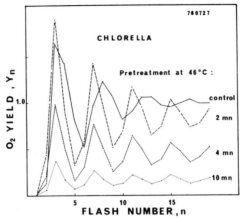

Fig. 2. Oxygen yield in sequences of flashes for heat-treated *Chlorella*. Y_n is the O_2 yield at flash number n. Each sample was incubated for various times (indicated) at 46°C and then returned at room temperature (25°C) for measurement. (Maison and Lavorel, 1977).

contribute photochemical misses (see [25]) provided the state Q^- is rapidly randomized between centers in the time interval between two flashes$^{(f)}$, which implies a rapid equilibrium of a negative charge between the acceptors of different centers through the plastoquinone pool. Such a rapid exchange does not seem to be possible according to the results of Velthuys and Amesz [29].

According to the results in Table 1, we have to conclude either that the simple scheme of photochemical misses is not general or, if photochemical misses do occur, that the correlation when it is direct is only apparent (*e.g.* the agents in Table 1 inactivate some centers *and* decrease the probability of photochemical misses in other centers). This unsatisfactory conclusion led us to investigate the possibilities offered by conservative misses.

Models with conservative misses

I shall first make a theoretical remark. Misses - or, more generally, zero order transition $S_i \longrightarrow S_i$ - in a n-state, step-by-step-model is not the obligate cause of

TABLE 1

Correlation between the changes in steady-state O_2 yield Y_{ss} and damping parameter σ_1 induced by various treatments or conditions in chloroplasts and green algae

Treatment or condition	Material (1)	Sign and percentage of change		Reference
		σ_1	Y_{ss}	
1. Inverse correlation				
- lack of acceptor	chloroplasts	+	- (*)	[25]
- Hg^{2+}	chloroplasts	+	- (*)	[26]
- *Scenedesmus* mutant n° 8	algae	+	- (*)	[25]
- NaN_3	algae	- 46	+ 36	[18]
- *Chlamydomonas* mutant F1 5	algae	+	- (*)	[23]
2. Direct correlation				
- CCCP 10^{-5}M	algae	- 56	- 6	[5]
- DNB $5 \cdot 10^{-4}$M	algae	- 32	- 45	[7]
- DCMU $3 \cdot 10^{-7}$M	algae	- 19	- 38	[6] [13]
- BQ $3 \cdot 10^{-4}$M	algae	- 42	- 40	[3]
- Heat : 46°C, 2 min	algae	- 64	- 15	[19]

(*) The changes were not calculated because both σ_1 and Y_{ss} vary rapidly along the flash sequence.

damping. Actually, the only requirement for damping is that in the corresponding transition matrix there be more than one diagonal carrying non-zero elements. This point is illustrated by the comparison of the transition matrices of several models shown in Fig. 3 which all predict damped $Y^{(n)}$ suites. Notice in particular that Joliot's original model (Fig. 3(a)) has no miss coefficient (1st diagonal) (and in fact is the first instance of a "conservative" model). Thus in the above models, it is more appropriate to speak of the multiplicity of transitions rather than of disorder as the cause of damping. This remark

(b)
$$\begin{bmatrix} \alpha_0 & 0 & \gamma_2 & \beta_3 \\ \beta_0 & \alpha_1 & 0 & \gamma_3 \\ \gamma_0 & \beta_1 & \alpha_2 & 0 \\ 0 & \gamma_1 & \beta_2 & \alpha_3 \end{bmatrix}$$

(a)
$$\begin{bmatrix} 0 & 0 & 1 & 0 \\ p & 0 & 0 & (1-p) \\ (1-p) & 0 & 0 & p \\ 0 & 1 & 0 & 0 \end{bmatrix}$$

(c)
$$\begin{bmatrix} \gamma & (1-\gamma)q & 0 & (1-\gamma) \\ (1-\gamma) & \gamma & 0 & 0 \\ 0 & (1-\gamma)(1-q) & \gamma & 0 \\ 0 & 0 & (1-\gamma) & \gamma \end{bmatrix}$$

Fig. 3. Transition matrices of several models for the O_2 yield in sequence of flashes. (a) Joliot *et al.* (1969), (b) Kok *et al.* (1970), (c) Mar and Govindjee (1972) (model C).

points to the very large number of *a priori* conceivable models which could at least partly explain the data. And this casts some doubt on the success of the whole enterprise!

Whatever it be, we were led to consider a class of models slightly different in their structure from the foregoing. I did not stress enough the point rather obvious that the models considered so far were more or less non-cooperative as far as the contribution of different centers to the charge storage was concerned (Kok's model being absolutely non-cooperative). Actually this restriction can be partly relaxed, *i.e.* cooperativity in the above defined sense cannot be excluded since, as we shall see, it allows to define models with as good a predictive value as the non-cooperative ones[g].

One first model we proposed earlier [16] is model C depicted in Fig. 4 (in a simplified version). The idea was to allow a lateral (possible free) carrier C to reversibly store a + charge. It is easily seen that such a model produces transitions exactly mimicking photochemical

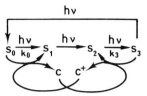

Fig. 4. A model with lateral carrier C able to reversibly store a + charge. In the depicted simplified version, a free diffusing species C would pick up a charge on S_3 and feed it back to the system S_0 [16].

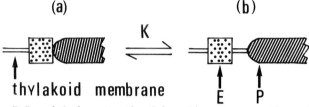

Fig. 5. The E-P model (see text). Schematic cross-section of the thylakoid membrane. E, the O_2 evolving system, and P, the PS II end of the photosynthetic chain, may either be associated (a), thus allowing + charge transfer from P to E or not (b). K is the equilibrium constant of the dissociation process. In this model, the miss coefficient α is equal to the fraction of free E and the damping coefficient σ_1 depends on both K and the ratio of the total amount of E to that of P. (*cf* Lavorel, 1976b).

"misses" and "double hits" but that in particular the apparent misses are strictly conservative, *i.e.* no photochemical loss whatever is incurred. C might be considered also as an accessory donor to PS II.

We then proposed [15] the so-called E-P model (Fig. 5) with the specific aim of explaining the observed direct correlation between Y_{ss} and σ_1 (*cf*. Table 1). It is an extension of model C towards more cooperativity. In model C, some exchange of + charges is allowed between different chains by the agency of the C carrier. A more radical step is taken in the E-P model: the OES (called E) is considered as an entity kinetically distinct from the reaction center (called P). E might be bound to P (thus allowing an exchange of + charge) or not. Aside from this, E is identical with the OES in Kok's model (4 states, etc). To be sure neither E nor P are free diffusing species: they may be thought of as intrinsic proteins (or protein complexes) embedded in the thylakoid membrane and endowed with lateral motion due to the fluidity of the lipid part of the membrane[h]. An important property of the model is that no strict stoichiometry between E and P is assumed, *i.e.* the number of active E and P may independently vary (*e.g.* due to experimental conditions). This property explains very simply the possibility of a direct correlation between Y_{ss} and σ_1. Assume that for some reason the number of active E is decreased, P being not changed. Since there is less E, the flash yield Y_{ss} will decrease. However, due to the relative excess of P, a larger fraction of the active E will get a + charge at each flash. Hence there is less disorder in the progression of states and σ_1 is also decreased. This mechanism is verified in a simulation of the model shown in Fig. 6 (compare with the heat-treatment, Fig. 2). Notice that, in spite of the damping being less, nevertheless the actual photochemical misses are more important (the excess light activated P disappear by back-reaction, *i.e.*, in the usual symbolism: $P^+Q^- \longrightarrow PQ$). It is interesting that an inverse correlation between Y_{ss} and σ_1 is also predicted in the symmetrical case where, instead of P, E happens to be in relative excess.

The above model has been extended in another sense. It was implicitly assumed that E could rather freely move with respect to P (conceivably a less mobile system!) so that each E might combine with any P. A more realistic assumption would be that E during the allowed time (*i.e.* between two successive flashes) could only combine with its nearest P neighbours (due to the dense packing of particles on the thylakoid surface or to the viscosity of the membrane). What would be the consequence of such a restricted mobility? The situation is depicted in Fig. 7 in the form of a game: initially, an equal number of E and P (male and female) alternate in a square lattice so that each E (P) has four nearest neighbours P (E); then

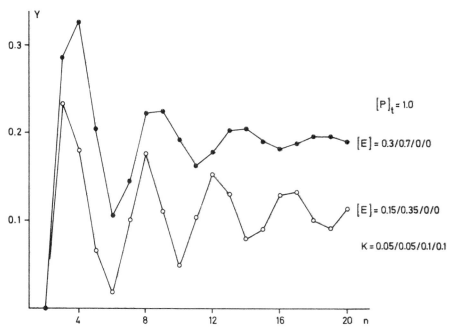

Fig. 6. Simulated Y_n suites according to the E-P model. The two sequences shown demonstrate the effect of varying only the total amount of E, $(E)_t$.●, $(E)_t=1.0$, initial distribution: $(E_0)/(E_1)/(E_2)/(E_3) = 0.3 / 0.7 / 0 / 0$. X, $(E)_t=0.5$, initial distribution: $(E_0)/(E_1)/(E_2)/(E_3) = 0.15 / 0.35 / 0 / 0$. in both cases, $K_0/K_1/K_2/K_3 = 0.05/0.05/0.1$ (see Fig. 5) and $(P)_t = 1.0$.

the elements E and P are allowed to associate randomly (get married) with the constraint that only nearest neighbours can do so; when all possible associations have been realized, it is found on the average that a certain percentage of the elements have not associated (bachelors). This simple topological constraint thus generates misses (photochemical in this case). It is interesting - but possibly coincidental - that σ_1 turns out to be equal to ~0.36, quite close to the value currently found in chloroplasts (~0.4).

The mobility (E-P) model is supported by independent evidences.

1. There is at least one well-documented case of bi-equimolecular reaction in the field of photosynthesis,

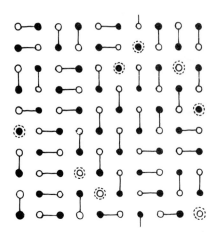

Fig. 7. Simulation of the E-P model with restricted mobility (see text). White and black circles represent E and P on a square lattice. The result of a Monte Carlo simulation is shown: the paired or bound elements (solid line between two circles) have been produced by randomly selecting the positions and the nearest neighbours. Notice that at the end of the process (as shown) some elements (dashed circles) have not been able to get paired.

i.e. a reaction obeying the scheme:

$$A + B \longrightarrow C \quad \text{with } (A) = (B) \tag{9}$$

this is the decay of Q^-, photochemically produced in *Chlorella* or chloroplasts, in the presence of DCMU [1]. The bi-equimolecular scheme (eq. 8) has been quantitatively verified for this process both by fluorescence [1] and luminescence measurements [13]. Now such a scheme is strange and unexpected in the usual conception of independent photochemical chains with integrated - permanently and mutually bound - reactive components where as a rule processes should be first order$^{(i)}$. The explanation is quite simple in the frame of the mobility model, since an equal number of E^+ and PQ^- are formed in light and may only disappear by back reaction (DCMU present) according to the scheme eq. 8 (it is verified that the model with restricted mobility also allows for a bi-equimolecular back-reaction scheme).

2. According to Kok's model, if DCMU is added to a sample in S_3 state (two flashes after dark-adaptation), a subsequent illumination should produce either of the two following mutually exclusive outcomes:

- O_2 emission and blockage of Q^- (no fluorescence yield decay, no luminescence emission);
- no O_2 emission and decay of Q^- (fluorescence yield decay, luminescence emission).

Actually, in *Chlorella* both the O_2 emission and decay of Q^- were observed [5]. This result also follows from the mobility model in as much as PQ^- can back-react with an E^+ distinct from the one which has just produced an O_2 molecule.

There is however a strong counter-argument to the E-P model. The head-pretreatment should always result in directly correlated changes in Y_{ss} and σ_1, since the thermolability of the OES is a well-established fact. Actually, it was observed, contrary to the case of *Chlorella*, that this treatment induced an increase in σ_1 in isolated chloroplasts [17] and in *Anacystis* [31].

Whatever the validity of the E-P model, I believe it is worth keeping in mind the idea of mobility (restricted or not) of the membrane bound components of the photosynthetic apparatus as a possibly important (and necessary) aspect of its functioning.

Non-reciprocity (or Non-linear) Photochemical Misses

In the light of the foregoing, we are currently considering a further possibility. Up to this point, we have implicitly assumed that the reciprocity law was obeyed, i.e. that a short flash of light density E * was photochemically equivalent to a continuous illumination (making allowance for turn-over considerations) of intensity I and duration t such that:

$$I \cdot t = E * \qquad (10)$$

The reciprocity law has been well verified in photosynthetic systems up to moderate light intensities (see *e.g.* [8]), but it might not be so for flashes with high light density. Indeed, anomalous - non-linear - effects have recently been described concerning the fluorescence properties of photosynthetic materials submitted to high density of the order of 10^{14} $h\nu \cdot cm^{-2}$ or more during microsecond (or sub-microsecond) flashes. One is ascribed to the formation of a short-living carotenoid triplet [4], another to excitonic annihilation processes [21].

A striking effect recently reported by Vermeglio [30] deserves careful consideration. He studied the photochemical reduction of ubiquinone by isolated and purified bacterial reaction centers illuminated by flashes and demonstrated the sequential and alternate formation of semiquinone and quinone. The absorbancy changes (at appropriate wavelength) follow an oscillatory pattern of periodicity 2 which is also damped. Applying σ analysis to this pattern (with an order 2 matrix), we find $\sigma_1 \stackrel{\sim}{=} 0.1$. This result strongly suggests that the damping of a photochemical response in flash sequence could be a property of reaction centers in general - and not of the OES in particular - when they are forced to function at high light density. Such an effect - if optically non-linear - would not be in contradiction with the finding of quantum yield of 1 electron $h\nu^{-1}$ currently found at low light intensity.

Accordingly, we can imagine a type of model of non-linear photochemical misses following the kinetic scheme:

$$PQ + h\nu \longrightarrow (PQ)^* \qquad (11a)$$

$$(PQ)^* \longrightarrow P^+Q^- \qquad (11b)$$

$$(PQ)^* + h\nu \longrightarrow PQ \qquad (11c)$$

In this, (PQ)* is a transient (not yet stabilized) species of the reaction center. Its life-time would be such that under ordinary light intensities the "quenching" step eq. (11c) would have a negligible chance to occur.

In my laboratory, we have performed preliminary experiments comparing the O_2 yield in *Chlorella* given rectangular flashes (argon laser) of fixed light density but variable duration ranging from 50 ns to 10 µs. We find a systematically higher O_2 yield (10 to 30%) for the long flashes than for the short ones. Note that this increase is much larger than predicted from normal multiple hits resulting from the turn-over of centres during the flashes: Assuming a turn-over half-time of 400 µs, only a 2% increase in O_2 yield is expected for a 10 µs flash as compared to a very short one.

Conclusion

In this report it is shown that the damping of the oscillatory pattern observed in the classical experiment of O_2 emission in sequence of flashes might have a large number of conceivable origins. Damping mechanisms may *a priori*, be classified as photochemical or conservative, but their detailed nature can be very diverse and the present list is certainly not complete! We have seen that so far there is no unequivocal evidence in favour of one mechanism (I believe, however, that a mechanism of photochemical misses as in the original conception of Kok is very unlikely). It is perhaps a safe position to assume that several independent mechanisms contribute to the damping effect. It is clear nevertheless that its complete elucidation would help to understand the various kinetic factors which control the yield of the photosynthetic reaction centres.

Footnotes

(a) It is true though that some models assume that atomic oxygen might be an intermediate form, but they nevertheless rely on the idea of storage.
(b) For instance, giving two flashes to a dark-adapted

sample places the OES in S_3 state, etc.; this, however, is only an approximation because we always are dealing with mixtures of states, a serious limitation to any quantitative analysis of the system.

(c) As a consequence also an average $\bar{Y}^{(n)}$ [16] may be calculated from the foregoing which should be equal to the steady-state O_2 yield Y_{ss} and thus may be taken as an instantaneous relative measure of the concentration of active centres. To be more precise, $\bar{Y}^{(n)}$ might be influenced both by changes in number of active centres and in the transition coefficients along the flash sequence.

(d) Obviously we are dealing here with meta-stable states, therefore nothing can be said concerning intermediate states (such as S'_i in: $S_i + h\nu \longrightarrow S'_i \longrightarrow S_{i+1}$)

(e) This result is by no means trivial. One cannot strictly conclude from a periodicity of four in a sequence to a four-state structure of the model, as exemplified in the five-state model D of Mar and Govindjee [20], their Fig. 3(d) giving nevertheless sequences of periodicity close to 4.

(f) We recall that misses are random events: all centres have equal probability, according to the model, of not reacting. This is not the case if the state Q^- on any centre lasts longer than the flashing period.

(g) Co-operative models are as a rule more complicated. As far as I can see, they seem to imply time-varying matrix coefficients. The practical importance of this point (as an object of testing) has not been investigated yet.

(h) In the original proposal [15], E was assumed to be confined in the interior space of the thylakoid only reaching P and binding to it from the inside face of the membrane. The present proposal seems more likely. Concerning the present problem, the two proposals are formally identical.

(i) A general second order process (with (A) $\neq$ (B) - see eq. 8) would not be actually unexpected because of the known occurrence of diffusible species such as protons. However the requirement (A) = (B) is very stringent: the two species must be produced in pair by light, they must be confined and may exclusively disappear by reacting with each other.

References

1. Bennoun, P. (1970). *Biochim. Biophys. Acta* **216**, 357-363.
2. Delrieu, M.J. (1974). *Photochem. Photobiol.* **20**, 441-454.
3. Diner, B. and Joliot, P. (1976). *Biochim. Biophys. Acta* **423**, 479-498.
4. Duysens, L.N.M., van der Schatte Olivier, T.E. and den Haan, G.A. (1974). Abstracts of the VI. International Congress on Photobiology, p. 277, Bochum.
5. Etienne, A.L. (1974). *Biochim. Biophys. Acta* **333**, 320-330.
6. Etienne, A.L. (1974). Thesis, Université de Paris-Sud.
7. Etienne, A.L., Lemasson, C. and Lavorel, J. (1974). *Biochim. Biophys. Acta* **333**, 288-300.
8. Ikegami, I. and Katoh, S. (1973). *Plant Cell Physiol.* **14**, 837-850.
9. Joliot, P., Barbieri, G. and Chabaud, R. (1969). *Photochem. Photobiol.* **10**, 309-329.
10. Joliot, P., Joliot, A., Bouges, B. and Barbieri, G. (1971). *Photochem. Photobiol.* **14**, 287-305.
11. Kok, B., Forbush, B. and McGloin, M. (1970). *Photochem. Photobiol.* **11**, 457-475.
12. Lavorel, J. (1975). *Photochem. Photobiol.* **21**, 331-343.
13. Lavorel, J. (1975). Unpublished.
14. Lavorel, J. (1976). *J. Theor. Biol.* **57**, 171-185.
15. Lavorel, J. (1976). *FEBS Lett.* **66**, 164-167.
16. Lavorel, J. and Lemasson, C. (1976). *Biochim. Biophys. Acta* **430**, 501-516.
17. Maison, B. (1977). Personal communication.
18. Maison, B. and Etienne, A.L. (1977). *Biochim. Biophys. Acta* **459**, 10-19.
19. Maison, B. and Lavorel, J. (1977). In "Photosynthetic Organelles, Structure and Function" (S. Miyachi, S. Katoh, Y. Fujita and K. Shibata, eds), pp. 55-65, Japanese Society of Plant Physiologists, Tokyo.
20. Mar, T. and Govindjee (1972). *J. Theor. Biol.* **36**, 427-446.
21. Mauzerall, D. (1976). *Biophys. J.* **16**, 87-91.
22. Mauzerall, D. and Chivvis, A. (1973). *J. Theor. Biol.* **42**, 387-395.
23. Picaud, A. (1977). Thesis, Université de Paris-Sud.
24. Rabinowitch, E.I. (1945). "Photosynthesis and Related Processes", vol. I, Interscience Publishers, New York.
25. Radmer, R. and Kok, B. (1973). *Biochim. Biophys. Acta* **314**, 28-41.
26. Radmer, R. and Kok, B. (1974). *Biochim. Biophys. Acta* **357**, 177-180.
27. Sun, A.S.K. and Sauer, K. (1971).*Biochim. Biophys. Acta* **234**, 399-414.
28. Thibault, P. (1977). Personal communication.

29. Velthuys, B.R. and Amesz, J. (1974). *Biochim. Biophys. Acta* **333**, 85-94.
30. Vermeglio, A. (1977). *Biochim. Biophys. Acta* **459**, 516-524.
31. Vernotte, C. (1977). Personal communication.

BICARBONATE: ITS ROLE IN PHOTOSYSTEM II

GOVINDJEE and R. KHANNA

*Department of Physiology and Biophysics and Botany
University of Illinois, Urbana, Illinois (USA)*

Summary

The major role of the bicarbonate ion in isolated thylakoids is to accelerate the rate of the electron flow from the secondary electron acceptor R (or B) of the photosystem II to the plastoquinone pool (PQ). This conclusion is based on the measurements of chlorophyll a fluorescence after single flashes of light [5] and of absorbance changes of the plastoquinone system [18]. Independent biochemical measurements on the partial reactions confirm the conclusion that the major bicarbonate effect is located somewhere between the "primary" electron acceptor Q (or X-320) and the PQ pool [10]. The bottleneck reaction of the normal chloroplasts resupplied with bicarbonate is $\simeq 20$ ms (reoxidation of PQH_2). It is concluded that the bottleneck reaction of the bicarbonate-depleted chloroplasts is 100 to 200 ms, and this is located in the reoxidation of R^{2-}, thus explaining the 5-10 fold stimulation, by bicarbonate, of the steady-state saturation rates of the Hill reaction.

Introduction

In addition to its role as a substrate for carbohydrate production, CO_2 is required for the Hill reaction (see *e.g.* [21]). When chloroplasts are depleted of bicarbonate, the rate of the Hill reaction is inhibited; a large increase is, however, observed when bicarbonate is resupplied to such depleted samples. Earlier, Stemler and Govindjee [21-23] and Stemler et al. [20] believed that the absence of bicarbonate blocks the oxidizing (or water) side of pigment system II (PS II). Wydrzynski and Govindjee [30], on the basis of their data on chlorophyll a fluorescence transient, suggested a new site of bicarbonate action on the reducing side of PS II. Jursinic et al. [9] provided evidence for a major stimulatory effect of bicarbonate on the reoxidation of Q^- to Q as measured by the decay of chlorophyll a fluorescence yield after a saturating flash. Jursinic et al. [9] further concluded that there was no effect of bicarbonate on the electron flow from H_2O to the reaction center chlorophyll a of PS II (P 680).

In this paper we shall review recent experiments which show that the site of bicarbonate effect is between the "primary" electron acceptor Q and the plastoquinone (PQ) pool. A large stimulatory effect of bicarbonate on the reoxidation of the reduced secondary electron acceptor R^{2-} by the PQ pool appears to be the major effect of this anion in the thylakoid membranes.

A Few Characteristics of the CO_2 Effect

Effect of external pH on bicarbonate stimulation

The effect of bicarbonate as a function of external pH (5 to 9) shows a much larger stimulation in the Hill activity around pH 6 to 7, where the bicarbonate species predominates [10,21]. This could imply that the bicarbonate ion is the active species in stimulating the Hill reaction. However, the affinity of HCO_3^- to the membrane component(s) may be different at different pHs. No definite assertion can, therefore, be made regarding the active species involved.

Bicarbonate stimulation in the presence of uncouplers of photophosphorylation

Bicarbonate has also been shown to cause enhancement of photophosphorylation [1,16]. To distinguish between the direct stimulation of electron flow by HCO_3^- from any indirect stimulation produced by an enhancement of photophosphorylation by bicarbonate, Khanna et al.[10] measured the bicarbonate effect in the presence of several uncouplers of photophosphorylation. The data presented in Table 1 show that there are no changes in the rates of electron

TABLE 1

Bicarbonate effect in the presence of uncouplers of photophosphorylation: rates of O_2 evolution (after [10]).

	Electron transport, µequivalents/mg Chl per h		Ratio
	$-HCO_3^-$	$+10$ mM HCO_3^-	$+HCO_3^-/-HCO_3^-$
(1) Control	26 ± 3	171 ± 9	6.6
Plus NH_4Cl	24 ± 4	163 ± 7	6.8
(2) Control	22 ± 3	134 ± 14	6.1
Plus methylamine hydrochloride	23 ± 3	147 ± 18	6.4
(3) Control	22 ± 4	165 ± 15	7.5
Plus nigericin	21 ± 1	144 ± 5	6.9
(4) Control	13 ± 1	135 ± 4	10.4
Plus gramicidin-D	12 ± 1	125 ± 7	10.4

The 2 ml reaction mixture consisted of 50 mM phosphate buffer (pH 6.8), 100 mM sodium formate, 100 mM NaCl and 0.5 mM ferricyanide. Concentrations of uncouplers used were: NH_4Cl (1 mM), methylamine, HCl (0.3 mM), nigericin (5 µM) and gramicidin-D (1 µM). Spinach chloroplasts containing 33 µg chlorophyll (Chl) ml^{-1} suspensions were used. Average of three experiments is shown.

flow or in the bicarbonate stimulation in the presence of NH_4Cl, methylamine-HCl, nigericin or gramicidin-D. Our samples are uncoupled to begin with; this may be so because we use high concentration of salts in the depletion medium. In any case, these data clearly establish that the bicarbonate stimulation of Hill activity studied here is distinct from the effect of HCO_3^- on photophosphorylation.

However, we have not yet established if the HCO_3^- effect on photophosphorylation is through its effects on electron flow or through a direct effect on the coupling factor protein [13].

Decay of the Reduced Form of the "Primary" Electron Acceptor after Light Flashes

Bicarbonate effect on the chlorophyll a fluorescence yield

Measurement of the decay of chlorophyll a fluorescence yield, 50 μs to few ms after a light flash, suggests that the reoxidation of the reduced "primary" acceptor Q^- is inhibited by bicarbonate depletion [9]. The half time of decay of fluorescence is slowed down to about 3 ms in the CO_2-depleted sample. A five fold decrease in the half time of decay is observed when 10 mM bicarbonate is resupplied to the depleted sample (Fig.1). This data indicates that a bicarbonate action is on the dark reoxidation of Q^-

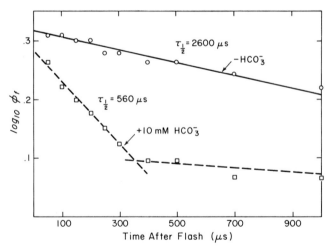

Fig. 1. Semilog. plot of the decay of chlorophyll a fluorescence yield (ϕf) after a saturing 10 ns 337 nm pulse with and without 10 mM bicarbonate. ϕf was measured with a variable delay analytic weak flash (General Electric Strobotac 1538-A; neutral density filters and Corning blue glass C.S. 4-96). Photomultiplier, EMI 9558 B protected with Wratten 2A and Schott RG-8 filters. *Lactuca sativa* chloroplasts were depleted of bicarbonate and resuspended in buffer as described by Wydrzynski and Govindjee [30]. Similar results were obtained with *Zea mays* chloroplasts (after [9]).

following a flash. Furthermore, the chlorophyll a fluorescence transients measured as a function of decreasing concentrations of bicarbonate are qualitatively similar to those observed with increasing concentrations of DCMU which imposes a block on the reducing side (see [30]). This also implies that bicarbonate depletion acts as a partial block of electron flow between Q and the intersystem carriers in much the same way as DCMU does at low concentrations.

Kinetics of absorption changes at 320 nm

Absorption changes at 320 nm indicate the formation and decay of semiquinone. Siggel *et al.*[18] studied the transition of quinone to semiquinone anion of the "primary" acceptor X-320 (Q) of PS II and its dark decay by monitoring absorption changes at 334 nm instead of at 320 nm (changes at 334 nm and 320 nm provided the same conclusions). Absorption change at 334 nm shows a very fast rise which is of the order of $\lesssim$ μs and a biphasic relaxation in dark with a half time of ∿500 μs for the faster component. The fast rise is attributed to the rapid photoreduction of X-320 and the 500 μs decay to the subsequent reoxidation of X-320$^-$. [X320$^-$ (or Q$^-$) donates its electron to the PQ pool *via* a two-electron acceptor species R (or B) [3,24].]

A semi-logarithmic plot of the time course of the absorption change at 334 nm, measured with repetitive flashes, shows two exponential phases of about equal magnitude with half-lifetimes ($t_{\frac{1}{2}}$) of 500 ± 100 μs and 7 ± 3ms in bicarbonate-depleted chloroplasts (Fig.2). The biphasic relaxation in these chloroplasts reflects heterogeneity of the sample introduced by the depletion procedure — about 30% of the total sample is unaffected by the procedure of depletion, another 30% exhibits slow kinetics; the rest of the signal (40%) was, however, totally inactivated, which is in agreement with the earlier conclusion

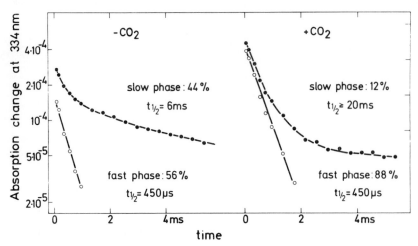

Fig. 2. Semilog. plots of the absorbance change at 334 nm induced by 20 μs repetitive flashes (indicating mainly Q^- and R^-, see text) for bicarbonate depleted ($-CO_2$) and reconstituted chloroplasts ($+CO_2$). *Spinacia oleracea* chloroplasts, which had been frozen and thawed, were depleted of bicarbonate as described earlier [21,30]. To obtain the $+CO_2$ sample, 20 mM bicarbonate was added. 512 flashes were averaged; the darktime between the flashes (td) = 250 ms; the electrical bandwidth was 10 kHz. The relative magnitude and half time of the two exponential phases are indicated (after [18]).

that about 30-40% of the reaction centre complex P680 is inactivated [9,20]. In the control and those chloroplasts reconstituted with bicarbonate, most of the signal decays with a t½ of ∼500 μs, but, there is a minor proportion (10-15%) of the signal which decays with a half-lifetime of ∼20 ms. This is due to an uncharacterized signal probably due to the formation of plastohydroquinone or the reduction of oxidized P700.

Site of Bicarbonate Effect is Between Q and the PQ Pool

Partial biochemical reactions

To establish the site of the bicarbonate action by independent biochemical means, Khanna *et al.* [10] measured the electron flow in partial reactions. This was achieved by using appropriate electron donor-acceptor combinations in conjunction with specific inhibitors of electron carriers. The electron transport chain was divided as shown below:

$$H_2O \to \overset{h\nu_2}{\to} Z \to P680 \to \overset{SiMo}{Q^-} \to R \to \overset{DBMIB}{PQ} \to cytf \to PC \to \overset{h\nu_1}{P700} \to X \to MV$$
$$O_2 \nearrow \qquad \qquad DCMU \quad DAD_{ox} \qquad DAD_{red}$$

where silicomolybdate (SiMo) accepts electrons from Q^- (see e.g. [31]), DCMU blocks electron flow from Q^- to R, oxidized diaminodurene (DAD_{ox}) accepts electrons somewhere before the plastoquinone pool [14], dibromothymoquinone (DBMIB) is an inhibitor of electron flow beyond the PQ pool [2], reduced diaminodurene (DAD_{red}) acts as an artificial electron donor to photosystem I [8], and methylviologen (MV) accepts electrons from X^-.

The bicarbonate effect in various partial reactions is shown in Table 2.

TABLE 2

Effect of Bicarbonate on Three Partial Reactions (after [10])

System (See scheme in the text)	Electron transport µequivalents/mg Chl per h	
	$-HCO_3^-$	$+10$ mM HCO_3^-
(1) H_2O to silicomolybdate	117 ± 16	108 ± 17
(2) H_2O to oxidized diaminodurene	12 ± 1	90 ± 2
(3) Reduced diaminodurene to methyl viologen	662 ± 12	673 ± 16

Chloroplasts containing 33 µg Chl/ml were illuminated in a continuously stirred reaction mixture (2 ml) containing 50 mM phosphate buffer (pH 6.8), 100 mM sodium formate, 100 mM NaCl and the indicated donor and acceptor system. These systems were: (1) H_2O — silicomolybdate (SiMo); 5 µM 3-(3'4'dichlorophenyl)1,1'dimethylurea (DCMU) and 25 µM SiMo. (2) H_2O — oxidized diaminodurene (DAD_{ox}); 0.5 mM DAD, 0.5 mM ferricyanide and 0.5 µM 2,5-dibromo-3-methyl-6-isopropyl-p-benzoquinone (DBMIB). (3) DAD_{red} — methyl viologen (MV); 50 µM MV, 0.5 mM DAD, 2.0 mM sodium ascorbate and 1 µM DCMU. When SiMo or DAD_{ox} was the electron acceptor, electron transport was observed as O_2 evolution. When MV was the acceptor, electron transport was followed as O_2 uptake. All data have been converted to µequivalents of O_2 (or µmoles of electrons) mg (chlorophyll) $Chl^{-1} h^{-1}$. Average of three experiments is shown.

H_2O to silicomolybdate system shows no significant stimulation of O_2 evolution upon the addition of 10 mM HCO_3^- to the bicarbonate depleted samples. This implies that there is no major site of HCO_3^- action on the oxidizing side of photosystem II. The absence of bicarbonate causes a reversible inactivation of up to 40% of the photosystem II reaction centers (see refs [9,20]). Thus we had expected a small HCO_3^- effect on this reaction. However, this inactivation can range from 5 to 40% [12]. The absence of the HCO_3^- effect in the $H_2O \rightarrow$ SiMo reaction may have been due to the low inactivation of reaction center II in our preparations.

In the H_2O to oxidized diaminodurene reaction, a 7- to 8-fold enhancement of O_2 evolution is observed when 10 mM $NaHCO_3$ is added to the depleted samples. Oxidized diaminodurene, perhaps, intercepts electrons before they reach the last molecule in the PQ pool. The complete electron transfer chain from H_2O to X shows the same effect as the system just described. Since DBMIB prevents the electron flow out of the reduced PQ pool, we conclude that HCO_3^- acts somewhere between Q and the PQ pool.

The above conclusion is further confirmed by the observation that there is no effect of HCO_3^- depletion in the photosystem I electron transport chain ($DAD_{red} \rightarrow MV$).

Absorbance changes at 265 nm in repetitive flashes

Absorbance changes at 265 nm are indicative of changes in Q, R and the PQ pool. In the discussion below, the following flow diagram of the intersystem electron carriers will be used:

$$PSII \rightarrow Q \xrightarrow{(A)} R \xrightarrow{(B)} PQ \xrightarrow{(C)} cyt\ f$$

The dark relaxation of the absorption change at 265 nm has a major slow phase ($t_{\frac{1}{2}} \sim 20$ ms) (see *e.g.* [17]). The 20 ms time represents the oxidation of plastohydroquinone (reaction C). The control chloroplasts of this study behaved slightly differently than the normal chloroplasts

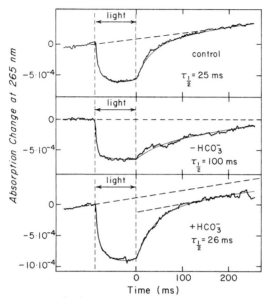

Fig. 3. Time course of the absorption change at 265 nm induced by 85 ms repetitive flashes (indicating mainly R^{--} and PQ^{--}, see text) for untreated (control), CO_2-depleted ($-CO_2$) and reconstituted ($+CO_2$) spinach chloroplasts (frozen and rethawed). The halftimes of dark relaxation are shown. Number of flashes, 64; darktime (td) = 5s; electrical bandwidth, 600 Hz. 720 nm background light ($\Delta\lambda$ = 15 nm) of 400 erg.cm^{-2}.s^{-1} intensity was used except in the case of the control (leading to a reduced amplitude in this case; after [18]).

as the slow phase had a half time of 25 ms [18]. CO_2-depletion retards this slow phase from 25 ms to about 100 ms (Fig. 3). This effect is reversible as the addition of 10 or 20 mM bicarbonate restored the 25 ms phase. This slow phase could represent either (1) the oxidation of PQH_2, if the relaxation of the signal is governed by reaction C or, (2) reaction B, if the relaxation is determined by the formation of PQ^{--} from R^{--} (accompanied by no or a minor absorption change) and the absorption change is brought about by the consecutive faster oxidation of the PQH_2 (reaction C). We suggest that the first possibility holds for control and reconstituted chloroplasts and the second holds for bicarbonate-depleted chloroplasts; the latter is consistent with the experiments on chlorophyll a fluorescence which show that a 150 ms time represents electron flow from R^{2-} to PQ pool in these depleted chloroplasts [5].

A retardation of the electron flow from R^{2-} to PQ explains the 5-10 times decreased rate of Hill reaction in the bicarbonate-depleted chloroplasts as the bottleneck reaction changes from a $t\frac{1}{2}$ of $\sim$20 ms to 100-200 ms.

Bicarbonate effect on the secondary electron acceptor R (or B)

Govindjee *et al.*[5] studied the effect of bicarbonate depletion on the electron transport from the primary acceptor, Q, to the plastoquinone pool, *via* the secondary electron acceptor R (the two electron carrier). They measured the fluorescence yield 150 ms after the flash as a function of flash number. Fig. 4 shows that this fluorescence yield reaches a high level in the CO_2- depleted chloroplasts after the third and succeeding flashes. The fluorescence yields 150 ms after the 1st and 2nd flashes are, however, low. In control chloroplasts and depleted

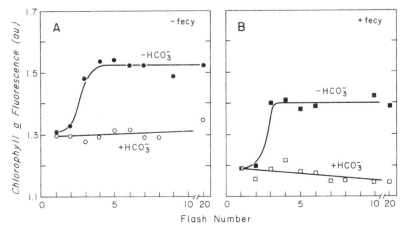

Fig. 4. Chlorophyll a fluorescence intensity ~150 ms after the last of a series of 3 µs saturating light flashes, spaced at 30 ms, as a function of the number of flashes. (au stands for arbitrary units.) Additions as indicated. Concentrations: Bicarbonate, 20 mM; chlorophyll, 20 µg-ml^{-1} of spinach chloroplast suspension; ferricyanide (FeCy), 20 mM (after [5]).

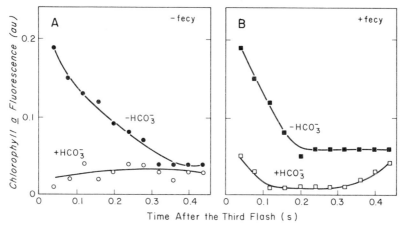

Fig. 5. Chlorophyll a fluorescence intensity after the third *minus* that after the second flash as a function of time. Additions as indicated; see the legend of Fig. 4 (after [5]).

chloroplasts resupplied with bicarbonate, the fluorescence yield is low after each of the light flashes. These results indicate that the fluorescence decay is slowed down from the third flash onwards in bicarbonate-depleted chloroplasts. The same results were obtained in the absence and the presence of ferricyanide. Thus, the observed effects were not influenced by the rate of electron acceptance by ferricyanide. Fig. 5 shows the kinetics of the slow fluorescence decay, plotted as the difference between the fluorescence intensity after the third minus that after the second flash, as a function of time. The halftime of this decay is of the order of 150 ms.

The above data are interpreted as follows. It is suggested that the absence of bicarbonate retards the electron flow from R^{2-} to the PQ pool. The first flash creates QR^-, the second flash QR^{2-}, and the third flash, if given fast enough, would form Q^-R^{2-} and the fluorescence yield would be high. Succeeding flashes would give the same result. The 150 ms, noted above, would then reflect the $t_{\frac{1}{2}}$ of the reduction of PQ by R^{2-} in CO_2-depleted chloroplasts.

It thus appears, from the decay of chlorophyll a fluorescence yield after the third flash and of the decay of absorbance changes at 265 nm after repetitive long flashes, that the bicarbonate depletion slows down the rate of PQ reduction ($t_{\frac{1}{2}}$ of 100-200 ms); this becomes the new bottleneck reaction. In control and reconstituted chloroplasts, the bottleneck reaction has a $t_{\frac{1}{2}}$ of ~20 ms (oxidation of PQH_2). Thus, the 5-10 fold inhibition of the electron flow in the absence of bicarbonate, or, else, the 5-10 stimulation of the Hill reaction by the addition of HCO_3^- to the depleted samples, can be finally explained.

Concluding Remarks

Warburg [27] believed that the requirement of CO_2 for the Hill reaction (which he had discovered [28]) was due to its involvement in the O_2 release in photosynthesis. This is the main rationale for the presentation of the present report at this symposium. The early work on the effect of CO_2 (or bicarbonate) on the Hill reaction was performed by several investigators (see refs [4,6,7,26,29]); Stemler and coworkers [19,24] and Metzner [12] have recently revived interest in this problem. Jursinic *et al.* [9] failed to find an effect of CO_2 on the oxygen evolution side of photosynthesis. The major effect of the bicarbonate ion has been shown here to be on the electron flow from the reduced form of the secondary electron acceptor (R^{2-}) to the plastoquinone (PQ) pool. The mechanism of this effect needs to be investigated. Perhaps, HCO_3^- is needed to place "R", which is a plastoquinone [15] in a proper conformation so that it can accept electrons from Q^- and donate electrons to the PQ pool in an efficient fashion. Kreutz [11] and Metzner [12] have suggested involvement of HCO_3^- in the O_2 evolution steps. However, all the recent experiments in our laboratory support a different site of HCO_3^- action. Thus, if there is any direct role of bicarbonate in the O_2 evolution, it still needs to be discovered.

References

1. Batra, P.P. and Jagendorf, A.T. (1965). *Plant Physiol.* **40**, 1074-1079.
2. Böhme, H., Reimer, S. and Trebst, A. (1971). *Z. Naturforschg.* **26b**, 341-352.
3. Bouges-Bocquet, B. (1973). *Biochim. Biophys. Acta* **314**, 250-256.
4. Good, N.E. (1963). *Plant Physiol.* **38**, 298-304.
5. Govindjee, Pulles, M.P.J., Govindjee, R., van Gorkom, H.J. and Duysens, L.N.M. (1976). *Biochim. Biophys. Acta* **449**, 602-605.
6. Heise, J. and Gaffron, H. (1963). *Plant Cell Physiol.* **4**, 1-11.
7. Izawa, S. (1962). *Plant Cell Physiol.* **3**, 221-227.
8. Izawa, S., Gould, J.M., Ort, D.R., Felker, P. and Good, N.E. (1973). *Biochim. Biophys. Acta* **305**, 119-128.
9. Jursinic, P., Warden, J. and Govindjee (1976). *Biochim. Biophys. Acta.* **440**, 322-330.
10. Khanna, R., Wydrzynski, T. and Govindjee (1977). *Biochim. Biophys. Acta,* in press.
11. Kreutz, W. (1974). In "On the Physics of Biological Membranes" (K. Colbow, ed.), pp. 419-429, Department of Physics, Simon Fraser University, Vancouver.
12. Metzner, H. (1975). *J. Theor. Biol.* **51**, 201-231.
13. Nelson, N., Nelson, H. and Racker, E. (1972). *J. Biol. Chem.* **247**, 6506-6510.
14. Ouitrakul, R. and Izawa, S. (1973). *Biochim. Biophys. Acta* **305**, 105-118.
15. Pulles, M.P.J., van Gorkom, H.J. and Willemsen, J.G. (1976). *Biochim. Biophys. Acta* **449**, 536-540.
16. Punnett, T. and Iyer, R.V. (1964). *J. Biol. Chem.* **239**, 2335-2339.
17. Siggel, U. (1976). *Bioelectrochem. Bioenerg.* **3**, 302-318.
18. Siggel, U., Khanna, R., Renger, G. and Govindjee (1977). *Biochim. Biophys. Acta,* in press.
19. Stemler, A. (1977). *Biochim. Biophys. Acta* **460**, 511-522.
20. Stemler, A., Babcock, G.T. and Govindjee (1974). *Proc. Natl. Acad. Sci. Washington* **71**, 4679-4683.
21. Stemler, A. and Govindjee (1973). *Plant Physiol.* **52**, 119-123.
22. Stemler, A. and Govindjee (1974). *Photochem. Photobiol.* **19**, 227-232.
23. Stemler, A. and Govindjee (1974). *Plant Cell Physiol.* **15**, 533-544.
24. Stemler, A. and Radmer, R. (1975). *Science* **190**, 457-458.
25. Velthuys, B.R. and Amesz, J. (1974). *Biochim. Biophys. Acta* **333**, 85-94.
26. Vennesland, B., Olson, E. and Ammeral, R.N. (1965). *Fed. Proc.* **24**, 873-880.
27. Warburg, O. (1964). *Ann. Rev. Biochem.* **33**, 1-14.
28. Warburg, O. and Krippahl, G. (1960). *Z. Naturforschg.* **15b**, 367-369.
29. West, J. and Hill, R. (1967). *Plant Physiol.* **42**, 819-826.
30. Wydrzynski, T. and Govindjee (1975). *Biochim. Biophys. Acta* **387**, 403-408.

31. Zilinskas, B.A. and Govindjee (1975). *Biochim. Biophys. Acta* **387**, 306-319.

PHOTOSYSTEM II ACTIVITY DEPENDS ON MEMBRANE-BOUND BICARBONATE

A. STEMLER

Carnegie Institution of Washington, Department of Plant Biology, Stanford, California, (U.S.A.)

The rate of electron flow through photosystem II depends on the presence of bicarbonate ions [1,6,9]. The nature of the interaction between HCO_3^- and the electron transport chain, however, remains in question. It may now be possible to approach this question, as it can be demonstrated that the bicarbonate ions involved with PS II are tightly bound to the thylakoid membranes.

Washed thylakoid membranes obtained from osmotically shocked maize chloroplasts can take up and bind HCO_3^- in the dark. This process can be shown by competitive binding experiments which use $H^{14}CO_3^-$ as a tracer. Chloroplast fragments were given $H^{14}CO_3^-$ along with increasing concentrations of unlabelled HCO_3^- to compete for binding sites with the $H^{14}CO_3^-$. The details of this procedure are published elsewhere [5]. The results of competitive binding experiments are usually presented by means of a Scatchard plot [4]. Fig. 1 is an example of such a plot.

In the case of a pure enzyme having a single binding site, a Scatchard plot will yield points on a straight line, the intercept of the abscissa giving the concentration of binding sites. Chloroplasts, binding HCO_3^-, demonstrate instead a curved Scatchard plot (Fig. 1). This result indicates that there are at least two binding sites

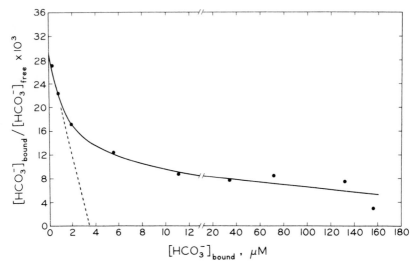

Fig. 1. Scatchard plot for the binding of HCO_3^- to chloroplast grana. The suspension contained 1.0 mg chlorophyll ml^{-1} (1.1 mM), 0.1 M sodium phosphate pH 6.5, 0.4 M NaCl, 1.2 µCi[^{14}C]NaHCO$_3$ ml^{-1}. The concentration of unlabelled NaHCO$_3$ was varied from 0.0 to 50 mM. The grana were incubated for 5 min at 30°C after addition of [^{14}C]HCO$_3^-$. Each of the last 4 points is an average of 8 separate sample measurements at different HCO$_3^-$ concentrations.

in terms of affinity for HCO_3^- on, or within, thylakoid membranes. The sharp drop in the left portion of the curve shows a small pool of binding sites having a relatively high affinity for HCO_3^-. A rough estimate of the size of this pool can be obtained by a line tangent to the upper section of the curve and intersecting the abscissa [2]. The dashed line so drawn in Fig. 1 intersects at 3.5 µM and this value represents the upper limit of the concentration of this binding site when the chlorophyll concentration is 1.1 mM. The ratio of bound HCO_3^- to chlorophyll molecules is, therefore, about 1:300. A more precise method [5] of determining the high-affinity pool size gave a slightly lower ratio, 1:380-400. Assuming one PS II reaction centre per 400 chlorophyll molecules, it is evident that one HCO_3^- is bound per reaction centre.

The actual size of the photosynthetic units in the chloroplast membranes was not measured, however, providing a possible source of error. It is the opinion of the author that one HCO_3^- bound per PS II reaction centre is, nevertheless, the most reasonable estimate, with 2 HCO_3^- bound per centre being the upper limit.

The long tail of the Scatchard plot (Fig. 1) which approaches the abscissa very obliquely to the right indicates that there also exists in grana another quite large pool of relatively low affinity HCO_3^- binding sites. The size of this pool is difficult to determine since it cannot be saturated because of the limited solubility of HCO_3^-/CO_2 at the experimental pH (6.5). A reasonable guess is that it is some large fraction of the chlorophyll concentration.

Because they have different binding properties, the two pools of HCO_3^- binding sites can be studied independently. If a small amount of $H^{14}CO_3^-$ is given to chloroplast membranes previously depleted of HCO_3^- with no competing cold HCO_3^-, most of the label will become attached to the high-affinity sites. If a large amount of cold HCO_3^- is given along with the $H^{14}CO_3^-$, the high-affinity sites will become saturated predominantly with cold HCO_3^-, and $H^{14}CO_3^-$ will then be attached primarily to low-affinity sites (which cannot be saturated). Moreover, the two pools differ in their ability to withstand repeated washing of the chloroplasts. HCO_3^- attached to the low-affinity sites can be removed with one or two washings in a solution buffered at pH 7.0. Once HCO_3^- is attached to the high-affinity sites, in contrast, it cannot be removed even after 3 or 4 washings. Removing HCO_3^- bound to low-affinity sites by washing does not suppress photosynthetic oxygen evolution and to date, no photochemical function can be assigned to this pool. Removal of HCO_3^- bound to the high-affinity sites (by the depletion procedure described below) does, on the other hand, markedly suppress oxygen evolution. The remaining discussion will concern only the high-affinity bound HCO_3^-.

The requirement of high-affinity bound HCO_3^- for maximum PS II activity is shown by the following experiments: Thylakoid membranes were given $H^{14}CO_3^-$ under conditions allowing binding to take place, then washed twice to remove all unbound and loosely bound $H^{14}CO_3^-$. The remaining $H^{14}CO_3^-$ was that taken up by the small number of high-affinity sites. The grana were then washed a third time in various media and collected by centrifugation. The dpm remaining in the pellet was determined as well as the ferricyanide-supported oxygen evolving activity of the chloroplasts, first in the absence, then in the presence of 12 mM added $NaHCO_3$. The effect of the third washing on both HCO_3^- removal and on oxygen-evolving ability could thus be observed.

The results are shown in Table 1. Control chloroplasts (line 1) were washed a third time in the same solution as the first two washings (see legend). Afterwards, these chloroplasts retained a measurable amount of $H^{14}CO_3^-$ and evolved oxygen normally without added HCO_3^-. When the third washing was in high salt medium at pH 5.0, *i.e.* "HCO_3^- depletion medium", the chloroplast pellet lost more than 99 per cent of its bound HCO_3^- (line 2). In these same chloroplasts oxygen evolution was suppressed by more than 90 per cent. When 12 mM HCO_3^- was added back, oxygen was increased more than 10-fold.

When chloroplast membranes were washed in a low salt medium containing 20 mM unlabelled $NaHCO_3$, they retained all their bound $H^{14}CO_3^-$ (line 3). This result indicates that the HCO_3^- bound in the high-affinity pool does not normally exchange with free HCO_3^-.

A very interesting result was obtained when 3-(3',4'-dichlorophenyl)-1,1-dimethylurea (DCMU) was given to the chloroplasts prior to washing with HCO_3^- depletion medium. In the presence of DCMU less than half of the bound HCO_3^-

TABLE 1

Wash Solution	dpm/mg chlorophyll ×10⁻¹	HCO₃⁻ removed % of control	μmol O₂/mg chlorophyll·h − HCO₃⁻	μmol O₂/mg chlorophyll·h + HCO₃⁻	$\left(\dfrac{+ HCO_3^-}{- HCO_3^-}\right)$
(1) 0.1 M sodium phosphate, pH 7.0 0.01 M NaCl 0.3 M sucrose	440	control	40.0	41.3	1.03
(2) 0.1 M sodium phosphate, pH 5.0 0.175 M NaCl 0.1 M sodium formate	< 3	< 99	2.6	38.7	14.9
(3) 0.1 M sodium phosphate pH 6.5 0.01 M NaCl 0.3 M sucrose 0.02 M NaHCO₃	433	2	48.5	47.0	0.96
(4) 10⁻⁵ M DCMU (added first) 0.1 M sodium phosphate pH 5.0 0.175 M NaCl 0.1 M sodium formate	249	44	-	-	-

Effectiveness of various wash media at removing tightly bound [^{14}C]HCO₃⁻ and inducing dependence of oxygen evolution on added HCO₃⁻. The chloroplasts were charged with [^{14}C]HCO₃⁻ by suspension in 1 ml medium containing 0.1 M sodium phosphate pH 5.8, 0.2 M NaCl, 168 μM [^{14}C]NaHCO₃ (1.2 μCi) and 2 mg Chl. After a 5 min incubation at 30°C, 6 ml of ice cold wash medium containing 0.1 M sodium phosphate pH 7.0 0.01 M NaCl, 0.3 M sucrose was added. After centrifugation the pellet was washed twice in 7 ml of the wash medium then a third time in the medium indicated. The dpm were determined in the final pellet and portions were measured for oxygen evolving ability in saturating light immediately after suspension in reaction mixture containing 0.1 M sodium phosphate pH 7, 0.175 M NaCl, 0.1 M sodium formate, 2 mM K₃[Fe(CN)]₆, 50 μg Chl·ml⁻¹. Initial rates were measured before and after injection of NaHCO₃ to 12.5 mM.

was removed (line 4) whereas more than 99 per cent was removed in the absence of DCMU (line 2). This result strongly suggests that the binding site of HCO_3^- and that of DCMU are very close.

The experiments mentioned thus far were done in the dark. Illuminating the chloroplasts failed to produce any noticeable increase in $H^{14}CO_3^-$ uptake. HCO_3^- binding is, therefore, a "dark" reaction. Experiments were also done to test whether or not bound HCO_3^- was released in the light. Chloroplast membranes were charged with $H^{14}CO_3^-$, then washed twice to remove all but the tightly bound label. They were then illuminated for 18 min in saturating light while oxygen evolution was monitored. At various time intervals, aliquots were drawn from the reaction mixture, centrifuged, and the pellet assayed for bound $H^{14}CO_3^-$. Results are depicted in Fig. 2. At the end of the illumination period, net oxygen evolution in the chloro-

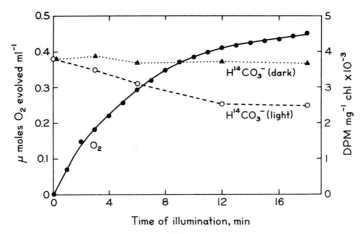

Fig. 2. Loss of tightly bound [^{14}C]HCO$_3$ from chloroplasts during oxygen evolution. Chloroplasts were charged with [^{14}C]HCO$_3^-$ as explained in the legend, Table 1. After a third washing the grana were suspended in reaction mixture containing 0.1 M sodium phosphate pH 7.0, 0.01 M NaCl, 0.3 M sucrose, 0.01 M K$_3$Fe(CN)$_6$, 162 µg Chl ml^{-1}, then illuminated. At times indicated, 1 ml of the reaction suspension was withdrawn from the illumination chamber, centrifuged, and the pellet measured for bound HCO_3^-. Similar results were observed when 20 mM unlabelled NaHCO$_3$ was added to the reaction mixture before illumination.

plasts was reduced to almost zero due to photoinactivation. Meanwhile the chloroplasts had lost only about 35 per cent of the $H^{14}CO_3^-$ bound in the small pool. Dark controls lost no $H^{14}CO_3^-$. When 20 mM unlabelled $NaHCO_3^-$ was added to the reaction mixture, the results obtained were identical. Again the chloroplasts lost only 35 per cent of their bound $H^{14}CO_3^-$ and evolved the same amount of oxygen.

These results indicate that while a small fraction of the tightly bound HCO_3^- is lost during prolonged illumination, there is no rapid exchange of the bound HCO_3^- with free HCO_3^- during oxygen evolution in the light. Rather, a bicarbonate ion joins each of the high affinity binding sites, activates electron flow, and remains in place even in the presence of a large excess of free HCO_3^-, while the reaction centre evolves many molecules of oxygen. The small amount of bound HCO_3^- which is lost in the light may be the result of membrane damage of a non-specific nature.

Knowledge of the membrane binding characteristics of HCO_3^- permits a quite different interpretation of earlier work done to decide if this ion, rather than water, could be the immediate source of photosynthetically evolved oxygen. Carbon dioxide splitting as the source of photosynthetically evolved oxygen was a theory to which Otto Warburg [8] was dedicated. More recently Metzner [3] has proposed that HCO_3^-, acting catalytically, is the immediate source of oxygen. In an attempt to settle this question, ^{18}O was employed as a tracer in a study which employed mass spectrometry. The work was done by the author in collaboration with Richard Radmer at the Martin Marietta Laboratories, Baltimore. At the time of its completion this experiment seemed to indicate clearly that HCO_3^- is not the source of photosynthetic oxygen.

The experiment was done as follows: HCO_3^--depleted chloroplasts were suspended in CO_2-free reaction mixture containing ferricyanide and placed in the reaction cell attached to the mass spectrometer (see ref.[7] for experi-

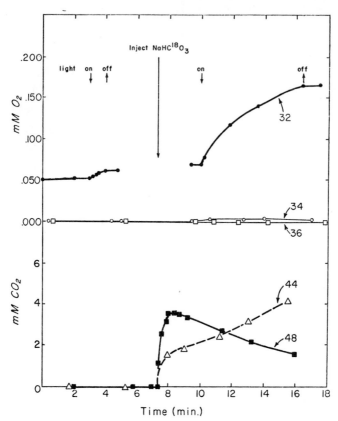

Fig. 3. ^{18}O content of dissolved carbon dioxide and photosynthetically evolved oxygen with time. The reaction mixture contained 0.175 M NaCl, 0.1 M Na formate, 0.05 M N-2-hydroxyethyl-piperazine-N-2-ethanesulfonic acid (HEPES) pH 7.5, 2 mM $K_3Fe(CN)_6$ and 100 μg chlorophyll/ml. Light from a projection lamp was filtered through 30 cm of water and a 3-110 filter (Corning) and focussed onto the reaction vessel. The intensity was about one-half of saturation. Mass numbers represented by each curve are indicated; 32 ($^{16}O_2$); 34 ($^{16,18}O_2$); 36 ($^{18}O_2$); 44 ($C^{16}O_2$); 48 ($C^{18}O_2$).

mental details). The chloroplasts were illuminated for one minute, $HC^{18}O_3^-$ was injected in a subsequent dark period and the chloroplasts were then reilluminated. The molecular masses observed were 32($^{16}O_2$), 34($^{16,18}O_2$), 36($^{18}O_2$), 44($C^{16}O_2$) and 48($C^{18}O_2$).

The results are shown in Fig. 3. Before adding $HC^{18}O_3^-$ depleted chloroplasts gave off only a small amount of O_2, seen as an increase in mass number 32. At 7.3 min, NaHC^{18}O$_3$ stock solution was injected into the reaction cell, giving rise to a large mass 48(C^{18}O$_2$) signal. This signal slowly declined as ^{18}O was exchanged for ^{16}O in the unlabelled H$_2$O. However, the rate of exchange was slow enough (indicating an absence of carbonic anhydrase) so that at the onset of the second illumination period (at 10 minutes) there still existed in solution more C^{18}O$_2$ than C^{16}O$_2$.

Oxygen evolution began again with the onset of the second light period. Because HCO$_3^-$ was present, the initial rate was about 5 times that seen during the first illumination. In other words, the usual HCO$_3^-$ effect was observed. We noted, however, that while most of the HCO$_3^-$ was labelled with ^{18}O at the time of illumination, the evolved O$_2$ was almost completely ^{16}O. The small amount of singly labelled O$_2$(^{16,18}O$_2$) which was evolved was consistent with the amount of ^{18}O in the water (as part of the HC^{18}O$_3^-$ stock solution, 20 μl of 97.2 per cent H$_2$^{18}O was injected so the final reaction mixture was about 2 per cent H$_2$^{18}O).

At first examination, the conclusions to be drawn from this experiment are unmistakable. Although HC^{18}O$_3^-$ markedly stimulated oxygen evolution, the oxygen evolved was almost exclusively ^{16}O$_2$. Apparently HCO$_3^-$ is not the immediate source of photosynthetic O$_2$. If we consider the membrane binding characteristics of HCO$_3^-$, however, such a conclusion is not nearly so compelling. The mass spectrometer was able to monitor the isotopic composition only of the free HCO$_3^-$, whereas that which stimulates oxygen evolution is now known to be a small pool tightly bound near the PS II reaction centre. Since this pool does not exchange with free HCO$_3^-$ in either light or dark, its isotopic composition at any given time cannot be measured directly. If, for the moment, we assume that this small pool of HCO$_3^-$ were the immediate source of oxygen, we might expect that after adding HC^{18}O$_3^-$ the first molecule of oxygen to emerge would likewise contain ^{18}O. CO_2 would be left behind (see Metzner ref. [3], for a more detailed scheme) and, before another oxygen could be evolved, the CO_2 would have to react with water to form HCO_3^- + H^+ once again. The water, however, would contain only ^{16}O so the newly formed HCO_3^- would contain substantially less ^{18}O than it did before the first molecule of oxygen was evolved. It is clear that having added $HC^{18}O_3^-$, only the first few molecules of O_2 given off by each reaction centre will contain ^{18}O. After that, the isotopic composition of evolved oxygen must increasingly reflect that of the solution water even if HCO_3^- were, in fact, the immediate source. To eliminate the ambiguity resulting from the existence of the tightly bound HCO_3^- pool, it will therefore be necessary to measure the isotopic composition of only the first few molecules of O_2 given off after $HC^{18}O_3^-$ is given to the chloroplasts. One hopes this will not meet insurmountable technical problems. A more worrisome consideration is that the oxygen exchange rate between tightly bound HCO_3^- and free water may be very fast. If so, even the first molecules of oxygen evolved after giving $HC^{18}O_3^-$ will reflect the isotopic composition of the suspension water. The occurrence of rapid isotope exchange would, unfortunately, be difficult to rule out.

The foregoing discussion may be summarized as follows: The electron flow rate through PS II depends on a small pool of tightly bound HCO_3^- which does not exchange with free HCO_3^- in the dark or while oxygen is evolved in the light. The site of HCO_3^- binding appears to be very close to that of the inhibitor DCMU. The use of ^{18}O as a tracer in mass spectrometry studies has not yet eliminated the possibility that membrane bound HCO_3^- may be the immediate source of photosynthetic oxygen.

References

1. Good, N.E. (1963). *Plant Physiol.* **38**, 298-304.

2. Klotz, I.M. and Hunston, D.I. (1971). *Biochemistry* **10**, 3065-3069.
3. Metzner, H. (1966). *Naturwissenschaften* **53**, 141-150.
4. Scatchard, G. (1949). *Ann. N. Y. Acad. Sci.* **51**, 660-672.
5. Stemler, A. (1977). *Biochim. Biophys. Acta* **460**, 511-522.
6. Stemler, A. and Govindjee (1973). *Plant Physiol.* **52**, 119-123.
7. Stemler, A. and Radmer, R. (1975). *Science* **190**, 457-458.
8. Warburg, O. (1964). *Ann. Rev. Biochem.* **33**, 1-14.
9. Warburg, O. and Krippahl, G. (1960). *Z. Naturforschg.* **15b**, 367-369.

RECENT STUDIES ON PHOTOSYSTEM II WITH THE MODULATED OXYGEN ELECTRODES

J. SINCLAIR, A. SARAI, T. ARNASON and S. GARLAND

*Department of Biology,
Carleton University, Ottawa (Canada)*

We have been studying the characteristics of the rate-limiting reaction which lies between the water-splitting act and the reaction centre of photosystem II, using the modulated oxygen electrode devised by Joliot [7]. I shall briefly describe this apparatus and how it can be used to monitor the rate-limiting reaction.

Fig. 1 shows a diagram of this apparatus and the associated electrical circuit. The electrode or polarograph was divided into three compartments by two stretched dialysis membranes. A flat shiny platinum electrode lined the bottom compartment and a reference silver/silver chloride electrode was located in the upper compartment. The platinum electrode was at an electric potential of -0.7 V

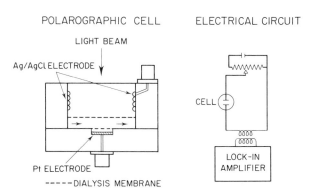

Fig. 1. Schematic diagram of modulated oxygen electrode of associated electrical circuit.

relative to the reference electrode and acted as an oxygen sink. The *Chlorella* cells or isolated chloroplasts were layered on top of the platinum electrode and were immersed in an experimental medium which was maintained at a constant chemical composition by the fresh medium which flowed through the central compartment.

The *Chlorella* suspensions were illuminated from above by a light whose intensity was modulated in a periodic manner. The chloroplasts responded to this light by producing waves of oxygen which diffused out of the cells to the platinum electrode with which they interacted to produce an alternating electric current in the measuring circuit. The phase lag (relative to the light) and the amplitude of this current were measured with the vector voltmeter, and these represented the experimental observations.

When the phase lag and amplitude of the oxygen signal were measured at different modulation frequencies it was found that the phase lag increased with the modulation frequency and the amplitude was attenuated. These observations were first made by Joliot, Hoffnung and Chabaud [8], and to explain them these workers proposed that there were two processes which influenced the experimental results in an appropriate way. These were firstly the diffusion process, whereby the oxygen waves travelled from the *Chlorella* to the electrode, and secondly the thermal reaction acting between photosystem II and the water splitting act, which rate-limited the production of oxygen. This reaction was postulated to have first order kinetics because the phase lag of the oxygen signal was found to be independent of the average intensity of the modulated light. As we shall see, this is a characteristic of a first order but not of a second order reaction. Joliot, Hoffnung and Chabaud [8] derived a pair of equations which described the influence of modulation frequency on the oxygen signal size and phase lag in terms of the diffusion process and the first order, rate-limiting reaction:

$$\ln\left[A \cdot \nu^{\frac{1}{2}}\right] = B_1 - \varepsilon \left[\frac{\pi \cdot \nu}{D}\right]^{\frac{1}{2}} - \tfrac{1}{2}\ln\left[1 + \left(\frac{2\pi\nu}{k}\right)^2\right] \qquad (1)$$

$$\phi = \frac{\pi}{4} + \varepsilon\left[\frac{\pi \cdot \nu}{D}\right]^{\frac{1}{2}} + \arctan\left[\frac{2\pi\nu}{k}\right] \qquad (2)$$

A is the O_2 signal amplitude and ϕ is the phase lag, B_1 is a proportionality factor which is a measure of the average rate of O_2 evolution, D is the diffusion coefficient of O_2 in water, ν is the modulation frequency, ε is the average distance between the closest O_2 sources and the platinum electrode, k is the rate constant of the rate-limiting reaction. It can be seen that there is no intensity-dependent term in the phase lag equation, as indicated above. The first two terms in both equations express the influence of diffusion while the last terms derive from the first order thermal reaction for O production. In every case we found that the values of k and ε which described the phase results, also gave a good description of the attenuation of the amplitude results with frequency. Fig. 2 demonstrates, how the experimental data fit these equations.

Having reviewed the methodology let us now examine the results which have been obtained with both *Chlorella* cells and isolated spinach chloroplasts. The rate constant k had a means value of 305 s^{-1} in *Chlorella* (3) and 215 s^{-1} in isolated spinach chloroplasts [1]. These values can be compared with estimates made by other workers for the rate-limiting reaction. Joliot, Hoffnung and Chabaud [8] produced a single value in their study of 820 s^{-1} although they used a different procedure for analysing their results. Etienne [5] using a method which involved the illumination of a flowing *Chlorella* suspension found a value of about 300 s^{-1}. Bouges-Bocquet [3] evaluated the rate constants for the dark steps in the 4-state Kok model of photosystem II, and the smallest rate constant had a value of about 580 s^{-1}. Thus there would not appear to be a very

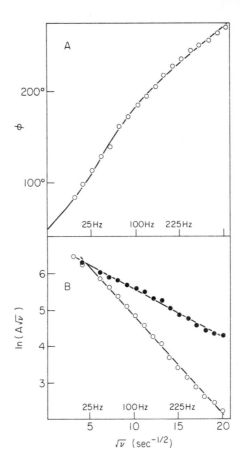

Fig. 2A. The phase lag between the current and the light modulation plotted against the square root of the modulation frequency (closed circles), obtained with a 35 μm layer of *Chlorella*. The solid line represents the curve generated by the equation

$$\phi = \phi_0 + \frac{\xi}{\sqrt{2D}} \sqrt{\omega} + \arctan\left(\frac{\omega}{k}\right), \text{ for } k = 330 \text{ s}^{-1}$$

$\xi = 3.0 \times 10^{-4}$ cm and $\phi_0 = 50°$. The standard deviation between the experimental and calculated points was 2.3°.

2B. The amplitude results of the same experiment as in (A) for the open circles $x = A\sqrt{\nu}$ and for the closed symbols $x = \frac{A\sqrt{\nu}\sqrt{k^2+\omega^2}}{k}$ with $k = 330$ s^{-1}; both are plotted against $\sqrt{\nu}$.

According to the theory described in the text, the upper line should only reflect the influence of oxygen diffusion while the lower line should also reflect the influence of a thermal rate-limiting reaction. From the slope of the upper line a value of 3.2×10^{-4} cm was derived for ξ.

close agreement on the value of the rate constant for the rate-limiting reaction. It is possible that part of the disparity between Bouges-Bocquet's results and our own can be explained by different conditions of measurement. Of necessity her observations were made on cells which had been in darkness and were in the process of adapting to the light whereas our measurements were made with cells which were completely light-adapted. It is commonly believed that when chloroplasts are illuminated, an electrical potential difference and a pH difference are created across the thylakoid membranes, and alterations in the local concentrations of magnesium ions, oxygen and carbon dioxide might also be expected. So perhaps the changed environment of the thylakoids affected the enzymes involved and led to the discrepancies noted above.

We have also studied the temperature dependence of k and an example of our results is shown as an Arrhenius plot (Fig. 3). The mean activation energy for *Chlorella* was 24.7 kJ·mole^{-1} [10] and 29.3 kJ·mole^{-1} for chloroplasts [1]. Joliot *et al.* [8] quoted a value of 32.2 kJ·mole^{-1} for *Chlorella* and Etienne [5] a value of 27.6 kJ·mole^{-1}. The linearity of our Arrhenius plot indicated that the same reaction was rate-limiting at all temperatures studied.

We have been using a first order reaction to describe the production of oxygen, i.e. a reaction in which the rate of O_2 evolution was proportional to the concentration of some unknown substance. The identity of this precursor was the objective of a series of experiments we performed using algae immersed in a medium in which the water had been replaced by deuterium oxide. We began these experiments by postulating that the precursor of the evolved oxygen was a hydroxyl group and that the rate-limiting reaction involved a rupture of the OH bond to separate the oxygen and hydrogen from each other. If this postulate were correct, we would have expected to see a large

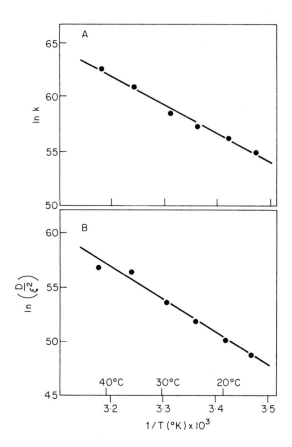

Fig. 3. (A) The natural logarithm of the rate constant, k, derived from the phase data, versus 1/T, obtained for a 35 μm layer of *Chlorella* cells. A least squares line is drawn through the points from which an activation energy of 21.8 kJ. mole^{-1} was calculated.

(B) The natural logarithm of D/ξ^2 from the same phase data, also plotted against 1/T. The least squares line through these results corresponds to an activation energy of 25.1 kJ. mole^{-1}.

decrease in the value of the rate constant k when the algae were immersed in deuterium oxide. This decrease in rate constant is known as the kinetic isotope effect or in this case simply the deuterium isotope effect. This effect can be accounted for by looking at the potential energy diagram for the reaction (Fig. 4). There is no alteration in the potential energy profile when we consider

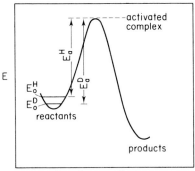

Fig. 4. Energy diagram showing the origin of the kinetic isotope effect for the rupture of OH and OD bonds in the differences of zero point energies.

the rupture of an OD bond instead of an OH bond but there is a change in the zero point energy, i.e. the lowest vibrational energy level, which is by far the most heavily populated energy level at room temperature. The zero point energy level for an OD bond lies significantly below that for an OH bond and consequently the activation energy is larger for an OD bond. This difference in activation energies could in theory make the rate constant for an OH bond rupture 10.6 times large as that for an OD bond. However, such a large ratio is seldom seen in chemical reactions where deuterium isotope effects normally lie between 3 and 7.

Fig. 5 shows what happened when we performed frequency scans on a layer of *Chlorella* cells in deuterium oxide and in water. There was a slight increase in the phase lag at any given frequency and a slight decrease in the amplitude. When the equations were fitted to those results we found that the rate constant in water was on average about 1.29 times as large as that in deuterium oxide [10]. The corresponding figure for spinach chloroplasts was 1.1 [1]. These were very low values for the ratio of rate constants, values which might well be due to effects of D_2O on enzyme

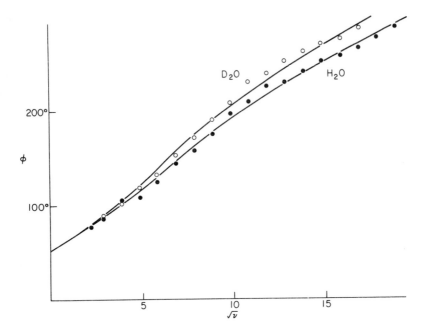

Fig. 5. The polarographic results when D_2O is substituted for water in the experimental medium, for a 6 µm layer of *Chlorella vulgaris* cells at 22°C. The open circles are the phase values obtained shortly after suspension of the algae in D_2O based buffer. Virtually the same results were obtained 1½ hours later. After one hour in water based solution, the phase values represented by the closed circles were obtained. The upper line is the best fit curve for D_2O, generated for $k=232$ s^{-1}, $p/\sqrt{D} = .136$ $s^{-\frac{1}{2}}$, and $\phi_0=0°$. The standard deviation between the experimental and theoretical points was 1.3°. The lower line is the best fit curve for the experiment in water for which $k=289$ s^{-1}, $p/\sqrt{D}= .129$ $s^{-\frac{1}{2}}$, $\phi_0=0°$ and the S.D.=.8°. P is the mean distance between the *Chlorella* cells and the platinum electrode.

structure for example. We have concluded that the deuterium isotope effect was absent here and that the reaction we studied probably did not involve the rupture of an OH bond.

We decided to pursue our search for a deuterium isotope effect associated with photosystem II by examining the effect which deuterium oxide had on the oscillating flash yields seen when *Chlorella* cells are illuminated by a series of saturating light flashes after a prolonged

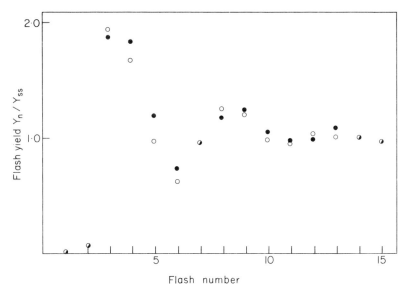

Fig. 6. The flash yield sequence in H_2O and D_2O. The normalized yield of oxygen plotted against the flash number for a sample of *Chlorella* cells suspended for 1 hour in D_2O (closed circles) and later in H_2O (open circles). The values of α and β from the computer fit of the Kok model were 0.22 and 0.017 respectively for cells in H_2O and 0.25 and 0.008 for cells in D_2O.

period of darkness [2]. In Fig. 6 we see the relative values of the flash yields for a sample of *Chlorella* plotted against flash number for a water based medium (open symbols) and for deuterium oxide based medium (closed symbols). When we fitted the Kok model to these results we found that α=0.22 in water and 0.25 in deuterium oxide and β=0.017 in water and 0.008 in deuterium oxide. Thus D_2O caused an increase in the number of misses, and a decrease in the number of double hits, effects which were reversible on returning to a water based medium. Figs 7-10 illustrate the effect of deuterium oxide on the progress of the 4 dark reactions associated with the Kok scheme of photosystem II reaction centres. We monitored these 4 reactions using the procedures developed by Bouges-Bocquet [3] and in each case the sample of *Chlorella* or chloroplasts was used to generate the results in both D_2O and H_2O. Fig. 7 shows the

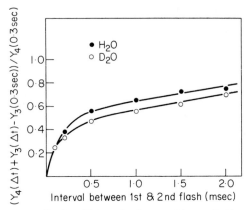

Fig. 7. Time course of the first dark reaction of the Kok scheme, $S_0^* \longrightarrow S_1$ in H_2O and D_2O. *Chlorella* cells were kept in the dark for 5 min, then subjected to a sequence of three flashes at 0.3 s intervals (the pre-illumination sequence). This was followed by an additional dark period of 3 min before the second sequence of 25 flashes. In the second sequence the interval between the first and second flash was variable. All other intervals were 0.3 s.

$$\gamma_0(\Delta t) = \frac{Y_4(\Delta t) + Y_3(\Delta t) - Y_3(0.3 \text{ s})}{Y_4(0.3 \text{ s})}$$ is plotted against Δt. The curves were fitted by eye. Y_n is the yield of the nth flash.

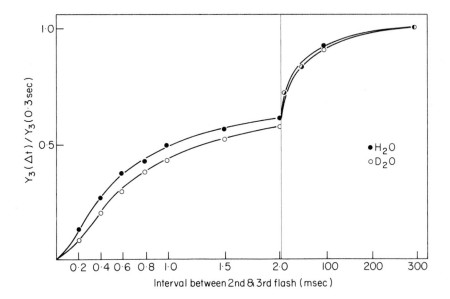

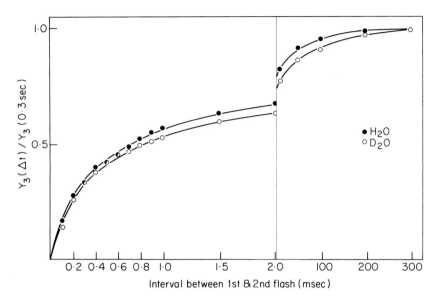

Fig. 9. The time course of the third dark reaction of the Kok scheme, $S_2^* \longrightarrow S_3$ in H_2O and D_2O. *Chlorella* cells were left in the dark for 5 min before the sequence of 25 flashes. Between the second and third flash a variable time interval Δt was used. The spacing between all other flashes was 0.3 s. The variation of the yield of the third flash $Y_3(\Delta t)/Y_3(0.3\ s) = \gamma_2(\Delta t)$ is plotted against Δt. The curves were fitted by eye.

decay of the S_0^* state to the S_1 state (the dark symbols in each of these four Figs give the results for water and the open symbols the results for deuterium oxide). The mean ratio of half times for D_2O and H_2O was 1.3. Fig. 8 depicts the S_1^* to S_2 transition and the mean ratio of half time was 1.0. Then we have S_2^* to S_3 with the mean half time ratio being 1.2 (Fig. 9) and finally S_3^* to S_0, obtained with spinach chloroplasts, with a mean half time ratio of 1.1 (Fig. 10). Thus the results obtained with this study

---

Fig. 8. (Opposite) The time course of the second dark reaction of the Kok scheme, $S_1^* \longrightarrow S_2$, in D_2O and H_2O. *Chlorella* cells which had been kept in the darkness for 5 min were subjected to a flash sequence of 25 flashes in which the interval Δt between the first and second flash were variable. The intervals between all other flashes was 0.3 s. The variation of the yield of the third flash $Y_3(\Delta t)/Y_3(0.3\ s) = \gamma_1(\Delta t)$ is plotted against Δt. Curves fitted by eye.

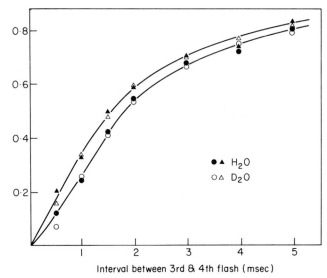

Fig. 10. Time course of the fourth dark reaction of the Kok scheme $S_3^* \longrightarrow S_0$ in D_2O and H_2O. Spinach chloroplasts, which were suspended in a medium containing 0.1 mM ferricyanide, were subjected to 5 min darkness followed by a flash sequence of 35 flashes in which the interval between the third and fourth flash was variable. All other intervals between flashes were 0.3 s.
$\frac{Y_7(\Delta t) - Y_7(0.3 \text{ s})}{Y_7(0.3 \text{ s}) - Y_6(0.3 \text{ s})} = x$ is plotted against Δt (circles). The triangles represent $\gamma_3(\Delta t) = 0.8 \ x \ (\Delta t) + 0.05 \ \gamma_0(\Delta t) + 0.18 \ \gamma_1(\Delta t)$. Curves were fitted by eye.

on saturating flash yields were consistent with those obtained with the modulated electrode, i.e. there was no kinetic isotope effect found whose magnitude was close to that which we expected.

We can offer three possible explanations of these results.

(1) The OH bond cleavage in photosystem II is an unusual reaction which does not give rise to a kinetic isotope effect. Chemists have found reactions where the expected kinetic isotope effect failed to appear because of an unusual degree of binding in the transition state. However, such reactions are extremely rare.

(2) The OH bond cleavage is extremely fast and does

not rate limit any of the reactions we have observed.

(3) There is no OH bond cleavage associated with photosystem II because the evolved oxygen comes from CO_2, a proposal which Metzner [9] has recently discussed.

In our work with isolated chloroplasts [1] we have taken advantage of the fact that the thylakoids are relatively accessible and have exposed them to a variety of conditions to see if any would change the value of k. The following treatments had no appreciable effect: 1. Varying the pH of the medium between 5.8 and 8.2; 2. adding 1 mM ammonium chloride to the medium as an uncoupler; 3. removing the 0.33 M sorbitol to drastically reduce the osmotic pressure of the medium; 4. varying the oxygen concentration between 10% and 30%.

The only treatment which changed the value of the rate constant was a decrease in the ionic strength of the

TABLE 1

The properties of the rate-limiting thermal reaction for O_2 evolution in Chlorella *and isolated spinach chloroplasts.*

Chlorella	Spinach Chloroplasts
1. It is a first order reaction	1. It is a first order reaction
2. The rate constant at 22°C is 305 ± 20 s^{-1}	2. The rate constant at 23°C is 215 ± 10 s^{-1}
3. The activation energy is 24.7 ± 2.1 kJ. mole^{-1}.	3. The activation energy is 29.3 ± 8.4 kJ. mole^{-1}.
4. The reaction probably does not involve the splitting of an O-H bond.	4. The reaction probably does not involve the splitting of an O-H bond.
$\frac{k_{H_2O}}{k_{D_2O}}$ = 1.29 ± 0.05	$\frac{k_{H_2O}}{k_{D_2O}}$ = 1.1 ± 0.1
5. The same reaction is rate limiting between 8°C and 42°C	5. The rate constant decreases reversibly at low salt concentration
	6. The rate constant is insensitive to the presence of 1 mM NH$_4$Cl, moderate changes in pH, the osmolarity of the solution.

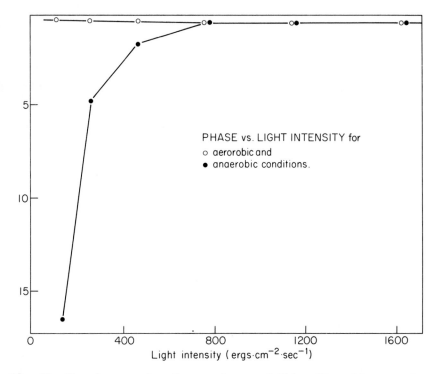

Fig. 11. The phase angle of a monolayer of *Chlorella* cells as a function of the modulated 650 nm light intensity. The results indicated by open symbols were obtained with a medium in equilibrium with air enriched with 10% CO_2; those for the closed symbols were obtained with a medium in equilibrium with a gas mixture containing 0.28% O_2, 10% CO_2 and the balance nitrogen, temperature 22°C.

medium. The latter normally contained 100 mM KCl, 1 mM $MgCl_2$, 5 mM $NaHCO_3$, 20 mM HEPES/NaOH buffer (pH 7.6), 0.33 M sorbitol. Removing the 100 mM KCl was sufficient to reduce the rate constant from about 215 s^{-1} to 110 s^{-1}. This change could be reversed by adding back ions, e.g., 25 mM $MgCl_2$ or 100 mM Na acetate. It is possible that this reduction of the rate constant was related to the changed thylakoid structure observed in low ionic strength media by Izawa and Good [6].

Table 1 summarizes the various characteristics which

have been ascribed to the rate-limiting reaction by these experiments.

Our most recent work was initiated by an observation which was disturbing. We found that under certain circumstances the phase lag of the oxygen signal from *Chlorella* cells varied with the intensity of the illuminating light. This was at variance with out earlier observations and those of Joliot, Hoffnung and Chabaud [8] and naturally put into question our use of a first order reaction to rate-limit oxygen production. I shall return to this problem later.

Under what conditions did the phase lag become dependent on the modulated light intensity? Fig. 11 shows two sets of phase lag measurements obtained with a monolayer of *Chlorella* cells for different intensities of 650 nm lgiht, at a temperature of 22°C. When the cells were

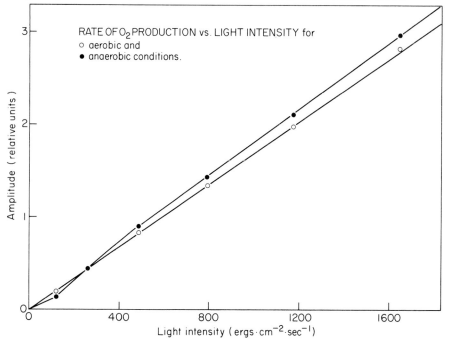

Fig. 12. The amplitude of the O_2 signal (which is proportional to the mean rate of O_2 evolution) measured at various intensities of modulated 650 nm light for the experiment shown in the previous figure.

immersed in a solution which was in equilibrium with air enriched with 10% CO_2, the results represented by open symbols were obtained. Under these circumstances there was virtually no phase change with intensity. But when the cells were immersed in a medium in equilibrium with a gas containing 10% CO_2, 0.28% O_2 and the balance nitrogen, the results indicated by the closed symbols were obtained. There was a very definite phase dependence on the light intensity - the phase lag being similar at low light intensities. Fig. 12 shows how the rate of O_2 evolution varied with light intensity in this experiment. While the rate in the low O_2 solution was higher than that in the high O_2 solution at high intensities, their relative magnitudes were reversed at low intensities. Thus as the light intensity was lowered in the low O_2 solution there was a decrease in the oxygen yield. Fig. 13 is also from

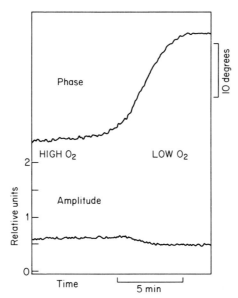

Fig. 13. The change of phase and signal amplitude as a function of time when the solution flowing past the cells was changed from one in equilibrium with air enriched with 10% CO_2 to one in equilibrium with 0.28% O_2, 10% CO_2 and the balance nitrogen. An upward phase deflection represents a decrease in phase lag. The intensity of 650 nm modulated light was 120 erg $cm^{-2} \cdot s^{-1}$ and the temperature was 22°C.

this experiment and shows in more detail what happened to the phase lag and rate of O_2 evolution when, at an intensity of 120 erg. cm^{-2} s^{-1}, the oxygen concentration was lowered. The upward deflection of the phase lag indicates a decrease in phase lag while the downward deflection of the amplitude indicates a decrease in the rate of oxygen evolution.

Having shown that at a sufficiently low oxygen concentration the phase lag and oxygen yield became intensity-dependent we investigated the effects of employing solutions in equilibrium with different mixtures of oxygen, nitrogen and carbon dioxide to discover to what level the oxygen content of the gas mixture had to be raised before the intensity dependency disappeared. We have found that if oxygen was present at the 1% level there was sometimes a small phase change with intensity and sometimes there was none at all and we believe that this is close to threshold O_2 concentration. It would of course be more satisfying if the actual oxygen concentration around the *Chlorella* cells could be measured but this is an extremely difficult task. The cells were sitting upon the platinum electrode which acted as an oxygen sink. The cells were producing oxygen, both those on the electrode and those on the perspex surface around the electrode. Oxygen was diffusing from the central compartment past the cells to the platinum electrode. These different elements produced a highly complex diffusion situation which we have not solved. Thus we were forced to rely upon an operational description of what the oxygen concentration was.

The results which we have described so far were all obtained at room temperature but we have found that if we raised the temperature the results we obtained changed quite significantly. Fig. 14 gives the results of intensity scans performed on a monolayer of *Chlorella* cells at two temperatures, 22°C (open symbols) and 36°C (closed symbols). The solution had been bubbled vigorously with nitrogen containing less than 10 ppm O_2 for one hour after which 2 mM sodium bicarbonate was added to the medium. The large symbols are the phase results which should be referred to the vertical axis to the right and the small symbols are the relative rates of O_2 evolution and should be referred to the vertical axis to the left. As the intensity of the modulated light was lowered below 1000 erg. cm^{-2} s^{-1} at 36°C, the phase lag began to decrease. At room temperature the phase lag change did not begin until the light intensity was reduced below 400 erg. cm^{-2} s^{-1}. Thus the threshold light intensity was a function of the temperature. It is also apparent that the shapes of the phase lag plots were different at the two temperatures. That at 36°C was approximately sigmoidal and suggested that the cells could exist in two different states, one at high intensities and one at very low intensities. When we examined the amplitude plots the non-linearity which was just observable at room temperature becomes much

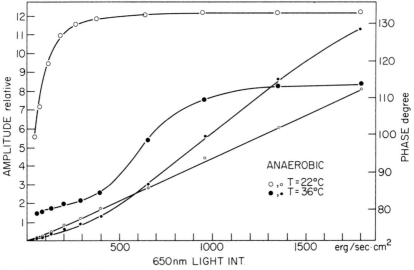

Fig. 14. The influence of temperature upon the phase (large symbols) and signal amplitude (small symbols) for a monolayer of *Chlorella* cells for various intensities of modulated 650 nm light. The solution had been bubbled with nitrogen for an hour after which 2 mM $NaHCO_3$ was added.

more pronounced at 36°C. Indeed the rate of O_2 evolution at 36°C was less than that at 22°C below intensities of 600 erg. cm^{-2} s^{-1}. This surprising result was obtained despite the temperature dependence of the oxygen electrode in which a given O_2 concentration would give rise to a larger electric current at 36°C than at 22°C. So the rate of O_2 evolution at 36°C was probably less than that at 22°C under aerobic conditions at intensities even higher than the 600 erg. cm^{-2} s^{-1} indicated here.

Since raising the temperature enhanced both the phase lag dependency on light and the reduced oxygen yield seen under low light intensities, we wondered whether elevating the temperature could also produce these phenomena under aerobic conditions. As can be seen on Fig. 15 both the decrease on phase lag and the reduced oxygen yield did appear at temperatures of about 34°C when the *Chlorella* cells were immersed in a solution in equilibrium with the air. Thus raising the temperature elevated the

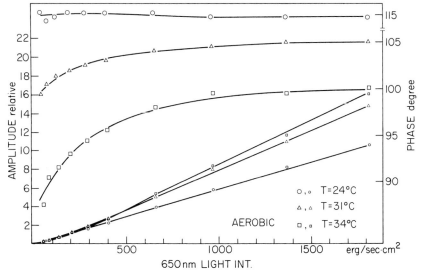

Fig. 15. The influence of temperature upon the phase (large symbols) and signal amplitude (small symbols) for a monolayer of *Chlorella* cells for various intensities of modulated 650 nm light under aerobic conditions.

concentration of oxygen necessary to abolish these intensity-dependent phenomena.

From our earliest experiments it was apparent that it was necessary to bubble nitrogen vigorously through the solution reservoir and flow the solution past the *Chlorella* cells for an hour or more before these phenomena could be detected. However, measurements of the O_2 concentration in the reservoir with a Clark oxygen electrode showed that the oxygen rapidly fell to a low value once nitrogen bubbling commenced. It therefore seemed that there might be a conditioning of the *Chlorella* which had to be completed before the phase change with intensity could be observed. We have since found that if we replaced a sample of *Chlorella* which was exhibiting these phenomena with a fresh sample of cells, the latter immediately manifested a phase lag which was intensity-dependent. We have therefore abandoned the idea of an extended conditioning process for the *Chlorella* and now believe that

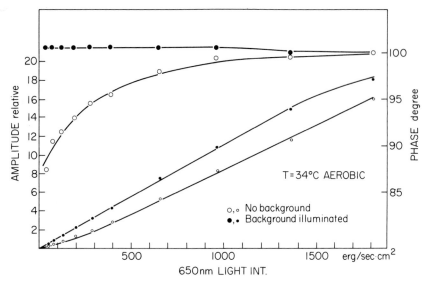

Fig. 16. The influence of a steady background 700 nm light (4030 erg. cm^{-2} s^{-1}) on the phase (large symbols) and signal amplitude (small symbols) for a monolayer of *Chlorella* cells for various intensities of 650 nm modulated light.

the long bubbling with nitrogen was necessary to remove oxygen from the walls of the apparatus and associated tubing.

Finally, we have studied the effect of adding a non-modulated background light of wavelength 700 nm to the modulated 650 nm light (Fig. 16). It should be realized that since our O_2 detection system responded only to the modulated production of oxygen, the direct effect of this background light was not observed and only its influence on the modulated oxygen production was detected. This was an experiment performed under aerobic conditions but at a temperature of 34°C. The open symbols give the phase and amplitude in the 650 nm modulated beam alone and it is apparent that there was a decreased phase lag as the intensity was lowered below 1000 erg. cm^{-2} s^{-1} and a distinct non-linearity in the amplitude plot indicating a

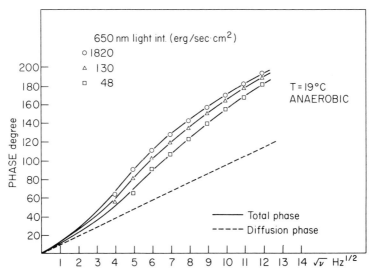

Fig. 17. The phase lag for a monolayer of *Chlorella* cells at different modulation frequencies for three intensities of 650 nm light under anaerobic conditions. The points represent the experimental results while the solid lines are the least fitting theoretical curves. The diffusive component of the theoretical curves did not alter with light intensity and is represented by the hatched line. The rate constant was 176 s^{-1} at 1820 erg. $cm^{-2} s^{-1}$, 249 s^{-1} at 130 erg. cm^{-2} s^{-1} and 369 s^{-1} at 48 erg. cm^{-2} s^{-1}.

reduced oxygen yield. When the background light was added (I_{700} = 4030 erg. cm^{-2} s^{-1}) the phase variation with intensity disappeared and the amplitude plot became linear. These results showed that if we speeded up photosystem I relative to photosystem II we could eliminate the intensity sensitivity of both the phase lag and the oxygen yield.

There appeared to be several possible explanations of these results. For example we considered that the enzyme catalysing the rate-limiting reaction might become labile under the conditions we have been using but that at a given light intensity it might still function with a specific value of k. We therefore performed frequency scans at 1820, 130 and 48 erg. cm^{-2} s^{-1} on a monolayer of *Chlorella* and fitted the equations derived by Joliot, Hoffnung and Chabaud [8] to the results (Fig. 17). We found that there was no change in the diffusive component of our best fitting curve as the intensity was altered

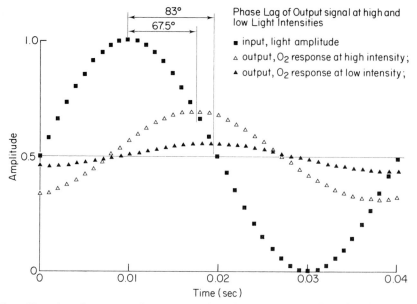

Fig. 18. The phase lag for a second order rate equation for O_2 production at high and low modulated light intensities.

and this is indicated by the straight line in Fig. 17 of phase lag against the square root of the frequency. But the value of k did vary, being 176 s^{-1} at 1820 erg. cm^{-2} s^{-1}, 249 s^{-1} at 130 erg. cm^{-2} s^{-1} and 369 s^{-1} at 48 erg. cm^{-2} s^{-1}. However, the theoretical equation linking the amplitude and the rate constant showed that this increase in k with decrease in intensity should also lead to an increase in oxygen yield. This was in direct contradiction with the observed fall in oxygen yield and therefore we concluded that a labile enzyme was not responsible for our results.

We then considered the possibility that when the O_2 concentration is lowered or the temperature is raised sufficiently, the rate-limiting reaction changes from the one we studied earlier to a second order reaction which we expected would give an intensity dependent phase lag for the O_2 signal. Using a computer we have solved the differential equations which describe a second order dependence of the rate of O_2 evolution on some substance X produced by photosystem II (Fig. 18). The large amplitude wave in the diagram represents a single oscillation of the light intensity. The small amplitude waves represent the resulting oscillations in O_2 production from a 2nd order reaction for intensities of 1820 and 130 erg. cm^{-2} s^{-1}. By measurement we find that the phase lag at the high intensity is 67.5° and that at low intensity is 83°, i.e. the phase lag increases as the light intensity decreases, which is, of course, the reverse of what we observed. We

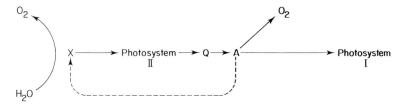

Fig. 19. Possible routes of electron transport about photosystem II.

were therefore forced to reject the hypothesis of a 2nd order reaction.

We now believe that to explain all our results within one theoretical framework, we must postulate that there are several electron transport routes operating within the chloroplasts of *Chlorella* and that the relative importance of these different routes can be altered by variations in the experimental conditions. Fig. 19 illustrates the model we have used to explain our results. Electrons released by the decomposition of water, flow to photosystem II and on to the redox pool A. Electrons leaving this redox pool can proceed either to photosystem I or, as proposed by Diner and Mauzerall [4], be used to reduce oxygen or be cycled back to the oxidizing site of photosystem II to reduce the oxidized form of the substance X. Under normal conditions we believe that the latter route is relatively unimportant but if the light intensity is lowered and the O_2 concentration is reduced it has a very significant effect on oxygen evolution.

The differential equations (3) and (4) were set up by Joliot, Hoffnung and Chabaud [8] when they proposed that the rate of oxygen evolution was linearly dependant on the concentration of the oxidized form of the substance X. The latter is assumed to be linearly dependant on the light intensity in a photochemical reaction. The light

$$\frac{dO_2}{dt} = kX^+ \qquad (3)$$

$$\frac{dX^+}{dt} = k_c I_o (1 - \sin \omega t) - kX^+ \qquad (4)$$

Solution $\qquad \phi = \tan^{-1} \left(\frac{\omega}{k}\right)$

$$\text{Amp.} = \frac{k_c I_o}{\sqrt{\frac{\omega^2}{k^2} + 1}}$$

$$\frac{dO_2}{dt} = kX^+ \tag{5}$$

$$\frac{dX^+}{dt} = k_c I_o (1 - \sin \omega t) - kX^+ - k_a A^- X^+ \tag{6}$$

Solution

$$\phi = \tan^{-1}\left(\frac{\omega}{k + k_a A^-}\right)$$

$$\text{Amp.} = \frac{kk_c I_o}{\sqrt{\omega^2 + (k + k_a A^-)^2}}$$

intensity is taken to vary sinusoidally with time. The amplitude and phase lag of the sinusoidal component of the rate of O_2 evolution which form part of the solution to these differential equations are shown. It can be seen that if the rate constant k is increased, the phase lag will decrease but the amplitude will increase.

The differential equations (5) and (6) describe our model which permits a backflow of electrons from the reduced form of A, A^- to X^+ by the introduction of the term $-k_a A^- X^+$ to the equation (6). It has been assumed in solving these equations that the value of A^- does not fluctuate in the steady state. The amplitude and phase lag of the sinusoidal component of the solution are shown below and it can now be seen that increasing the value of the product $k_a A^-$ will cause both the phase lag and the amplitude to decrease in value. Thus this model tells us that if the experimental conditions are such as to permit A^- to increase in value we should expect to see the intensity dependance of the phase lag and the oxygen yield, but if A^- is kept at a small value this intensity dependence should be absent. These ideas are consistent with our experimental results where the absence of O_2 which should lead to an increase in the value of A^- gave rise to the intensity-dependent results whereas the restoration of O_2 abolished them. Similarly the addition of background 700 nm light which should speed up the removal of electrons from A^-, also abolished the intensity-dependence of the phase lag and the O_2 yield. The influence of temperature on our results can be explained on the basis of our model if the ratio $\frac{k_a A^-}{k}$ increased with temperature. But it is not possible to say whether the influence of temperature is primarily exerted on the rate constants or on the value of A^-. We can make an estimate of the value of $k_a A^-$ from the phase results made during the frequency scans at three different light intensities. If we assume that A^- was zero at the highest intensity and close to maximum at the lowest intensity then $k_a A^-$ has a value of about 190 s^{-1}.

In conclusion we believe that the information we obtained about the rate constant k was valid because the backflow of electrons from A^- to X^+ was not a significant factor in those experiments and that this backflow only becomes important under special circumstances.

References

1. Arnason, T. and Sinclair, J. (1976). *Biochim. Biophys. Acta* **430**, 517-523.
2. Arnason, T. and Sinclair, J. (1976). *Biochim. Biophys. Acta* **449**, 581-586.
3. Bouges-Bocquet, B. (1973). *Biochim. Biophys. Acta* **292**, 772-785.
4. Diner, B. and Mauzerall, D. (1973). *Biochim. Biophys. Acta* 305, 329-352.
5. Etienne, A.L. (1968). *Biochim. Biophys. Acta* **153**, 895-897.
6. Izawa, S. and Good, N.E. (1966). *Plant Physiol.* 41, 544-552.
7. Joliot, P. (1965). *C.r. Acad. Sci. Paris* **260**, 5920-5923.
8. Joliot, P., Hoffnung, M. and Chabaud, R. (1966). *J. Chim. Phys.* **10**, 1423-1441.
9. Metzner, H. (1975). *J. Theor. Biol.* **51**, 201-231.
10. Sinclair, J. and Arnason, T. (1974). *Biochim. Biophys. Acta* **360**, 393-400.

MANGANESE AND CHLORIDE: THEIR ROLES IN PHOTOSYNTHESIS

GOVINDJEE, T. WYDRZYNSKI and S.B. MARKS*

*Departments of Physiology and Biophysics, and Botany, and *Chemistry, University of Illinois, Urbana, Illinois (U.S.A.)*

Introduction

Manganese [4] and chloride [9] ions have long been implicated to play important roles in oxygen evolution during photosynthesis. Chloride-free chloroplasts and thylakoid membranes depleted of bound manganese fail to evolve O_2, but are capable of electron flow from artificial donors (such as diphenylcarbazide, feeding electrons to an intermediate (Z) between the oxygen evolving intermediate (S) and the reaction centre II (P 680)) to Hill oxidants. There has, however, been no experimental observation to show functional changes in either manganese or chloride ions during oxygen evolution. Manganese and/or chloride ions could simply be parts of a structural component.

Joliot and coworkers and Kok and coworkers (see ref. [10]) suggested the existence of a charge-accumulating species (S), which according to Kok and coworkers, undergoes changes as follows (also see Mar and Govindjee, [11]):

$$S_0 \xrightarrow{h\nu} S_0' \longrightarrow S_1 \xrightarrow{h\nu} S_1' \longrightarrow S_2 \xrightarrow{h\nu} S_2' \longrightarrow S_3 \xrightarrow{h\nu} S_3' \longrightarrow S_4 \quad (1)$$

with $O_2 + 4H^+ \longleftarrow 2H_2O$ cycling back to S_0

where each S_{n+1} state is assumed to differ from the previous S_n state by the accumulation of an additional oxidizing equivalent. The prime states refer to the initial states created as a consequence of light reactions, and $S_n' \rightarrow S_{n+1}$ steps refer to the dark relaxation steps which lead to the recovery of the reaction centre complex in a state that can undergo a light reaction. This scheme predicts a periodicity of four in O_2 evolution/flash when chloroplasts are given a series of brief light flashes separated by dark periods long enough to allow $S_n' \rightarrow S_{n+1}$ reactions, but not long enough to allow deactivation of S states. The periodicity of 4 in the O_2 pattern was first observed by Joliot and coworkers (see ref. [10]), and the maximum was in the 3rd flash suggesting that dark-adapted chloroplasts start mainly from the S_1 state. There are no other steps in photosynthesis which oscillate with a periodicity of four. Thus, an oscillation of 4 is restricted to the O_2 evolving side of photosynthesis. (However, an oscillation with a period of two has been located on an intermediate R (or B) on the system I side of photosystem II [2,7,18].)

To our knowledge, only two other signals, besides O_2, oscillate with a period of 4 in brief flashes of light: (1) transverse (spin-spin) relaxation rates $(1/T_2)$ of water protons (not to be confused with H^+ evolved in the scheme shown above, but all protons whether as H^+ or in H_2O etc. in the sample which are related to bound Mn) [25]; and (2) an absorbance change in the near ultraviolet region [15]. The chemical identity of the latter signal is not yet known.

In this paper, we shall review our recent observations [20,22-24] which strongly support the thesis that water proton relaxation rates monitor changes in contribution of paramagnetic Mn(II) as related to the O_2 evolving mechanism. In addition, data will be presented to show the possible role of halogen ions (i.e. chloride) in O_2 evolution.

Proton relaxation measurements

In order to appreciate our conclusions, it is necessary

to provide a brief background of the measurements of the water proton relaxation rates $1/T_1$ (longitudinal or spin-lattice) and $1/T_2$ (transverse or spin-spin) [5,6]. When a sample is placed in a strong magnetic field, $\overline{H}_o$ parallel to the z axis (see Fig. 1 (a)), the individual nuclear magnetic dipoles tend to align with it and precess about the z axis. In a rotating frame of reference (x', y', z', Fig. 1 (b)) moving at the precessional frequency of the nuclear dipoles, the individual dipoles appear stationary to an observer rotating along with the frame in the x'y' plane. The vector sum of all nuclear dipoles gives the net magnetization, $\overline{M}$, of the spin system. Since more dipoles are aligned with $\overline{H}_o$ rather than against it, $\overline{M}$ appears parallel along the +z' direction (Fig. 1 (c)). If another field, $\overline{H}_1$, is applied orthogonally to $\overline{H}_o$ in the x' direction (see Fig. 1 (d)), $\overline{M}$ will move about $\overline{H}_1$. If $\overline{H}_1$ is app-

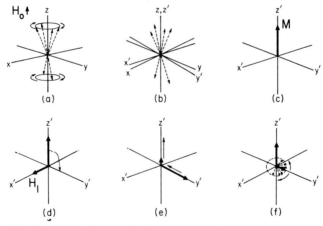

Fig. 1. Definition of $1/T_1$ and $1/T_2$ relaxation in the rotating frame. (a) Precession of nuclear spins (I=1/2) about the applied field, $\overline{H}_o$. $\overline{H}_o$ is defined along the z-axis. (b) Spins appear stationary in the rotating frame (x',y',z'). (c) Net magnetization, $\overline{M}$, of the nuclear spins. (d) Application of an rf field, $\overline{H}_1$, in the rotating frame. (e) After rf is turned off, $\overline{M}$ relaxes back to equilibrium condition. The growth of $\overline{M}$ along the +z axis is called longitudinal relaxation and is characterized by a time constant T_1 while the decay of $\overline{M}$ along the y axis is called the transverse relaxation characterized by a time constant T_2. (f) Contribution of the dephasing of the individual nuclear dipoles in the x'y' plane to $1/T_2$. (See ref. [5]).

lied for a long enough time $\overline{M}$ will tip to the x'y' plane (this is called a 90° pulse). After $\overline{H_1}$ is turned off, then $\overline{M}$ tends to return to its equilibrium position *via* relaxation mechanisms. The rate of build-up of $\overline{M}$ along the z' axis is defined as $1/T_1$ or the longitudinal relaxation while the decay of $\overline{M}$ along the y' axis (the direction of the detection system) is defined as $1/T_2$ or the transverse relaxation (Fig. 1 (e)). The decay of $\overline{M}$ along the y' axis arises from a dephasing of the individual nuclear dipoles in the x'y' plane (Fig. 1 (f)); thus, $1/T_2$ is always $\geq 1/T_1$.

In practice $1/T_1$ is measured by the inversion recovery method, where a 180°, τ, 90° rf pulse sequence is applied. In the first 180° pulse, $\overline{M}$ is inverted to the -z' direction. The succeeding 90° pulse then tips $\overline{M}$ to the y' axis, the direction in which the magnetization signal can be measured. The extent of build-up of $\overline{M}$ along the z' axis back to its equilibrium position is dependent upon the time, τ, between the 180° and 90° pulse. The slope of a plot of $\ln \frac{\overline{M_o} - \overline{M_z}(\tau)}{2\overline{M_o}}$ *versus* τ is proportional to $-1/T_1$, where $\overline{M_o}$ is the equilibrium value of $\overline{M}$ and $M_z(\tau)$ is the value of the z component of M at a time τ.

The measurement of $1/T_2$ is somewhat more complex in that field inhomogeneities in $\overline{H_o}$ also lead to dephasing in the x'y' plane and thus obscure the true $1/T_2$ rate. The Carr-Purcell-Meiboom-Gill pulse sequence overcomes errors due to field inhomogeneities. This method can be described as a 90°, τ, 180°, 2τ, 180°, 2τ...pulse sequence. The first 90° pulse tips $\overline{M}$ to the +y' direction (Fig. 1). $\overline{M}$ starts to decay because of $1/T_2$ processes and field inhomogeneities. The succeeding 180° pulse inverts the dipoles to the -y' axis. The dipoles which dephase due to field inhomogeneities tend to refocus back along the y' axis while the dephasing due to $1/T_2$ processes continues. By applying a series of 180° pulses until $1/T_2$ relaxation is complete, errors due to field inhomogeneities are essentially overcome and the decay of the magnetization signals along the y' axis gives the true $1/T_2$. The Meiboom-Gill phase modification of the H_1 pulse after the initial 90° pulse is used to overcome pulse width errors. The slope of a plot of $\ln \frac{\overline{M}_z(\tau)}{M_o}$ versus τ gives $-1/T_2$.

Influence of paramagnetic ions on proton relaxation rates

Paramagnetic ions are known to influence strongly both $1/T_1$ and $1/T_2$ [5]. Neglecting the contributions outside the first coordination sphere of paramagnetic ion-H_2O complexes, the paramagnetic contribution to the water proton relaxation rates ($1/T_{1,2(P)}$) is given as:

$$\frac{1}{T_{1,2(P)}} = \frac{\rho q}{T_{1,2(M)} + \tau_m} \tag{2}$$

where, ρ is the mole fraction of paramagnetic ion (i.e., [Mn]/55.6 for Mn-HOH complex), q is the number of bound nuclei (i.e., $6 \geq q \geq 0$ for Mn-HOH), $T_{1,2(M)}$ are the relaxation times for the bound nuclei and τ_m is the lifetime of the nuclei bound to paramagnetic species. The $1/T_{1,2(M)}$ are related to the distance (r) between the paramagnetic ion and the nuclei, the nuclear Larmor frequency (ω_I), and the dipolar correlation time (τ_c) by the following simplified Solomon-Bloembergen equation:

$$\frac{1}{T_{1(M)}} = \frac{2}{15} \frac{S(S+1)(\gamma_I)^2 g^2 \beta^2}{r^6} \left[\frac{3\tau_c}{1+\omega_I^2 \tau_c^2}\right] \tag{3a}$$

$$\frac{1}{T_{2(M)}} = \frac{1}{15} \frac{S(S+1)(\gamma_I)^2 g^2 \beta^2}{r^6} \left[4\tau_c + \frac{3\tau_c}{1+\omega_I^2 \tau_c^2}\right] \tag{3b}$$

where, S is the total electron spin, γ_I is the nuclear magnetogyric ratio, g is the electronic g factor, β is the Bohr magneton, and other symbols are as defined above (The hyperfine (contact) contribution has been neglected in the above equations.)

The $1/\tau_c$ itself is a function of the molecular processes which modulate the dipolar fields:

$$\frac{1}{\tau_c} = \frac{1}{\tau_s}\frac{1}{\tau_m} + \frac{1}{\tau_r} \tag{4}$$

where, τ_s = electronic spin relaxation time, τ_m = lifetime of the paramagnetic ion water complex and τ_r =rotational correlation time. In macromolecular and membrane systems, $1/\tau_r$ is negligible and thus $1/\tau_s$ and/or $1/\tau_m$ will govern the relaxation rates. When τ_s dominates all other relaxation processes, $1/T_1$ and $1/T_2$ follow a predicted frequency dependence [5] which is given by the Bloembergen-Morgan equation:

$$\frac{1}{\tau_s} = B \left[\frac{\tau_v}{1+\omega_s^2\tau_v^2} + \frac{4\tau_v}{1+4\omega_s^2\tau_v^2} \right] \tag{5}$$

where, ω_s is the electronic Larmor frequency, B is a constant related to the zero field splitting parameters and τ_v is the correlation time for the modulation of the zero field splitting by molecular collisions.

Proton relaxation rates and manganese

(1) Conditions which are known to remove most of the bound manganese from thylakoid membranes (NH_2OH-EDTA treatment [14], TRIS-washing [26], TRIS-acetone washing [27], heat treatment + EDTA) decrease $1/T_1$ of chloroplasts significantly. This is shown in Table 1. (EDTA itself has some effect because it binds paramagnetic ions.) These data indicate that the proton relaxation rates monitor bound Mn.

(2) As one replaces Mn with Mg according to Chen and Wang [3], O_2 evolution decreases. However, O_2 evolution becomes nil even when about 1/3 of Mn is still in the chloroplast. Figs 2 (left) and (right) show that as O_2 evolution decreases due to lowered Mn, both $1/T_1$ and $1/T_2$ also decrease but then they attain a fixed background value at lowered rates of O_2 evolution. The details of

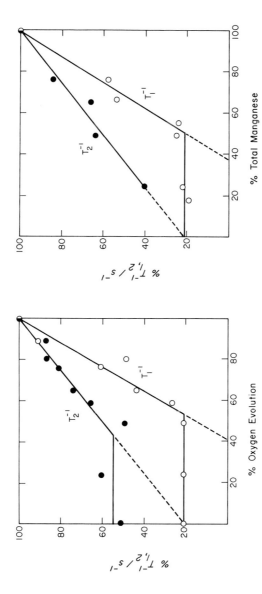

Fig. 2. (Left). Plot of percent $1/T_1$ and $1/T_2$ *versus* percent O_2 activity. (Right). Plot of percent $1/T_1$ and $1/T_2$ *versus* percent Mn content of the membrane. Mn extraction was achieved by incubating pea chloroplasts for 2 h in dark at 4°C in HEPES buffer medium (50 mM HEPES, pH 7.5, 400 mM sucrose and 10 mM NaCl) containing MgCl$_2$ at the following [Mg]/[Chl] ratios: 0, 167, 332, 667, 2,500, and 10,000. After incubation the chloroplasts were centrifuged and the pellet resuspended to 3 mg Chl/ml in HEPES buffer medium. Mn content was determined by neutron activation analysis. For the control the manganese content was 0.62±0.03 μg Mn/mg Chl and the rate of O_2 evolution was 120 μmoles O_2/mg Chl-hr. (After refs. [20,23].)

TABLE 1

Effect of various treatments which alter the environment of native bound manganese on the $1/T_1$ of chloroplast membranes. (After ref. [20].)

Condition	T_1^{-1}/s^{-1}
Spinach	
(a) Washed control	1.04
(b) NH_2OH-EDTA washed	0.54
(c) TRIS washed	0.41
+ 10^{-4} M EDTA	0.74
Peas	
(a) Washed control	0.82
(d) TRIS-acetone washed	0.21

(a) Spinach or pea chloroplasts were washed in HEPES buffer medium (50 mM HEPES, pH 7.5, 400 mM sucrose and 10 mM NaCl) and resuspended to a final concentration of 3 mg Chl/ml. $1/T_1$ was measured at 27 MHz, at 25°C.

(b) Chloroplasts were incubated in HEPES buffer medium containing 2 mM $MgCl_2$, 1 mM EDTA and 3 mM NH_2OH for 20 min in dark at 25°C, centrifuged and resuspended in HEPES buffer medium, according to Ort and Izawa [14].

(c) Chloroplasts were incubated in 0.8 M TRIS, pH 8.2, for 20 min in dark at 4°C, centrifuged and resuspended in HEPES buffer medium, according to Yamashita and Butler [26].

(d) Chloroplasts were incubated in 0.8 M TRIS, pH 8.2, 20% acetone (v/v) for 20 min at 4°C, centrifuged and resuspended in HEPES buffer medium, according to Yamashita and Tomita [27].

these results are not easily understood but two points are clear (i) in the region of 50-100% Mn content, there is a linear relationship between $1/T_1$, $1/T_2$, O_2 evolution and Mn content; and (ii) there are some background $1/T_1$ and $1/T_2$ signals.

(3) Fig. 3 (a) shows $1/T_1$ as a function of frequency of the rf pulse from several different samples of chloro-

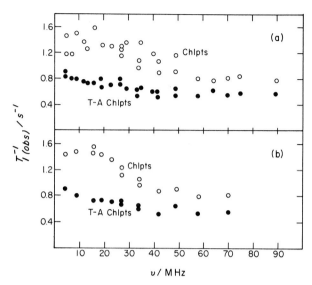

Fig. 3. Frequency dependence of $1/T_{1(obs)}$ for dark-adapted chloroplast membranes (Chlpts). (a) Uncorrected observed rates for several different samples. (b) Observed rates normalized to the same Mn concentration. Samples of pea chloroplasts contained 3 mg Chl/ml. T-A Chlpts refers to TRIS-acetone washed chloroplasts in which the loosely bound fraction of Mn is removed. $1/T_1$ was measured at room temperature (23-25°C). (After refs. [20,23]).

plasts. We can see that the scatter in data is very large. However, when corrections are made for differences in [Mn], much of the scatter is smoothed out (Fig. 3 (b)). The reduction in the scatter upon correction for [Mn] again shows the relationship between $1/T_1$ and [Mn].

Relationship of proton relaxation rates to Mn[II]

For normal chloroplasts in Fig. 3 (b), a broad peak in the 10-25 MHz region is apparent whereas only a flat curve is obtained for TRIS-acetone washed chloroplasts in which the large pool of Mn is removed. The broad peak in the control curve indicates that the dipolar correlation time governing the rates is itself frequency dependent and hence is dominated by electronic relaxation [5]. Since this frequency dependence is absent in TRIS-acetone (T-A)

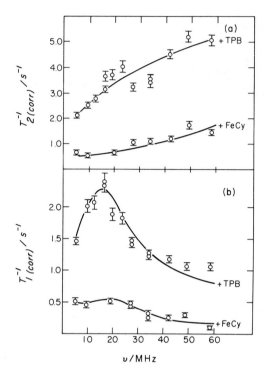

Fig. 4. "Best fit" theoretical curves to the frequency dependence of the relaxation rates for dark-adapted chloroplast membranes in the presence of redox reagents. (a) $1/T_2$ relaxation; (b) $1/T_1$ relaxation. $1/T_{1,2(corr)} = [1/T_{1,2(obs)} - 1/T_{1,2(T-A)}]$. T-A = TRIS-acetone washed chloroplasts. Rates were measured at room temperatures and normalized to the same Mn concentration at each frequency. Either 5 mM TPB or 50 mM FeCy was used. Pea chloroplasts at 3 mg Chl/ml were used. (After refs. [20,23]).

washed chloroplasts, this contribution must arise from the large pool of Mn. Analysis of Fig. 3 (b) (see refs [8,23]) led us to the conclusion that $1/T_1$ and $1/T_2$ monitor mainly Mn[II]. This analysis was extended to chloroplasts treated with the reductant tetraphenylboron (TPB) and the oxidant ferricyanide (FeCy). These results are shown in Fig. 4. Both $1/T_1$ and $1/T_2$ were corrected by subtracting off the value of $1/T_1$ and $1/T_2$ for TRIS-acetone washed chloroplasts to obtain the contributions of

TABLE 2

"Best Fit" NMR parameter values for chloroplast membranes and other Mn(II) systems. (After ref. [20].)

	$\tau_v/s \times 10^{12}$	$\tau_m/s \times 10^8$	$B/(\text{rad s}^{-1})^2 \times 10^{-19}$	$\tau_s/s \times 10^9$
Chlpts	20	2.2	0.90	1.11
Chlpts+TPB	20	1.0	0.9	1.11
Chlpts+FeCy	9.3	0.1	1.1	1.96
Aqueous Mn(II)	(a) 2.1	2.3	10	0.95
Mn(II)-carboxy-peptidase A	(b) 7±0.6	0.25±0.3	3.1±0.4	0.92
Mn(II)-pyruvate-kinase	(a) 6	0.5±0.1	1.46	2.28
	(b) 14±4	0.4±0.04	0.8±0.1	1.78

(a) Bloembergen and Morgan [1].
(b) Navon [13].
(c) Reuben and Cohn [16].

the large pool of Mn. The $1/T_2$ increases with frequency, whereas $1/T_1$ shows a peak in the 10-20 MHz range (open circles) as has been observed for control chloroplasts. The solid lines are theoretical lines obtained by fitting the various NMR parameters to theoretical equations. A master equation was obtained by combining equations (5),(4), and (3). Then, at each frequency, B, τ_v and τ_m were varied until curves $1/T_1$ and $1/T_2$ were obtained which fit the experimental points (note that $1/\tau_r$ was assumed to be negligible in our case); an outersphere contribution had to be also included to obtain the "best" fits shown by solid lines in Fig. 4.

Table 2 gives the calculated parameter values. τ_v ranged from 9.3×10^{-12}s (FeCy) to 20×10^{-12}s (TPB), B from 1.1×10^{19} rad·s^{-1} (FeCy) to 1.0×10^{19} rad·s^{-1} (TPB), and τ_m from 0.1×10^{-8}s (FeCy) to 1×10^{-8}s (TPB). These values are not too different from those of control chloroplasts and are in the same order of magnitude as for other Mn[II] systems [1,13,16,19]. We note that the correlation time (τ_s) for Mn[II] calculated to be ~10^{-9}s from Eq. 5 (at zero field) is two orders of magnitude larger than for Mn[III] and high spin iron [5]; hence, these ions have only a small effect on the proton relaxation rates in comparison to Mn[II[.

Proton relaxation rates and oxygen evolving mechanism

Wydrzynski *et al.* [25] discovered the oscillations in $1/T_2$ as a function of flash number shown in Fig. 5 (top) for spinach chloroplasts. Fig. 5 (bottom) shows the O_2

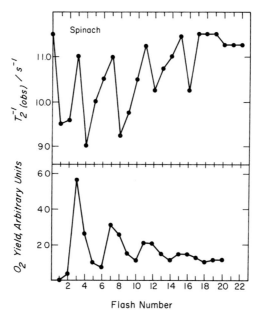

Fig. 5. $1/T_{2(obs)}$ and O_2 yield flash patterns for spinach chloroplast membranes. (Top) $1/T_2$ relaxation; (bottom) O_2 yield. $1/T_2$ was measured at 27 MHz, 25°C, on sample aliquots in HEPES buffer medium, pH 7.5, at 3 mg Chl/ml while O_2 was measured on sample aliquots at 1 mg Chl/ml. Flash procedure: Flashes spaced 2 s apart were given in sequence from 1 to 22 flashes. The Carr-Purcell-Meiboom-Gill pulse train was initiated at the last flash in the sequence. A 7 min dark period was allowed between each flash sequence. The same flash procedure was used to measure the O_2 yield. Light flashes were obtained from a xenon strobe flash lamp. The results in the top figure are from ref. [25] and in the bottom figure from ref. [20].

flash pattern for spinach chloroplasts under identical conditions. The following similarities in the two curves are observed: (a) peaks at 3rd, 7th, etc. flashes; (b) periodicity of four; and (c) damping of oscillations at about the 17th flash. However, there are several dissimilarities: (a) high dark $1/T_2$ *versus* no O_2 in darkness; (b) upward trend in $1/T_2$ pattern and downward trend in O_2; (c) minima are at 4th, 8th etc. flashes in $1/T_2$ and at 6th, 10th etc. flashes in O_2; and (d) when $1/T_2$ increases from 4th to 5th to 6th flashes, O_2 yield decreases. Clearly, the two ($1/T_2$ and O_2) do not monitor exactly the same thing. However, the periodicity of 4 establishes the relationship of $1/T_2$ to the water side of pigment system II. The differences could be due to several reasons

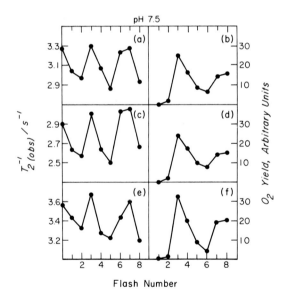

Fig. 6. $1/T_{2(obs)}$ and O_2 yield flash patterns at pH 7.5. (a), (c) and (e) are the $1/T_2$ relaxation and (b), (d) and (f) are the O_2 yield for three different samples. $1/T_2$ was measured at 27 MHz, 25°C, on sample aliquots in HEPES buffer medium at 2 mg Chl/ml while O_2 was measured on sample aliquots of 1 mg Chl/ml. $1/T_2$ and O_2 yield were measured after each flash in a pulse sequence of 2 μs saturating flashes obtained from a pulsed dye laser (λ, 590 nm). Flashes were spaced 4 s apart to allow for complete $1/T_2$ relaxation.(After refs. [20,22].)

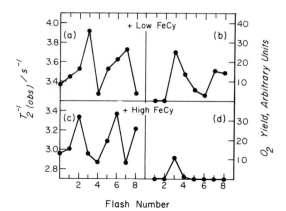

Fig. 7. $1/T_{2(obs)}$ and O_2 yield flash patterns in the presence of ferricyanide (FeCy). (a) [FeCy]/[Chl] = 0.222; (b) [FeCy]/[Chl] = 22.2. Flash procedure and conditions for pea chloroplasts were as described in the legend of Fig. 6. (After refs. [20,22].)

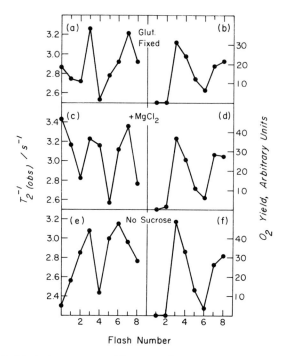

Fig. 8. (For legend see opposite page.)

including (a) $1/T_2$ monitors the individual S states, whereas O_2 only monitors the $S_4 \to S_0$ transition; and (b) the $1/T_2$ unrelated to Mn not involved in O_2 evolution, which has not been subtracted, influences the $1/T_2$ but not O_2 results. We believe that exploitation of the differences between $1/T_2$ and O_2 yield should lead to a better understanding of the O_2 evolving mechanism.

Fig. 6 shows $1/T_2$ and O_2 yield as a function of flash number in 3 different samples of pea chloroplasts at pH 7.5. (Data for chloroplasts at pH 6.7 were presented at an earlier conference [8].) Here, again, the similarities between $1/T_2$ and O_2 are (a) peaks are at the 3rd flash in both cases; (b) an approximate periodicity of 4 is observed in both cases. And, the differences are (a) $1/T_2$ has a minimum at the 5th flash, whereas O_2 has the usual minimum at the 6th flash; (b) the 1st and 2nd flashes decrease $1/T_2$ from its high dark level where O_2 shows the opposite trend; and (c) the $1/T_2$ for the 3rd and 7th flashes are about the same, but O_2 shows a clear decrease at the 7th flash. For the same pea chloroplasts at pH 6.7, there was no O_2 in the 2nd flash under our illumination (brief 2 μs laser flash, $\lambda=590$ nm). Thus, O_2 in the second flash, observed at pH 7.5, should not have been due to a technical double hit.

The question of high dark $1/T_2$

(1) Various conditions were shown to lower the high dark level of $1/T_2$ without significantly interfering with the main patterns of O_2 and $1/T_2$ (Figs 7 and 8). This includes (a) treatment of chloroplasts with low concen-

Fig. 8. $1/T_{2(obs)}$ and O_2 yield flash patterns after various treatments which affect membrane conformations. (a), (c) and (e) are $1/T_2$ relaxations, (b), (d) and (f) are O_2 yields. (a) and (b) chloroplasts were fixed in glutaraldehyde (98.5% fixation) according to procedure given in ref. [28]; (c) and (d), $[MgCl_2]/[Chl] = 2.22$; (e) and (f) chloroplasts suspended in medium with no sucrose. Flash procedures and conditions for pea chloroplasts in Fig. 6 were used. (After refs. [20,22].)

tration of ferricyanide (Fig. 7 (a)); (b) treatment with low osmoticum (absence of sucrose, see Fig. 8 (e)); this treatment does cause the second peak to be after the 6th flash; and (c) fixation of chloroplasts with glutaraldehyde [28] (Fig. 8 (a)). Addition of $MgCl_2$ ($[MgCl_2]/[Chl]$ = 2.22) did not decrease the $1/T_2$ in darkness.

(2) Certain conditions which abolish O_2 evolution, e.g., addition of DCMU and very high concentrations of ferricyanide, also decreased $1/T_2$ in dark, the former eliminating $1/T_2$ flash pattern in light [8,20] and the latter retaining it but changing it to show peaks at 2nd and 6th flashes (Fig. 7 (b)).

(3) Unpublished results of R. Khanna in our laboratory show that in blue-green alga *Phormidium luridum*, $1/T_2$ in dark is low. Thus, the high dark $1/T_2$ in isolated chloroplasts is not a universal feature of all the samples, although it still needs to be explained when it exists.

A model to explain $1/T_2$ flash pattern

We have begun to explain, at least, qualitatively $1/T_2$ flash pattern in terms of a manganese model [8,12,21]. Using Kok et al.'s model (see ref. [10]) as a starting point, the concentration of $[S_0]$ and $[S_1]$ in darkness, α(misses) and β (double hits) were calculated from O_2 data. At pH 6.7, $[S_0]$ = 0.30, $[S_1]$ = 0.70, α = 0.10 and β = 0 were obtained. From these data, the concentration of S_0, S_1, S_2 and S_3 were calculated after each flash. Then, each S state was assigned a certain weighting factor which we assume is proportional to Mn[II] contribution and $1/T_2$ was calculated from the net sum of the contributions of all S states after each flash. The calculated $1/T_2$ was normalized at flash 3. The "best" fit curves for calculated $1/T_2$ for chloroplasts at pH 6.7 gave the model to be 2, 1, 1, 3 for S_0, S_1, S_2, S_3 states with the qualification that S_0 in darkness has a contribution of 4, the numbers indicating the contribution of Mn[II] to each S state.

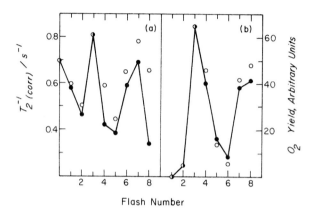

Fig. 9. Theoretical fit to the $1/T_2$ and O_2 yield flash patterns at pH 7.5. (a) $1/T_2$ relaxation and (b) O_2 yield. Closed circles are experimental data points [$1/T_{2(corr)}$], open circles are theoretical points. For the theoretical fit, $[S_0] = 0.30$, $[S_1] = 0.67$, $[S_2] = 0.03$, $\alpha = 0.10$, $\beta = 0$, and weighting factors for S_0, S_1, S_2 and $S_3{}^+$ are 2,1,2,1 on the first cycle, 1,3,2,1, on succeeding cycles, except that $S_0 = 4$ in the dark. (After refs. [20,22].)

The high value of S_0 in darkness was rationalized in view of Velthuys's [17] suggestion that a special dark-adapted state S_{-1} may exist in isolated chloroplasts.

Fig. 9 shows the experimental data (solid points) for $1/T_2$ and O_2 patterns for a particular preparation of chloroplasts at pH 7.5. Since no technical double hits exist with our laser flashes (no O_2 in the 2nd flash for chloroplasts at pH 6.7), we assumed that starting states are $[S_0] = 0.30$, $[S_1] = 0.67$, $[S_2] = 0.03$, $\alpha = 0.10$ and $\beta = 0$. The first cycle $1/T_2$ is fitted with a model (open circles in Fig. 9 (a)) in which the weighting factors are 2,1,2,1 for S_0, S_1, S_2, and S_3 (quite different from the previous fit at pH 6.7 [8]), *but* in the remaining cycles the contributions *are* 2,1,1,3 and S_0 in dark is 4 (as before). Qualitatively, the fit is quite good except for the 8th flash where the model gives a much higher value than observed experimentally. Obviously, a finer "tuning" of the model remains to be accomplished.

It is not difficult to generate qualitatively similar (i.e., with a periodicity of 4), but quite different curves, e.g., peaks at 2nd and 6th flashes, as observed when O_2 evolution is inhibited in the first cycle with TPB, NH_2OH or CCCP [8], or curves with peaks at 4th and 8th flashes. A low dark level is simple to generate by eliminating the additional condition that S_0 in dark has a very high Mn[II] contribution. Also, minima can be

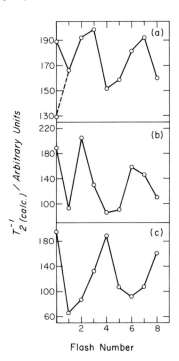

Fig. 10. Several theoretical patterns that qualitatively fit various experimental $1/T_2$ flash patterns. (a) $[S_0] = 0.30$, $[S_1] = 0.70$, $\alpha = 0.10$, $\beta = 0$; weighting factors for S_0, S_1, S_2 and S_3 are: 2,1,2,2, except $S_0 = 4$ in dark (solid line) or $S_0 = 2$ in dark (dashed line). (b) $[S_0] = 0.30$, $[S_1] = 0.70$, $\alpha = 0.10$, $\beta = 0$; weighting factors are: 0,1,1,3 except $S_0 = 4$ in dark (c) $[S_0] = 1.00$, $\alpha = 0.10$, $\beta = 0$; weighting factors are: 0,1,2,3 except $S_0 = 4$ in dark. The pattern in (a) is similar to the flash pattern for spinach chloroplasts (Fig. 5) and glutaraldehyde fixed chloroplasts (Fig. 8a); the dashed line in (a) is similar to that for pea chloroplasts with low FeCy (Fig. 7a). The pattern in (b) is similar to the flash pattern for pea chloroplasts containing either NH_2OH, TPB or CCP (see ref. [8]). And, the pattern in (c) is similar to the flash pattern for peas at pH 4 (Fig. 22b in ref. [20]). (After refs. [20,22].)

shifted from 5th to 4th flash, and so on.

Fig. 10 (a) shows a curve generated with 2,1,2,2, pattern with $S_0=4$ in dark (solid line) and $S_0=2$ (as in light, dashed line); the minimum is at the 4th flash as in spinach chloroplasts (Fig. 5) and the dark $1/T_2$ is low as in pea chloroplasts with low concentration of ferricyanide (Fig. 7 (a)). Fig. 10 (b) shows a $1/T_2$ curve with 0,1,1,3 pattern except that $S_0=4$ in dark. Such a curve is experimentally obtained when O_2 evolution is inhibited with TPB (in the first cycle), with CCCP or with low concentration of NH_2OH (see refs. [8,20,22]). Fig. 10 (c) was generated by assuming that $[S_0] = 1.00$ with 0,1,2,3 pattern except that $S_0=4$ in dark. Such a curve in $1/T_2$ is experimentally observed for chloroplasts at pH 4.0 when there is no O_2 evolution (see ref. [20]).

The above discussion indicates that using different weighting factors for S_0, S_1, S_2, S_3, almost all of our light effects on $1/T_2$ can be generated. What does this mean? We suggest that each specific S state when it is monitored in the time scale of 200-300 ms after the flash is either not in the same average redox state of manganese, or that differences in the number of exchangeable protons in bound water cause the apparent Mn[II] contribution to be different. However, what is clear is that in a great majority of cases, S_3 appears more reduced than S_2 in contrast to Kok's picture. Thus, we have suggested [8,12,21] that H_2O may be reducing the S states before the time scale of NMR measurements. This is consistent with the suggestion that H^+ may be released in steps prior to the S_4-S_0 transition. Such H^+ releases have now been shown by Fowler (prsonal communication) and Crofts and coworkers (personal communication).

Chloride depletion and ^{19}F NMR

It can be shown that chloride can be replaced by fluoride. This permits us to use ^{19}F as a tool to look at the

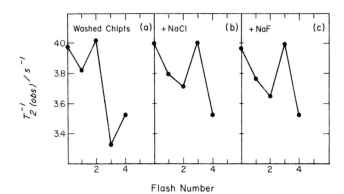

Fig. 11. $1/T_{2(obs)}$ flash pattern in the presence of NaCl and NaF. (a) Chloroplasts washed free of chloride; (b) [NaCl]/[Chl] = 4.44; (c) [NaF]/[Chl] = 4.44. Flash procedures and conditions for pea chloroplasts in Fig. 6 were used except NaCl was eliminated from the medium. (After ref. [20].)

question of the role of chloride in photosynthesis. Fig. 11 shows that chloride-depleted chloroplasts have their $1/T_2$ peak on the 2nd flash (in fact, 3rd flash gives the lowest $1/T_2$); such chloroplasts do not evolve O_2. Addition of either NaCl or NaF restores the $1/T_2$ pattern, i.e., maximum on the 3rd flash; also the Hill reaction is restored. Thus, fluoride replaces chloride well. Using such chloroplasts, ^{19}F spectra were measured. The $1/T_2$ values, calculated from bandwidths of the resonant peak, are shown in Table 3. Addition of chloroplasts to buffer enhances $1/T_2^*$, addition of reductant TPB to chloroplasts increases $1/T_2^*$, and removal of manganese by TRIS-acetone washing brings back the $1/T_2^*$ to that of the buffer control. These results parallel the measurements on water proton relaxation. This establishes a relationship between manganese related to O_2 evolution to the fluoride (and thus chloride). Whether this means that chloride is a ligand to manganese involved in O_2 evolution and thus may help stabilize the higher oxidation states of Mn (Mn[III]?) remains to be shown.

TABLE 3

^{19}Fluorine relaxation measurements of chloroplast membranes[a]
(After ref. [20].)

Condition	T_2^{*-1}/s^{-1}
[b] Buffer	13.7
[c] Chlpts	34.0
Chlpts + 5 mM TPB	60.2
TRIS-acetone washed Chlpts	15.6

[a] ^{19}F spectra were measured on a Jeol FX-60 Fourier Transform NMR spectrometer. Relaxation rates were calculated from the half bandwidth ($\Delta\nu$) of the resonance peak, $1/T_2^* = \pi\Delta\nu$.

[b] HEPES buffer medium containing 20 mM NaF was used.

[c] Dark-adapted pea chloroplasts, 3 mg Chl/ml, were washed free of Cl$^-$ and suspended in buffer medium containing NaF.

Summary and concluding remarks

(1) Water proton relaxation rate (PRR) is an excellent monitor of manganese bound to thylakoid membranes as (a) removal of Mn by TRIS-acetone washing or by NH$_2$OH-EDTA treatment dramatically decreases PRR; (b) PRR is approximately linearly related to [Mn] in chloroplasts with different [Mn] obtained by replacing Mn with Mg.

(2) The major species of Mn which influences PRR is Mn[II] according to an analysis of the dependence of PRR on the frequency of the rf pulse.

(3) The $1/T_2$ (transverse relaxation) rates of thylakoids as a function of flash number show a periodicity of 4 suggesting their correlation to the O$_2$ evolving mechanism.

(4) The particularly high dark $1/T_2$ is eliminated by (a) treatment with ferricyanide; (b) glutaraldehyde fixation; and (c) low osmoticum. This dark level is explained on the basis that in isolated chloroplasts, a very reduced state (high Mn[II] contribution) may exist (for details

see ref. [20]).

(5) A theoretical model, which assumes that in pea chloroplasts, suspended at pH 7.5, the Mn[II] contribution for S_0, S_1, S_2 and S_3 is 2,1,2,1, respectively, in the first flash sequence and then 2,1,1,3 in the succeeding cycles except that $S_0=4$ in the dark, fits in quite well with the experimental data. A model assuming a 0,1,1,3 Mn[II] contribution with $S_0=4$ in dark explains $1/T_2$ flash pattern under conditions when O_2 evolution has been inhibited by CCCP, NH_2OH or TPB.

(6) ^{19}F relaxation rate ($1/T_2^*$) was shown to behave in parallel with the PRR suggesting that chloride (which was replaced by fluoride) might play a role in stabilizing the Mn complex involved in O_2 evolution.

Acknowledgements

We are thankful to Drs P.G. Schmidt and H.S. Gutowsky for their collaboration in this research. Details of the experiments reviewed here will be published jointly with our coworkers in separate publications [21-23]. T.W. was supported by USPHS Training Award (GM728301 Sub. Proj-604), G. by a NSF grant (PCM 76-111657) and SBM by a grant to HSG (MPS 23-0498).

Footnote:

Abbreviations:

β, Bohr magneton; B, constant in Bloembergen-Morgan equation containing the value of the resultant electronic spin and the zero field splitting parameters; γ, nuclear magnetogyric ratio; $\overline{H}_o$, external applied magnetic field; $\overline{M}$, net magnetization of a spin system; $\overline{M}_o$, equilibrium magnetization; ν, frequency (Hz); ρ, mole fraction of ligand nuclei bound; PRR, proton relaxation rate; q, number of ligand nuclei bound; rf, radio frequency; S, total electron spin; $1/T$, longitudinal or spin-lattice relaxation rate, corrected with buffer medium rate; $1/T_2$, transverse or spin-spin relaxation rate, corrected with buffer medium rate; $1/T_{1,2(M)}$, relaxation rates at the bound site; $1/T_{1,2(obs)}$, observed relaxation rates; $1/T_{1,2(T-A)}$, relaxation rates of TRIS-acetone washed chloroplast membranes; τ_c, dipolar correlation time; τ_m, chemical exchange lifetime; τ_r, rotational correlation time; τ_s, electron spin relaxation time; τ_v, correlation time for the modulation of zero field splitting, ω_I, nuclear Larmor frequency; ω_s, electronic Larmor frequency; all other abbreviations see list on page xxx.

References

1. Bloembergen, N. and Morgan, L.O. (1961). *J. Chem. Phys.* 34, 842-850.
2. Bouges-Bocquet, B. (1973). *Biochim. Biophys. Acta* 314, 250-256.
3. Chen, K.Y. and Wang, J.H. (1974). *Bioinorg. Chem.* 3, 339-352.
4. Cheniae, G.M. (1970). *Ann. Rev. Plant Physiol.* 21, 467-490.
5. Dwek, R.A. (1973). "Nuclear Magnetic Resonance (NMR) in Biochemistry: Application to Enzyme Systems", Clarendon Press, Oxford.
6. Farrar, T.C. and Becker, E.D. (1971). "Pulse and Fourier Transform NMR: Introduction to Theory and Methods", Academic Press, New York.
7. Govindjee, Pulles, M.P.J., Govindjee, R., van Gorkom, H.J. and Duysens, L.N.M. (1976). *Biochim. Biophys. Acta* **449**, 602-605.
8. Govindjee, Wydrzynski, T. and Marks, S.B. (1977). *In* "Bioenergetics of Membranes" (L. Packer, ed.), Elsevier/North Holland Biomedical Press B.V., Amsterdam.
9. Izawa, S., Heath, R.L. and Hind, G. (1969). *Biochim. Biophys. Acta* **180**, 388-398.
10. Joliot, P. and Kok, B. (1975). *In* "Bioenergetics of Photosynthesis" (Govindjee, ed.), 387-412, Academic Press, New York.
11. Mar, T. and Govindjee (1972). *J. Theor. Biol.* 36, 427-446.
12. Marks, S.B., Wydrzynski, T., Govindjee, Schmidt, P.G. and Gutowsky, H.S. (1977). Paper presented at a conference on "Cellular Function and Molecular Structure: A Symposium on Biophysical Approaches to Biological Problems", Columbia, Missouri.
13. Navon, G. (1970). *Chem. Phys. Lett.* 7, 390-394.
14. Ort, D.R. and Izawa, S. (1974). *Plant Physiol.* 53, 370-376.
15. Pulles, M.P.J., van Gorkom, H.J. and Willemsen, J.G. (1976). *Biochim. Biophys. Acta* 449, 536-540.
16. Reuben, J. and Cohn, M. (1970). *J. Biol. Chem.* 245, 6539-6546.
17. Velthuys, B.R. (1976). Ph.D. Thesis, The State University, Leiden.
18. Velthuys, B.R. and Amesz, J. (1974). *Biochim. Biophys. Acta* 376, 162-168.
19. Villafranca, J.J., Yost, F.J. and Fridovich, J. (1974). *J. Biol. Chem.* 249, 3532-3536.

20. Wydrzynski, T. (1977). Ph.D. Thesis, University of Illinois, Urbana, Ill.
21. Wydrzynski, T., Govindjee, Marks, S.B., Schmidt, P.G. and Gutowsky, H.S. (1977), in preparation.
22. Wydrzynski, T., Marks, S.B., Govindjee, Schmidt, P.G. and Gutowsky, H.S. (1977), in preparation.
23. Wydrzynski, T., Marks, S.B., Schmidt, P.G., Govindjee and Gutowsky, H.S. (1977), in preparation.
24. Wydrzynski, T., Zumbulyadis, N., Schmidt, P.G. and Govindjee (1975). *Biochim. Biophys. Acta* **408**, 349-354.
25. Wydrzynski, T. Zumbulyadis, N., Schmidt, P.G., Gutowsky, H.S. and Govindjee (1976). *Proc. Natl. Acad. Sci. Washington* **73**, 1196-1198.
26. Yamashita, T. and Butler, W.L. (1968). *Plant Physiol.* **43**, 1978-1986.
27. Yamashita, T. and Tomita, G. (1974). *Plant Cell Physiol.* **15**, 252-266.
28. Zilinskas, B.A. and Govindjee (1976). *Z. Pflanzenphysiol.* **77**, 302-314.

KINETICS OF DEACTIVATION OF THE CHARGED STATE FORMED UPON ILLUMINATION ON THE OXIDIZING SITE OF PHOTOSYSTEM II IN THE PRESENCE OF DCMU IN *CHLORELLA* AFTER EXTRACTION OF MEMBRANE-BOUND MANGANESE BY NH_2OH

G. VIERKE

Institut für Physikalische Biochemie und Kolloidchemie, Universität Frankfurt, 6000 Frankfurt 71 (BRD)

Abstract

The kinetics of deactivation of the positive charge stored on the oxidizing site of photosystem II in the light in the presence of DCMU was studied in dark-adapted *Chlorella* after extraction of two thirds of the membrane-bound manganese by NH_2OH. Two independent methods were used: determination of the kinetics of the back reaction of photosystem II from luminescence measurements and investigation of the kinetics of formation of Q^- by way of fluorescence induction measurements.

The positive charge formed in extracted *Chlorella* in the light was found to exhibit kinetic properties that are characteristic for the S_2 state in untreated *Chlorella*.

(1) The kinetics of the back reaction is accelerated upon addition of DCMU.

(2) The kinetics of the back reaction is slowed down in the presence of NH_4Cl.

(3) The kinetics of formation of Q^- in the light in the presence of 200 µM DCMU is the same in dark-adapted untreated *Chlorella* and in NH_2OH-extracted *Chlorella*.

As the kinetics of accumulation of Q^- and the kinetics of the back reaction strongly depend on the nature of the S state formed upon illumination, these results provide some evidence for the formation of the S_2 state even after extraction of membrane-bound manganese.

Introduction

It is well known that the presence of membrane-bound manganese is essential for maintaining the oxygen evolution capacity of the photosynthetic membrane [9,13,14]. Photosynthetic oxygen evolution is thought to be brought about by the concerted action of single oxygen evolving centers that are conceived of as operating largely independently from each other [15,26]. Loss of membrane-bound manganese in a center leads to the destruction of its oxygen evolution capacity. The function of manganese in photosynthetic oxygen evolution, however, is still unknown. It is not clear at present, therefore, whether the S_4 state of the oxygen evolving centers simply loses its ability to oxidize water upon removal of manganese from its binding sites in the membrane, or whether the S states cannot be formed any more.

Two hypotheses on the role of manganese in photosynthetic oxygen evolution have been taken into consideration as yet. The first one states that higher oxidation states of manganese are involved in the accumulation of positive charges of the S states of the oxygen evolving system [5,7,8,21,33,35,36]. Bound water is then thought to be present in a manganese-coordinated form. The strong linear correlation of oxygen evolution [9,11,14] and of the S state dependent water proton spin-spin relaxation rate [45,46] with the manganese binding sites argues in favour of this assumption. Furthermore, it is well known that manganese is able to take on stable higher oxidation states when bound to suitable ligands [5,44].

The second hypothesis claims that the presence of manganese is essential for maintaining the correct structure of the water binding site only [41]. This assumption is supported by the observation that in manganese-deficient chloroplasts and algae the structure of the photosynthetic membrane is subjected to gross alterations [10,24,34].

A decision between these two hypotheses has not been reached as yet. One way of getting a decision is to examine whether S states are also formed in photosynthetic organisms after extraction of membrane-bound manganese. If the first hypothesis were correct, release of manganese from the photosynthetic membrane should lead to the destruction of the S states. If the second hypothesis were true, S states could still be formed in the absence of manganese though the ability to oxidize water is lost. However, it could be that the formation of the S states depends on the fulfilment of structural requirements that can only be met when the natural structure of the photosynthetic membrane is maintained.

This consideration leads to the following conclusion: when S states are still formed in extracted thylakoids, higher oxidation states of manganese can *not* be involved in photosynthetic oxygen evolution, but when S states are not formed in the absence of manganese, no clear-cut decision between the two hypotheses can be reached.

To decide unequivocally whether S states are formed after extraction of manganese is not straightforward, since the most powerful tool for the study of the S states — the measurement of flash-induced oxygen evolution — cannot be used in extracted photosynthetic organisms. One approach to the solution of this problem is to examine whether S state dependent properties of photosystem II maintain their S state dependence even after extraction of membrane-bound manganese.

This paper concentrates on the study of two S state dependent properties of photosystem II in *Chlorella*: the kinetics of the back reaction and the kinetics of formation of Q^-.

The back reaction of photosystem II involves electron back transfer from the primary electron acceptor Q^- to the

S states of the oxygen evolving system [6,22,27]. The kinetics of the back reaction as measured by the time course of reoxidation of Q^- strongly depends on the nature of the S state involved as substrate [22]. In *Chlorella* luminescence in the millisecond and second time region originates from the back reaction [27,28]. The kinetics of the back reaction in the presence of DCMU can be determined, therefore, from the luminescence decay curve [30, 43]. This offers the possibility to study the properties of the S states by way of luminescence measurements. The kinetics of formation of Q^- is represented by the kinetics of area growth over the fluorescence rise curve in the presence of DCMU [31]. The kinetics is biphasic in chloroplasts and green algae. The two first order rate constants strongly depend on the S states [31]. The S state dependence of the fluorescence rise curve has also been noticed by Etienne [22].

In this article the kinetic properties of the charged state formed upon illumination in dark-adapted NH_2OH extracted *Chlorella* in the presence of DCMU were compared with those exhibited by the S_2 state in untreated *Chlorella*. It was observed that the kinetic behaviour of these two charged states was strikingly similar.

Theory

The kinetics of the back reaction in the presence of DCMU can be evaluated from luminescence measurements in terms of the partial light sum

(1) $$\mathcal{L}(t) = \int_0^t L(t')dt'$$

and the total light sum

(2) $$\mathcal{L}_{tot} = \int_0^\infty L(t)dt.$$

Luminescence intensity $L(t)$ is a complex function of $[Q^-]/[Q_0]$ [30] but the integrated luminescence intensities $\mathcal{L}(t)$ and $\mathcal{L}_{tot}$ are quite simple functions of the relative concentration of Q^- (see below). This face offers the possibility to calculate $[Q^-](t)$ directly in terms of $\mathcal{L}(t)$ and $\mathcal{L}_{tot}$.

By combining the equations 10, 12, and 16 of Mar and Roy [30] it is seen that $L(t)$ can be represented by

$$(3) \qquad L = -\varphi \frac{S}{1 - S\phi_t} \frac{dc}{dt}$$

S is the exciton yield of the back reaction, φ is the live fluorescence quantum yield, and ϕ_t is the efficiency of exciton trapping by an open photosystem II trap. The concentration of Q^- is denoted by c in this treatment and should not be confused with the rate constant C of the back reaction (see eqn. 12). Both φ and ϕ_t depend on c.

The partial light sum is then given by

$$(4) \qquad \mathcal{L}(t) = \int_c^{c_0} \frac{\varphi(c) S}{1 - S\phi_t(c)} dc$$

This expression can be considerably simplified because S is a very small number. The value of S can be estimated from

$$(5) \qquad \Phi_{DF} = \varphi \cdot S$$

Φ_{DF} denotes the quantum yield of delayed light emission. Since Φ_{DF} has been shown to be 4×10^{-6} [40] and $\varphi = 4 \times 10^{-2}$ in *Chlorella* in strong light [32], S is equal to 10^{-4}.*

Because of this low value of S the product $S\phi_t$ is always

---

*The same order of magnitude for S was obtained from an approximate experimental determination of S from the number of photons emitted L_0 and the rate of the back reaction at time $t = 0$ using the equation $L_0 = -\varphi S (dc/dt)_0$. The rate of the back reaction was determined from measurements of Etienne [22].

much smaller than one. Expanding $(1-S\phi_t)^{-1}$ in eqn. 4 and neglecting the terms in S^2 and higher powers gives

$$\mathcal{L}(t) = S \int_c^{c_o} \varphi(c)dc \tag{6}$$

In the presence of DCMU the fluorescence quantum yield is given by [17]

$$\varphi = \varphi_0 + (\varphi_\infty - \varphi_0) \frac{(1-p)\frac{c}{c_o}}{1-p\frac{c}{c_o}} \tag{7}$$

φ_0 and φ_∞ are the fluorescence quantum yields when $[Q^-] = 0$ and $[Q^-] = [Q_0]$, respectively. p is the mean probability that an exciton which hits a closed trap will be transferred to another centre. For *Chlorella* p is 0.45 [17].

Evaluation of the integral of eqn. 6 yields

$$\mathcal{L}(t) = S\varphi_0 c_o \left\{ 1 - \frac{c}{c_o} + \left(\frac{\varphi_\infty}{\varphi_0} - 1\right)\frac{1-p}{p}\left[\frac{c}{c_o} + \frac{1}{p}\ln\left(1 - p\frac{c}{c_o}\right) - 1 - \frac{\ln(1-p)}{p}\right]\right\} \tag{8}$$

The total light sum is

$$\mathcal{L}_{tot} = S\varphi_0 c_o \left\{ 1 - \left(\frac{\varphi_\infty}{\varphi_0} - 1\right)\frac{1-p}{p}\left(1 + \frac{\ln(1-p)}{p}\right)\right\} \tag{9}$$

On expanding $\ln(1-p\frac{c}{c_o})$ and ignoring the third and higher terms in $p\frac{c}{c_o}$ one obtains for the difference of $\mathcal{L}_{tot}$ and $\mathcal{L}(t)$

$$\mathcal{L}_{tot} - \mathcal{L}(t) = S\varphi_0 c_o \left\{ \frac{c}{c_o} + \frac{1}{2}\left(\frac{\varphi_\infty}{\varphi_0} - 1\right)(1-p)\left(\frac{c}{c_o}\right)^2 \right\} \tag{10}$$

Dividing this expression by $\mathcal{L}_{tot}$ leads to an equation for the determination of c/c_o which solely depends on parameters that can be determined experimentally. The solution of this equation is

$$\text{(11)} \qquad \frac{c}{c_o} = \frac{[1 + 2ABN(t)]^{\frac{1}{2}} - 1}{A}$$

$$\text{(11a)} \qquad A = \left(\frac{\varphi_\infty}{\varphi_0} - 1\right)(1-p)$$

$$\text{(11b)} \qquad B = 1 - \left(\frac{\varphi_\infty}{\varphi_0} - 1\right)\left(1 + \frac{\ln(1-p)}{p}\right)\frac{1-p}{p}$$

$$\text{(11c)} \qquad N(t) = 1 - \frac{\mathcal{L}(t)}{\mathcal{L}_{tot}}$$

p can be determined from the fluorescence rise curve (Fig. 10) according to the method of Delosme [17] (p = 0.49 for extracted *Chlorella*). Because of the existence of a dead fluorescence component [16] the ratio φ_∞/φ_0 cannot be evaluated from the fluorescence rise curve in the presence of DCMU. The value of φ_∞/φ_0 was chosen such that the kinetics of the back reaction calculated from eqn. 11 is in accordance with the kinetics obtained from measurements of the dark decay of fluorescence induction [1,22]. This is possible because both techniques measure the time course of the concentration of Q^-. Figure 1 shows that $\varphi_\infty/\varphi_0 = 5$ is a proper choice for this ratio in *Chlorella*.

After integrating the luminescence decay curve the kinetics of the back reaction can now be conveniently calculated from eqn. 11.

Mar and Roy [30] have shown that the kinetics of the back reaction can be described in terms of the two rate constants C and D by the expression

$$\text{(12)} \qquad [Q^-] = \frac{[Q_0]e^{-ct}}{1 + D[1 - e^{-ct}]}$$

After evaluating the kinetics of the back reaction from eqn. 11 the parameters D and C may be determined by plotting the expression

$$\text{(13)} \qquad \ln\left[\frac{\frac{[Q_0]}{[Q^-]} + D}{D + 1}\right]$$

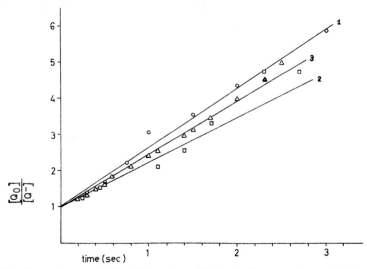

Fig. 1. Comparison of the kinetics of the back reaction calculated from luninescence measurements with that obtained from the kinetics of the dark decay of fluorescence induction. ○ Curve 1: Data from Bennoun [32]; □ Curve 2: Data from Etienne [23] at pH = 6.5; △ Curve 3: Data from luminescence measurements at pH = 6.75 with $\varphi_\infty/\varphi_0 = 5$, T = 24°C, chlorophyll content: 55 µg/ml.

against time. The plot yields a straight line for the correct value of D. The rate constant C is given by the slope of the straight line (see eqn. 12).

Because of $S\varphi_t \ll 1$ the rate constants D and C of the back reaction can now be represented in a more convenient form by simplifying the expressions given by Mar and Roy [30]

$$(14) \qquad D = n\frac{\Omega_1}{\Omega_0}$$

$$(15) \qquad C = \nu\Omega_0 W^\alpha \exp\left[-\frac{\Delta H^o}{kT}\right]$$

$$(15a) \qquad W = \frac{\sum_A P_A[A^-]_i + \sum_C P_C[C^+]_o}{\sum_A P_A[A^-]_o + \sum_C P_C[C^+]_i}$$

Here Ω denotes the partition function which determines the entropy of activation and Ω_0 and Ω_1 are the first two terms of the development in the series of Ω. n is the number of nearest neighbour reaction centres. Eqn. 14 states that the entropy of activation increases when the rate constant D decreases.

ΔH^0 represents the enthalpy of activation, W is an expression that originates from the Goldman equation. It describes the influence of a change in membrane potential induced by ions on the luminescence back reaction. P_A denotes the permeability of the anion A^- and P_C is the permeability of the cation C^+. The index i indicates the inside volume of the thylakoids and o the outside. α is the polarization constant of the membrane. ν is the frequency factor of the vibration of Q^- in its initial site.

It should be noted that the theory presented here also applies to chloroplasts though somewhat different values of p and φ_∞/φ_0 have to be used.

Materials and Methods

Culture of Chlorella, *preparation of the control sample, and determination of chlorophyll*

Chlorella fusca was cultivated as described by Soeder et al. [39]. *Chlorella* cells were always taken from a synchronous culture at the same time and were used immediately.

As extraction of *Chlorella* with NH_2OH requires six washings in total, untreated *Chlorella* used as control sample was also washed six times with phosphate buffer (67 mM, pH = 6.75) after separation of the algae from the nutrition medium by centrifugation. This was done in order to eliminate differences in the luminescence properties of untreated and extracted cells possibly induced by centrifugation.

Chlorella cells were then incubated with DCMU (K & K Lab. USA, recrystallized twice from benzene) in the dark for 10 min. The suspension was continuously stirred.

When the effect of NH_4Cl on the kinetics of the back

reaction was studied the following incubation procedure was used: *Chlorella* was at first incubated with 100 mM NH_4Cl in the dark for 30 min. Then DCMU was added and the sample was kept in the dark for additional 10 min. The suspension was stirred during the whole exposure time to NH_4Cl and DCMU.

Chlorophyll was determined spectrophotometrically in a 1 cm cuvette using the expression $E_{684} = 0.19[Chl]$ (chlorophyll concentration in µg/ml). No correction for scattered light was made. The dependence of the absorbance values at 684 nm (E_{684}) from the chlorophyll concentration was established by measuring the chlorophyll content of the sample according to the method of Ziegler and Egle [47] after extraction of chlorophyll with an acetone-water mixture.

Preparation of NH_2OH extracted Chlorella

NH_2OH extracted *Chlorella* was prepared according to the following procedure [11,12,14]: after centrifugation of the algae from the nutrition medium they were washed twice with phosphate buffer (67 mM, pH = 6.75). This was done in order to remove traces of heavy metal ions which catalyze the autoxidation of NH_2OH [13]. Then the algae were concentrated up to a chlorophyll content of 200 µg/ml and incubated with 10 mM NH_2OH in phosphate buffer (67 mM, pH = 6.75) at room temperature in the dark for 20 min.

After incubation with NH_2OH the algae were washed four times with buffer in the dark. The efficiency of the removal of NH_2OH by this procedure was controlled by luminescence measurements. NH_2OH is known to quench luminescence even at very low concentrations [4,25] because it acts as electron donor to photosystem II, thereby suppressing the luminescence back reaction. By measuring the intensity of 5 ms luminescence several times successively it was observed that in the presence of more than 10 µM NH_2OH luminescence intensity was diminished more and more till a stationary level was attained. This effect, however, was not noticed in extracted *Chlorella* which was washed four times. This shows that the washings were sufficient to remove NH_2OH to such an extent that luminescence intensity is no longer affected.

The efficiency of NH_2OH extraction was controlled by measurements of oxygen evolution, of the F_p amplitude of variable fluorescence, and of the concentration of EPR-detectable free manganese in the suspension. In extracted *Chlorella* cells the base rate of oxygen evolution was usually found to be 20-25% of the control rate. The F_p amplitude of variable fluorescence, however, was abolished. NH_2OH extraction was further observed to be accompanied by the release of free manganese into an EPR-detectable state (Fig. 2).

Incubations of extracted *Chlorella* with NH_4Cl and DCMU was carried out in the same way as described for untreated *Chlorella*.

Apart from a short exposure to dim daylight (several seconds) during handling of the sample extracted *Chlorella* cells were kept strictly in the dark to avoid photoreactivation.

Oxygen measurements and EPR measurements

Measurement of the rate of oxygen evolution was carried out with the Hansatech oxygen electrode unit (Hansatech, England). EPR measurements were performed in a flat cell at room temperature using the Varian E-12 spectrometer.

Measurement of luminescence

Luminescence measurements were carried out 5 ms after excitation with monochromatic light at 478 ± 10 nm. The excitation intensity was 6 mW/cm^2 measured at the surface of the cuvette. The emitted light was measured in the direction of the excitation beam with the EMI photomultiplier 9659B (extended S 20 response) which was cooled down

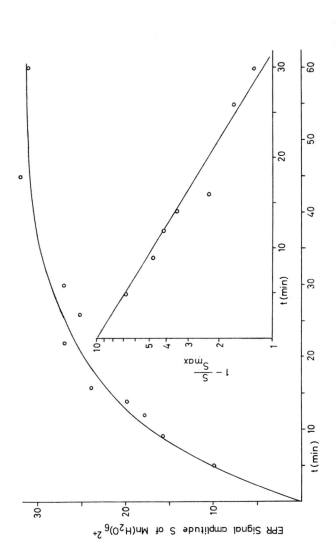

Fig. 2. Kinetics of release of membrane-bound Mn upon extraction of *Chlorella* with NH_2OH as measured by EPR spectroscopy. T = 25°C, pH = 6.75; Chlorophyll content: 1200 μg/ml; concentration of NH_2OH; 1 mM. The ordinate represents the signal amplitude of the sixth line in the EPR spectrum of free Mn(II) in arbitrary units. *Insert*: First order analysis of the kinetics of release of manganese.

to -40°C by a stream of cold nitrogen. The photomultiplier was protected from stray light by a cut-off filter 665 nm (WG 665, 10 mm, Schott & Gen., F.R.G.) thus permitting the measurement of the whole spectrum of luminescence. The multiplier signal was fed to a rapid DC amplifier (GV 9031, EGB, F.R.G.) and then was recorded by a light beam galvanometer recorder (Lumiscript 150-B, Hartmann & Braun, F.R.G.).

The sample was placed in the centre of a cylindrical shutter with two openings arranged at an angle of 85 degrees so that the sample was either illuminated with exciting light and the emission window was closed or after rotating the shutter by an electrical pulse within 5 ms luminescence was measured with the excitation window closed.

The temperature of the cuvette (made of quartz glass 'Suprasil', Heraeüs, F.R.G.) was regulated by a TUK thermostat (MWG Lauda, F.R.G.) and was measured by a calibrated copper-constantan thermocouple. The chlorophyll content of the sample was always kept below 50 µg/ml in order to avoid distortion of the luminescence decay curve resulting from reabsorption of luminescence.

In the presence of DCMU no luminescence induction effects were observed in *Chlorella* at times greater than 1 s after excitation. The preillumination time was 30 s unless otherwise noted.

Integration of the luminescence decay curve was done by weighing the area below L(t).

Fluorescence induction measurements

Fluorescence induction measurements were made with monochromatic exciting light at 478 ± 10 nm (excitation inensity: 2.7 mW/cm^2) using essentially the same apparatus as for luminescence measurements. The photomultiplier was shielded from the exciting light beam by a filter combination consisting of a cut-off filter (WG 665, 3mm, Schott &

Gen., F.R.G.) and an interference filter 689 nm. Fluorescence was measured, therefore, at 689 nm. The time resolution was 200 µs (bullet driven shutter).

Before each measurement the *Chlorella* suspension was gassed with air for 5 min in the dark. Then 200 µM DCMU was added. After 5 s incubation with DCMU in the dark the fluorescence induction curve did not change any more in accordance with the experimental results reported by Etienne [22].

Experimental Results

Kinetics of the release of membrane-bound manganese upon incubation with NH_2OH

It has been demonstrated that in isolated chloroplasts manganese can be released from the photosynthetic membrane upon incubation with NH_2OH [11]. Though the effect of NH_2OH extraction on the oxygen evolving site of photosystem II in algae has been studied extensively [12-15], it has not been shown as yet that manganese actually leaks out of the membrane. This is a crucial point of this investigation, however.

After treatment of chloroplasts with TRIS buffer membrane-bound manganese is set free into an EPR-detectable state [3]. The spectrum of the Mn(II) compound formed was identical to that of $Mn(H_2O)_6^{2+}$. The effect of NH_2OH on *Chlorella* in the dark was studied, therefore, by EPR spectroscopy. An EPR signal due to free manganese was observed to develop gradually with time. The kinetics of this process is shown in Fig. 2. It is seen that the kinetics of removal of membrane-bound manganese by NH_2OH is first order (Fig. 2, insert). This result is in accordance with that obtained by Cheniae and Martin [12] by way of oxygen evolution measurements.

Kinetics of the back reaction in NH_2OH extracted Chlorella

After extraction of manganese by NH_2OH luminescence in the millisecond and second time region in *Chlorella* is inhibited as compared to untreated *Chlorella* but is not suppressed completely (Fig. 3). This shows that at least one positive charge must be stored on the oxidizing site of photosystem II. This was also observed in TRIS-treated chloroplasts [42].

Calculation of the kinetics of the back reaction from the luminescence decay curves (Fig. 3) according to eqn. 11 and evaluation of the kinetics according to eqn. 12 reveals that the kinetics of the back reaction is slowed down in the absence of membrane-bound manganese (Fig. 4). Thus the charged state in extracted *Chlorella* is somewhat more stable than the S_2 state in untreated *Chlorella*.

The entropy of activation of the back reaction remains unchanged upon extraction of manganese because the rate constant D has the same value in extracted and untreated cells (see eqn. 14). Hence the enhanced stability of the charged state is solely due to the decrease of the rate constant C, thus indicating that the enthalpy of activation and/or the membrane potential is slightly modified in extracted cells.

The total light sum in extracted *Chlorella* is observed to be smaller by the factor 10 than in untreated *Chlorella*. This shows that the exciton yield S of the back reaction is strongly diminished in the absence of membrane-bound manganese (see eqn. 9).

Effect of DCMU on the kinetics of the back reaction

It has been shown that DCMU accelerates the deactivation reaction of S_2 in untreated *Chlorella* [6]. The same effect was noticed for the back reaction as derived from luminescence measurements (Fig. 5). Fig. 6 shows that this is also observed in extracted *Chlorella*. In this experiment dark-adapted cells were preilluminated for 5 s only

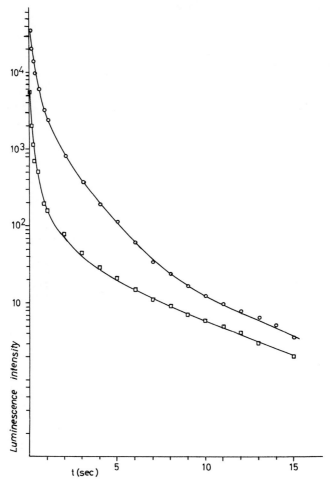

Fig. 3. Luminescence decay curves in untreated *Chlorella* and in NH_2OH extracted *Chlorella* at T = 24°C and pH = 6.75 in the presence of 20 μM DCMU. ○ Untreated *Chlorella* (chlorophyll content; 55 μg/ml); □ NH_2OH extracted *Chlorella* (chlorophyll content; 35 μg/ml). Luminescence intensity is given in arbitrary units. The rates of oxygen evolution were 181 μM O_2/mg Chl·hr for untreated *Chlorella* and 47 μM O_2/mg Chl·hr for extracted *Chlorella*.

in order to avoid photoreactivation to a greater extent. The consequence is that the entropy of activation of the back reaction is much smaller (D = 40) than in cells that have been illuminated long enough to create the stationary light-adapted state of the photosynthetic membrane.

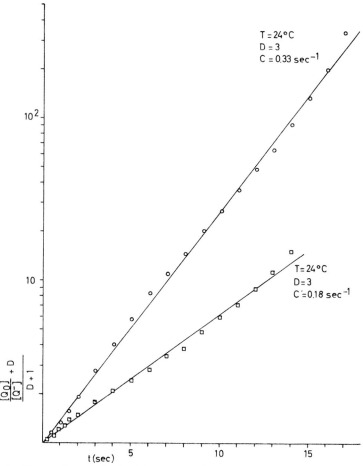

Fig. 4. Comparison of the kinetics of the back reaction in untreated and NH$_2$OH extracted *Chlorella* at T = 24°C and pH = 6.75 in the presence of 20 μM DCMU. ○ Untreated *Chlorella*, □ extracted *Chlorella*. Other data as in Fig. 3.

Effect of NH$_4$Cl on the kinetics of the back reaction

In *Chlorella* addition of NH$_4$Cl leads to the appearance of a slow phase in the deactivation reaction of the S$_2$ state [18]. In chloroplasts the life time of S$_2$ was also observed to be enhanced in the presence of NH$_4$Cl [41]. Likewise the kinetics of the back reaction as studied by luminescence measurements in the presence of DCMU in untreated *Chlorella* and in NH$_2$OH extracted *Chlorella* (Fig. 7)

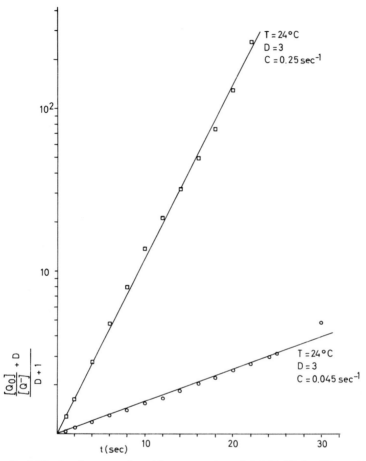

Fig. 5. Effect of non-saturating amounts of DCMU (0.5 μM) on the kinetics of the back reaction in untreated *Chlorella* at pH = 6.4. ○ *Chlorella* without DCMU; □ *Chlorella* with 0.5 μM DCMU. Chlorophyll content: 40 μg/ml. The buffer solution contained 50 mM phosphate and 100 mM KCl.

was slowed down when NH_4Cl was added. These results show that the deactivation of S_2 and of the charged state in extracted *Chlorella* is strongly affected by NH_4Cl whereas S_1 does not interact with NH_3 [41].

The entropy of activation is slightly increased in extracted *Chlorella* in the presence of NH_4Cl contrary to what is observed in untreated *Chlorella* (Fig. 7).

The exciton yield is enhanced in both cases when NH_4Cl

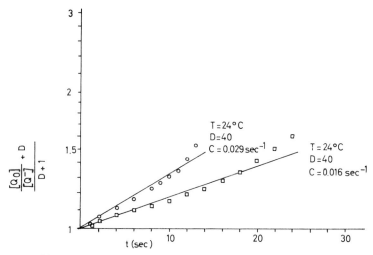

Fig. 6. Effect of non-saturating amounts of DCMU on the kinetics of the back reaction in extracted *Chlorella* at pH = 6.4. □ Extracted *Chlorella* without DCMU after 5 s preillumination before measurement. ○ Extracted *Chlorella* with 0.5 μM DCMU after 5 s preillumination. Other data as in Fig. 5.

is present. The enhancement factor is 1.9 in extracted *Chlorella* and 1.5 in untreated *Chlorella*.

Kinetics of formation of Q^- in NH_2OH extracted Chlorella

The effect of extraction of membrane-bound manganese on the kinetics of the back reaction was further studied using a more indirect but independent approach the measurement of the kinetics of the accumulation of Q^- in the light in the presence of DCMU according to the method of Melis and Homann [31].

In dark-adapted DCMU treated *Chlorella* the concentration of Q^- is solely determined by its photochemical production and by the competing reoxidation reaction $Q^- \rightarrow S_2$. However, the effect of the back reaction can only be seen distinctly when its rate is comparable to the rate of photochemical production of Q^-. Evaluating the data reported by Etienne [23] on the fluorescence rise kinetics in *Chlorella* in the presence of 50 μM DCMU and of 2 mM NH_2OH which donates electrons to the oxidizing site of photosystem

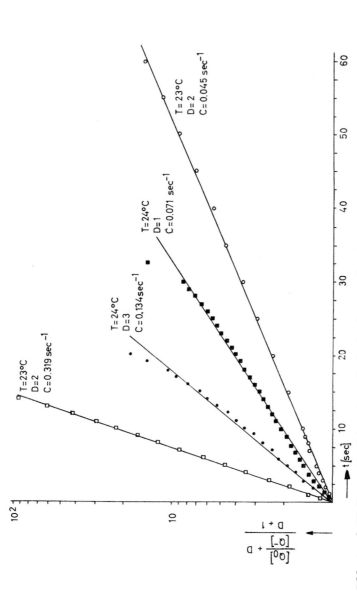

Fig. 7. Effect of NH_4Cl on the kinetics of the back reaction in untreated and NH_2OH extracted *Chlorella* at pH = 8 in the presence of 20 µM DCMU. □ Untreated *Chlorella* + 20 µM DCMU; ○ Untreated *Chlorella* + 20 µM DCMU + 100 mM NH_4Cl; ● Extracted *Chlorella* + 20 µM DCMU; ■ Extracted *Chlorella* + 20 µM DCMU + 100 mM NH_4Cl. Chlorophyll content: 40 µg/ml.

II [2,12,20] thereby suppressing the back reaction demonstrates that the back reaction actually seriously affects the kinetics of formation of Q^-. Both k_α and k_β are strongly decreased in the presence of NH_2OH (Fig. 8.).

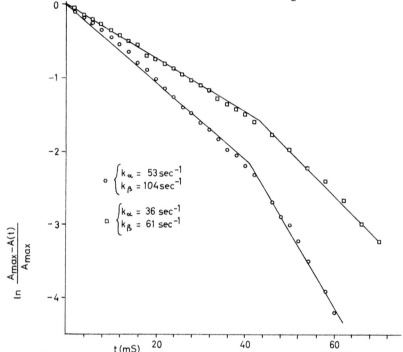

Fig. 8. Evaluation of the fluorescence rise curve in NH_2OH extracted *Chlorella* in the presence of 50 μM DCMU and 2 mM NH_2OH at pH = 6.5 according to the method of Melis and Homann [27]. Data are taken from Etienne [41]. □ *Chlorella* with 50 μM DCMU; ○ *Chlorella* with 50 μM DCMU and 2 mM NH_2OH after 20 min incubation. A(t) and A_{max} denote the area over the fluorescence rise curve at time t and at t = ∞ respectively. k_α and k_β represent the slopes of the two straight lines.

It has been reported by Etienne [22] that the life time of S_2 is extremely short (900 ms) in the presence of high concentrations of DCMU (400 μM). Under these conditions destruction of the S states upon extraction of manganese should severely affect the kinetics of formation of Q^-.

However, there was no difference for the k_α and k_β values in dark-adapted untreated *Chlorella* and in NH_2OH extracted *Chlorella* when NH_2OH is carefully washed out

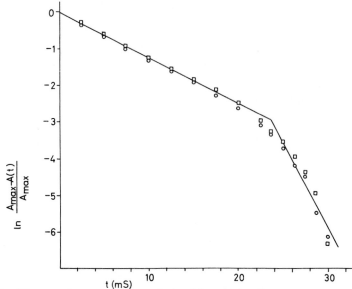

Fig. 9. First order analysis of the kinetics of accumulation of Q^- in untreated and NH_2OH extracted *Chlorella* in the presence of 200 μM DCMU. □ Untreated *Chlorella*; ○ Extracted *Chlorella*; T = 25°C; pH = 6.75. Data are taken from Fig. 10 and are evaluated according to the method of Melis and Homann [27]. $A(t)$ and A_{max} denote the area over the fluorescence rise curve at time t and at t = ∞, respectively. k_α and k_β are the two first order rate constants and are given by the slopes of the straight lines.

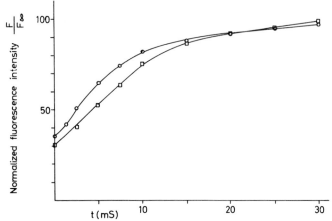

Fig. 10. Fluorescence rise curves in untreated and NH_2OH extracted *Chlorella*. ○ Untreated *Chlorella*; □ Extracted *Chlorella*. Excitation wavelength: 478 nm; excitation intensity: 2.7 mW/cm²; Emission wavelength: 689 nm; Chlorophyll concentration: 30 μg/ml; Temperature: 25°C; pH = 6.75; Time resolution: 200 μs.

after incubation (Fig. 9), though the fluorescence rise curves were slightly different (Fig. 10).

Discussion

The ground state of the oxygen evolving system of dark adapted *Chlorella* is represented by a distribution of the stable states S_0 and S_1 (30% S_0 and 70% S_1, [29]). In the presence of DCMU the S_2 state is then predominantly formed in the light.

It is well-known that the life time of S_2 in *Chlorella* is decreased upon addition of DCMU [6] whereas it is enhanced in the presence of NH_4Cl [18]. These effects have been ascribed to binding of these substances to S_2 [6, 18].

The S_1 state appears to be entirely different in nature as compared to the excited S states. S_1 does not interact with NH_3 [41] and the ADRY agent ANT 2s [37]. Since DCMU can be considered to be an ADRY agent as well [38] it is highly unlikely that it affects the stability of S_1.

Characteristic differences between the S_1 and S_2 state have also been observed with the kinetics of fluorescence induction in the presence of DCMU. The first order rate constants k_α and k_β of the kinetics of formation of Q^- are highest for S_1 and lowest for S_2 [31]. After extraction of membrane-bound manganese a positive charge is formed on the oxidizing site of photosystem II in the light both in the presence and absence of DCMU. Of course, this charged state may in principle be entirely different in nature to any of the S states. But surprisingly it shows kinetic properties that are considered to be characteristic for the double-charged S_2 state in untreated *Chlorella*: an enhanced reduction rate when DCMU is added (Fig. 6) and a decreased reduction rate in the presence of NH_4Cl (Fig. 7). Additionally, the kinetics of formation of Q^- which strongly depends on whether a single-charged state like S_1 or a double-charged state like S_2 is formed upon illumination does *not* change in dark-adapted extracted *Chlorella* as compared to untreated *Chlorella* which is in its S_2 state (Fig. 9).*

The acceleration effect of DCMU and the retardation effect of NH_3 on the kinetics of the back reaction is not as strongly pronounced as in untreated *Chlorella*. This difference and other differences, e.g. the decrease of exciton yield and the rate constant C and the increase of the entropy of the back reaction in the presence of NH_4Cl, must probably be attributed to changes in the molecular structure of the photosynthetic membrane after extraction of manganese.

The results presented here demonstrate that the charged state in extracted *Chlorella* exhibits properties that are characteristic for S_2 but not for S_1. However, this cannot be interpreted with certainty at present to indicate that the S_2 state is also formed after extraction of membrane-bound manganese though the results of this paper provide some evidence for it. This is because, first, it has not been demonstrated beyond doubt as yet that NH_3 and DCMU actually interact specifically with the S states e.g. through binding to these states though there is experimental evidence supporting this view. And, second, additional properties of the charged state must be investigated before its identity with the S_2 state can be considered to be proven.

Acknowledgements

I wish to thank Prof. A. Ried for the culture of *Chlorella fusca*, Mrs. Ute Bergmann and Mr. Mathias Schmidt

---

*It has been hypothesized that the existence of two first order phases in the formation of Q^- might be indicative of a sequential two-electron reduction of Q [31]. A two-quantum conversion has also been considered by Doschek and Kok [19]. If this hypothesis turns out to be true, it would demonstrate that at least two positive charges can be generated on the oxidizing site of photosystem II even after extraction of membrane-bound manganese.

for technical assistance, and the Deutsche Forschungsgemeinschaft for financial support.

References

1. Bennoun, P. (1970). *Biochim. Biophys. Acta* **216**, 357-363.
2. Bennoun, P. and Joliot, A. (1969). *Biochim. Biophys. Acta* **189**, 85-94.
3. Blankenship, R.E. and Sauer, K. (1974). *Biochim. Biophys. Acta* **357**, 252-266.
4. Bonaventura, C. and Kindergan, M. (1971). *Biochim. Biophys. Acta* **234**, 249-265.
5. Boucher, L.J. (1972). *Coord. Chem. Rev.* **7**, 289-329.
6. Bouges-Bocquet, B., Bennoun, P. and Tabury, J. (1973). *Biochim. Biophys. Acta* **325**, 247-254.
7. Calvin, M. (1965). *Rev. Pure Appl. Chem.* **15**, 1-10.
8. Calvin, M. (1976). *Photochem. Photobiol.* **23**, 425-444.
9. Cheniae, G.M. (1970). *Ann. Rev. Plant Physiol.* **21**, 467-498.
10. Cheniae, G.M. and Martin, I.F. (1968). *Biochim. Biophys. Acta* **153**, 819-837.
11. Cheniae, G.M. and Martin, I.F. (1970). *Biochim. Biophys. Acta* **197**, 219-239.
12. Cheniae, G.M. and Martin, I.F. (1971). *Plant Physiol.* **47**, 568-575.
13. Cheniae, G.M. and Martin, I.F. (1971). *Biochim. Biophys. Acta* **253**, 167-181.
14. Cheniae, G.M. and Martin, I.F. (1972). *Plant Physiol.* **50**, 87-94.
15. Cheniae, G.M. and Martin, I.F. (1973). *Photochem. Photobiol.* **17**. 441-459.
16. Clayton, R.K. (1969). *Biophys. J.* **9**, 60-76.
17. Delosme, R. (1967). *Biochim. Biophys. Acta* **143**, 108-128.
18. Delrieu, M.J. (1976). *Biochim. Biophys. Acta* **440**, 176-188.
19. Doschek, W.W. and Kok, B. (1972). *Biophys. J.* **12**, 832-838.
20. Ducret, J.M. and Lavorel, J. (1974). *Biochem. Biophys. Res. Comm.* **58**, 151-158.
21. Earley, J.E. (1973). *Inorg. Nucl. Chem. Lett.* **9**, 487-490.
22. Etienne, A.L. (1974). *Biochim. Biophys. Acta* **333**, 320-330.
23. Etienne, A.L. (1974). *Biochim. Biophys. Acta* **333**, 497-508.
24. Heath, R.L. (1973). In "International Review of Cytology" (G.H. Bourne, J.F. Danielli and K.W. Jeon, eds), Vol. 34, pp. 49-97, Academic Press, New York.
25. Joliot, P., Joliot, A., Bouges, B. and Barbieri, C. (1971). *Photochem. Photobiol.* **14**, 287-305.
26. Kok, B., Forbush, B. and McGloin, M. (1970). *Photochem. Photobiol.* **11**, 457-475.
27. Lavorel, J. (1975). In "Bioenergetics of Photosynthesis" (Govindjee, ed), pp. 225-314, Academic Press, New York.
28. Lavorel, J. (1975). *Photochem. Photobiol.* **21**, 331-343.
29. Lavorel, J. (1976). *FEBS Lett.* **66**, 164-167.
30. Mar, T. and Roy, G. (1974). *J. Theor. Biol.* **48**, 257-281.
31. Melis, A. and Homann, P.H. (1975). *Photochem. Photobiol.* **21**, 431-437.
32. Murty, N.R., Cederstrand, C.N. and Rabinowitch, E. (1965). *Photochem. Photobiol.* **4**, 917-921.

33. Olson, J.M. (1970). *Science* **168**, 438-446.
34. Possingham, J.V., Vesk, M. and Merser, F.V. (1964). *J. Ultrastruct. Res.* **11**, 68-75.
35. Renger, G. (1970). *Z. Naturforschg.* **25b**, 966-971.
36. Renger, G. (1973). *Europ. J. Biochem.* **27**, 259-269.
37. Renger, G. (1973). *Bioenergetics* **4**, 491-505.
38. Renger, G. (1973). *Biochim. Biophys. Acta* **314**, 113-116.
39. Soeder, C.J., Schulze, G. and Thiele, G. (1967). *Arch. Hydrobiol. Suppl.* XXXIII, 127-169.
40. Tollin, G., Fujimori, E. and Calvin, M. (1958). *Biochemistry* **44**, 1035-1047.
41. Velthuys, B.R. (1975). *Biochim. Biophys. Acta* **396**, 392-401.
42. Velthuys, B.R. and Amesz, J. (1974). *Biochim. Biophys. Acta* **333**, 85-94.
43. Vierke, G. (1977). *Photochem. Photobiol.*, submitted.
44. Vierke, G. and Müller, M. (1975). *Z. Naturforschg.* **30c**, 327-332.
45. Wydrzynski, T., Zumbulyadis, N., Schmidt, P.G. and Govindjee (1975). *Biochim. Biophys. Acta* **408**, 349-354.
46. Wydrzynski, T., Zumbulyadis, N., Schmidt, P.G., Gutowsky, H.S. and Govindjee (1976). *Proc. Natl. Acad. Sci. Washington* **73**, 1196-1198.
47. Ziegler, R. and Egle, K. (1965). *Beitr. Biol. Pflanzen* **41**, 11-37.

PROPERTIES AND FUNCTION OF TWO MANGANESE-CONTAINING PROTEINS FROM *DUNALIELLA* CHLOROPLASTS

E. VON KAMEKE and K. WEGMANN

*Institut für Chemische Pflanzenphysiologie
der Universität, Tübingen (F.R.D.)*

The significance of manganese for the photosynthetic oxygen liberation was first recognized by Pirson in 1937 [19] and further established by Warburg and Krippahl [31]. Tanner et al. [29] observed that the Mn^{++} EPR signal disappeared during illumination of photosynthetic material - in the presence of CO_2 - and reappeared in the dark. Cheniae and Martin [2] distinguished between a tightly bound manganese fraction and a loosely bound one. However, till now little is known about the binding site(s) for manganese and its molecular function. In order to study the manganese in its native protein-bound state we began to isolate manganese-proteins from chloroplasts. 54Mn as a tracer was used during the development of a suitable isolation method which enabled the rapid quantitative determination in the fractions. In a later stage of the experiments manganese was determined by atom absorption spectroscopy. The unicellular phytoplankter *Dunaliella tertiolecta* was used as the starting material because of its rapid growth and the simple disintegration of the cells by osmotic shock. After 48 h 88% of the $^{54}MnCl_2$ had been taken up by the cells. Numerous experiments led to the optimized isolation and separation procedure represented in Fig. 1. The essential steps are the selective extraction of the chloroplast membranes by a low concentra-

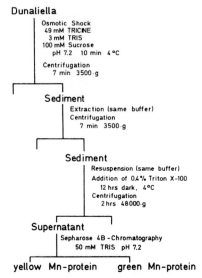

Fig. 1. Isolation scheme of manganese proteins from *Dunaliella* chloroplasts.

tion of Triton X-100 without sonification, and the subsequent column chromatography on a Sepharose 4B column. The procedure yields a yellow manganese protein and a green one.

The yellow manganese protein has a molecular weight (m.w.) of about 600 000 as determined by Sephadex column and TLC, and ultracentrifugation; the value was a bit

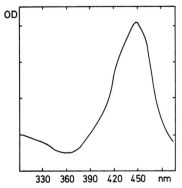

Fig. 2. Absorption spectrum of the yellow manganese protein in aqueous solution.

smaller in the SDS gel electrophoresis (580 000). The absorption spectrum (Fig. 2) indicates the probable presence of a carotenoid in a bound state. Its identification by absorption spectroscopy in chloroform, n-hexane and ethanol according to Hager and Meyer-Bertenrath [8], and by TLC in adsorption and distribution mode demonstrated neoxanthin besides a trace of β-carotene. The function of the neoxanthin is still unknown. Isoelectric focusing on granulated gel according to Radola [20] showed the isoelectric point of the yellow protein at pH 6.8. The amino acid composition is given in Table 1. A molecular ratio

TABLE 1

Amino acid composition of the isolated manganese proteins

Amino acid	yellow Mn-protein	green Mn-protein
	(mol-%)	
Cys	0.0	0.2
Asp	16.2	12.2
Thr	4.9	5.4
Ser	4.9	4.9
Glu	4.8	5.9
Pro	4.1	5.5
Gly	9.4	11.5
Ala	9.5	10.4
Val	13.0	8.0
Met	0.8	0.8
Ile	4.2	5.6
Leu	8.5	9.9
Tyr	2.4	2.5
Phe	4.2	4.7
His	2.5	2.1
Lys	6.5	5.7
Arg	4.1	4.7

Mn : protein : neoxanthin of 1.17:1:20.83 (average from 10 samples) was calculated from the analytical data. The EPR spectrum at -160°C (Fig. 3) does not contain the typical Mn^{++} lines. As the manganese content is known by ^{54}Mn labelling and AAS, it could either be present in a higher oxidized state or in strong complex bonding. The EPR spectrum, however, showed the presence of high and low spin Fe^{III} signals at g = 4.30 and 2.26 resp. The EPR spectrum

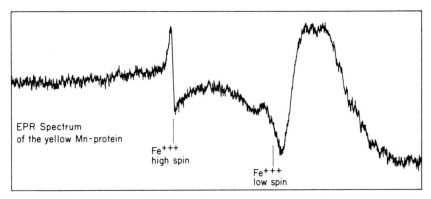

Fig. 3. EPR Spectrum of the yellow manganese protein from *Dunaliella* chloroplasts at -160°C. Varian X-band Spectrometer E 3, 10 mW, H_0 = 2000 G, magnetic field width 4000 G.

led us to the suspicion that a superoxide dismutase could be involved. Indeed, the microtest for SOD [1] gave positive results. Manganese- and iron-containing superoxide dismutases were already known from prokaryotes and organelles (for a survey see [5]). Chloroplastic superoxide dismutases have been investigated by Lumsden and Hall [15,16] and by Elstner and Heupel [3]. However, those SOD had m.w. between 40 000 and 80 000 and they did not contain carotenoids. When we isoelectrically focused the yellow protein after having it added prior to the gel polymerization, the SOD activity split off the large molecule and had then a m.w. of approx. 80 000; it did not contain the neoxanthin.

The green manganese protein had a molecular weight of approx. 500 000 estimated by Sephadex column and TLC, 470 000 by SDS gel electrophoresis. In the ultracentrifuge it had a m.w. of 950 000 indicating dimerization of the protein molecules. The absorption spectrum (Fig. 4) shows the presence of the chlorophyll band and the probable presence of a carotenoid. Further analysis of the chlorophylls by spectroscopy [18] and thin layer chromatography established chlorophylls *a* and *b*. The carotenoid was identified as above by optical spectroscopy in different

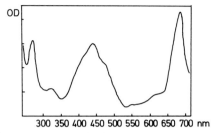

Fig. 4. Absorption spectrum of the green manganese protein in aqueous solution.

solvents and TLC. It is violaxanthin besides a very small amount of lutein. Violaxanthin is one of the intermediates of the xanthophyll epoxide cycle [7,25]. Therefore we checked the presence of the other components of the epoxide cycle, however, we never detected antheraxanthin or zeaxanthin in preparations from differently preconditioned cell material. The amino acid composition is given in Table 1. The molecular composition of the green protein is Mn : protein : chl.a : chl.b : violaxanthin = 3.33 : 1 : 7.5 : 5 : 2 (average from 10 samples). Isoelectric focusing led to the loss of approx. half the Mn content, accompanied by the cleavage of the protein into two subunits banding at pH 5.4 and pH 5.0, their m.w. being 260 000 and 240 000 resp. The chl.a/b ratios were 1.6 and 1.0, thus one of the subunits did not contain chlorophyll b. The EPR spectrum (Fig. 5) exhibits the Mn^{++} lines, which decrease to some extent during illumination, while the radical signal at g = 2.0025, which is well known in photosynthetic material, appears in the light. This signal was not observed in the presence of 10^{-3} M methylviologen.

Manganese-containing proteins from chloroplasts have also been isolated by other investigators. Lagoutte and Duranton [14] have obtained a 25 000 m.w. manganese protein from *Zea mays* chloroplasts, which obviously is not identical with our isolates; it does not contain any pigments. Holdsworth [10] reported the isolation of a high molecular weight manganese protein from the marine diatom

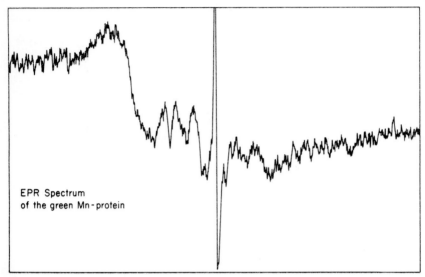

Fig. 5. EPR Spectrum of the green manganese protein from *Dunaliella* chloroplasts at -160°C in the light. Varian X-band, Spectrometer E 3, 10 mW, H_o = 3320 G, magnetic field width 2000 G.

Phaeodactylum tricornutum, which contained fucoxanthin as the carotenoid component. Latest data [11] give a m.w. of 850 000 and the presence of chlorophylls *a* and *c*, and fucoxanthin. Fucoxanthin, a structural relative of neoxanthin, and chlorophyll *c* are characteristic for diatoms. It remains, however, to be clarified, whether the 850 000 m.w. protein is an analogue to our protein, composed of a green and a yellow subunit.

The green protein was tested for photochemical activities. Spectrophotometric methods for the determination of photosystem I and II activities were negative. Oxygen measurements with a Clark type electrode showed photochemical activity, however not the expected oxygen evolution but an oxygen consumption. The reaction started in the light without any additives. The rate of oxygen uptake is proportional to the light intensity between 10^2 and 10^5 erg·cm^{-2}·s^{-1}. Most of our measurements were carried out with white light in order to obtain high light intensity, however, we made sure that red light was satisfactory for driving the reaction. By the addition of catalase in control experiments we could show that oxygen was reduced to H_2O_2. Within one minute after the addition of the enzyme an oxygen burst was observed, followed by almost no further apparent oxygen consumption. The photochemical oxygen reduction to H_2O_2, known as the Mehler reaction, always has been ascribed the photosystem I. Thus, it might be possible that we have obtained a different Mehler-type reaction associated with photosystem II. The subunits obtained by isoelectric focusing both are photochemically inactive, although in the presence of Mn^{++}. About 50% of the original activity were recovered by the complexation of the two subunits together with 10^{-3} M Mn^{++}.

Why does the green manganese protein not evolve oxygen? Possibly only the sum of two simultaneously running reactions is seen:

$$H_2O + Acc \longrightarrow Acc \cdot H_2 + 1/2\ O_2$$
$$Acc \cdot H_2 + O_2 \longrightarrow Acc + H_2O_2$$
$$\overline{H_2O + 1/2\ O_2 \longrightarrow H_2O_2}$$

Of course, these equations do nothing express about the reaction mechanism, only about the stoichiometry. If this is true, we should observe oxygen evolution after the addition of a proper hydrogen acceptor. The addition of 10^{-4}M $K_3[Fe(CN)_6]$ or $5 \cdot 10^{-4}$ M Janusgreen indeed reduced the oxygen consumption to 51 or 22% resp., but caused no oxygen evolution.

In a number of experiments (examples given in Table 2) the participation of the green protein in the early electron transport chain was studied by the addition of electron carriers as well as inhibitors and ADRY reagents as defined by Renger [21-24]. Our results may tentatively be considered in a scheme (Fig. 6) of the early electron transport in photosystem II based on investigations by Kimimura and Katoh [13], expanded for the Y_3 by the work

TABLE 2

Oxygen consumption by the green manganese protein from Dunaliella tertiolecta *chloroplasts. 10^5 erg·cm^{-2}·s^{-1} white light. Values given in percent of control. 100% represents consumption of 300 μmoles oxygen·mg Chl.$^{-1}$·h^{-1}.*

Green manganese protein	100
+ 10^{-4} M $K_3[Fe(CN)_6]$	51
+ 10^{-3} M NH_2OH	45
+ $5·10^{-3}$ M NaN_3	42
+ $5·10^{-4}$ M Janusgreen	22
+ $5·10^{-4}$ M Salicylaldoxime	170
+ $5·10^{-4}$ M CCCP	445
+ $5·10^{-4}$ M CCCP + 10^{-3} M $MnCl_2$	111
+ yellow Mn-protein (SOD) 20% on a protein basis	63

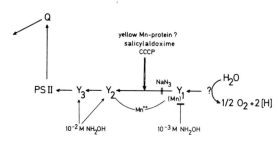

Fig. 6. Proposed electron transport scheme between H_2O and PS II.

of Siegenthaler [26], who observed an inhibitory site for unsaturated C_{18}-fatty acids between Y_2 and PS II. Mn^{++} probably can shunt the electron transport between Y_1 and Y_2 and thus abolish the effect of ADRY reagents. It seemed to us in some experiments that the yellow Mn protein may act as an ADRY reagent, and therefore it is expected to be closely associated with the green protein on the thylakoid membrane. We do not yet know the real nature of Y in the scheme and whether individuals Y_1, Y_2 and Y_3 exist. Nevertheless, we expect the protein we isolated from *Dunaliella* chloroplasts to be at least part of the Y system. Our efforts focus to experiments planting the isolated protein(s) on the surface of artificial lipid vesicles.

The role of SOD in photosynthesis remains to be explained. According to Epel and Neumann [4] the superoxide radical is formed in photosystem I of isolated chloroplasts. The superoxide radical formation in illuminated spinach chloroplasts has been established by EPR spectroscopy after spin trapping by DMPO [9]. No data are available excluding the superoxide radical formation in photosystem II. Therefore we are going to prepare spin trapping experiments with our protein preparations. A detection of $HO_2\cdot$ would support our idea that oxygen is liberated by the cleavage of a bound peroxycarbonate molecule previously formed by photochemical reaction at a manganese site, and by the subsequent reaction of the superoxide radical with SOD. The idea is consistent with the well-established participation of CO_2 [6,12,17,27,28,30,32], the valency change of the Mn indicated by the diminished EPR signal in the light, the role of the SOD in chloroplasts, and it overcomes theoretical difficulties with the energy requirement of a water-splitting reaction.

Acknowledgements

Thanks are due to the Fonds der Chemischen Industrie for a grant to E. von Kameke, and to the Deutsche Forschungsgemeinschaft for financial support.

References

1. Bohnenkamp, W. and Weser, U. (1975). *Hoppe Seyler's Z. physiol. Chem.* **356**, 747-754.
2. Cheniae, G.M. and Martin, I.F. (1969). *Plant Physiol.* **44**, 351-360.
3. Elstner, E.F. and Heupel, A. (1975). *Planta* **123**, 145-154.
4. Epel, B.L. and Neumann, J. (1973). *Biochim. Biophys. Acta* **325**, 520-529.
5. Fridovich, I. (1975). *Ann. Rev. Biochem.* **44**, 147-159.
6. Govindjee, Stemler, A.J. and Babcock, G.T. (1974). *In* "Proceedings of the Third International Congress on Photosynthesis" (M. Avron ed.), Vol. I, pp. 363-371, Elsevier Scientific Publishing Company,

Amsterdam, Oxford, New York.
7. Hager, A. (1975). *Ber. dtsch. Bot. Ges.* **88**, 27-44.
8. Hager, A. and Meyer-Bertenrath, T. (1967). *Planta* **76**, 139-168.
9. Harbour, J.R. and Bolton, J.R. (1975). *Biochem. Biophys. Res. Comm.* **64**, 803-807.
10. Holdsworth, E.S. (1974). Abstracts of the 1st Australian Symposium on Biochemistry and Physiology of Marine Algae, p. 11.
11. Holdsworth, E.S. and Juzu, H.A. (1977). Abstracts of the IV. International Congress on Photosynthesis, Reading.
12. Jursinic, P., Warden, J. and Govindjee (1976). *Biochim. Biophys. Acta* **440**, 322-330.
13. Kimimura, M. and Katoh, S. (1972). *Plant Cell Physiol.* **13**, 287-296.
14. Lagoutte, B. and Duranton, J. (1975). *FEBS Lett.* **51**, 21-24.
15. Lumsden, J. and Hall, D.O. (1974). *Biochem. Biophys. Res. Comm.* **58**, 35-41.
16. Lumsden, J. and Hall, D.O. (1975). *Biochem. Biophys. Res. Comm.* **64**, 595-602.
17. Metzner, H. (1975). *J. Theor. Biol.* **51**, 201-231.
18. Metzner, H., Rauh, H.-J. and Senger, H. (1965). *Planta* **65**, 186-194.
19. Pirson, A. (1937). *Z. Bot.* **31**, 193-267.
20. Radola, B.J. (1973). *Biochim. Biophys. Acta* **295**, 412-428.
21. Renger, G. (1971). *Z. Naturforschg.* 26b, 149-153.
22. Renger, G. (1972). *Biochim. Biophys. Acta* **256**, 428-439.
23. Renger, G. (1973). *Biochim. Biophys. Acta* **314**, 390-402.
24. Renger, G., Bouges-Bocquet, B. and Büchel, K.-H. (1973). *Bioenergetics* **4**, 491-505.
25. Siefermann, D. (1971). Thesis, Universität Tübingen.
26. Siegenthaler, P.A. and Horakova, J. (1974). *In* "Proceedings of the Third International Congress on Photosynthesis" (M. Avron, ed.), Vol. I, pp. 655-664, Elsevier Scientific Publishing Company, Amsterdam, Oxford, New York.
27. Stemler, A. and Govindjee (1973). *Plant Physiol.* **52**, 119-123.
28. Stemler, A. and Govindjee (1974). *Plant Cell Physiol.* **15**, 533-544.
29. Tanner, H.A., Brown, T.E., Eyster, C. and Treharne, R.W. (1960). *Biochem. Biophys. Res. Comm.* **3**, 205-210.
30. Warburg, O. and Krippahl, G. (1956). *Z. Naturforschg.* 11b, 718-727.
31. Warburg, O. and Krippahl, G. (1967). *Biochem. Z.* **346**, 429-433.
32. Wydrzynski, T. and Govindjee (1975). *Biochim. Biophys. Acta* **387**, 403-408.

MOLECULAR OXYGEN, LIGHT AND METALLOPORPHYRINS

J.-H. FUHRHOP

*Gesellschaft für Biotechnologische Forschung mbH and
Institut für Organische Chemie A der Technischen
Universität, Braunschweig (F.R.G.)*

Introduction

Chlorophyll *a* has been proposed as a photocatalyst in the light-driven oxidation of water to oxygen [5]. One pathway would be addition of water to a chlorophyll cation radical (or to chlorophyll in an excited state), removal of hydrogen peroxide from a pair of oxidized chlorophyll molecules, and catalytic decomposition of this peroxide to molecular oxygen.

net reaction: $2H_2O - 2\ominus \longrightarrow H_2O_2 + 2H^+$

Such a reaction, however, has not been observed yet. A possible reason for these failures is the complexity of the chlorophyll *a* molecule. Indeed, the vinyl group at C-3, the methine bridges (in particular C-20), the isocyclic ring carbons C-13' and C-15', and the fifth co-ordination site of the central magnesium ion are reactive sites for hydroxylations from which a hydrogen peroxide could be split off. However, only rapid decomposition of the chlorophyll chromophore is observed with nucleophiles, oxygen and light. This paper summarizes the relevant *in vitro* reactions of metalloporphyrins and discusses some mechanisms by which catalytically active chromophores may be protected.

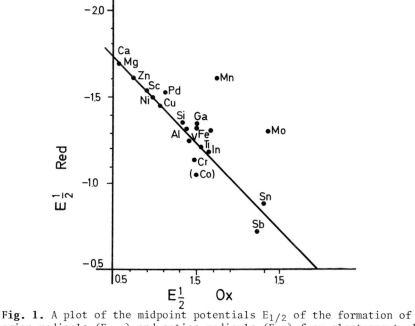

Fig. 1. A plot of the midpoint potentials $E_{1/2}$ of the formation of anion radicals (E_{Red}) and cation radicals (E_{Ox}) from electroneutral metalloporphyrins. The figure demonstrates, that a metalloporphyrin which is easy to oxidize is difficult to reduce and *vice versa*.

π-Radicals and their dimers

The fundamental reaction of the porphyrin chromophore in its metal complexes is the reversible π-radical formation with oxidants [15]. Chemical oxidation of magnesium complexes is particularly favourable [13].

Although metallochlorins produce oxidation potentials which are 300 mV lower than those of the corresponding metalloporphyrins (*e.g.* magnesium octaethylchlorin, 120 mV *vs* SCE) [10] the midpoint potential of chlorophyll is comparably high (550 mV *vs* SCE) [15]. This indicates a strong electron withdrawing effect of the isocyclic ring.

The vinyl side-chain and the isocyclic ring undergo rapid irreversible reactions in the π-radical state which finally lead to a complete destruction of the chlorophyll chromophore. Porphyrins form stable dimers in many solvents ($\Delta H \simeq 40$ kJ·mol^{-1}). This tendency is fortified in the case of their π-radicals. The π-π'dimer of zinc octaethylporphyrin radical, for example, exhibits an enthalpy of 72 kJ·mol^{-1} in methanol/chloroform [17]. An intense charge-transfer band (oscillator strength $\simeq 0.7$) around 900 nm is typical for such dimers. The structure of the dimer, given in Fig. 2, has been evaluated from the EPR spectra of dimers with paramagnetic copper ions in the porphyrin centre. From the D-values of the corresponding triplet state a copper-copper distance of 4.2 Å is obtained, which is only consistent with a parallel approach of the porphyrin planes [30].

The above mentioned two chemical prerequisites of the modified Franck hypothesis are therefore inherent in porphyrin chemistry: radical and dimer formation.

Addition of nucleophiles

π-Cation radicals of metalloporphyrins react with nitrogen-containing nucleophiles such as NO_2^- [3], pyridine [2] and imidazole [4] to produce *meso*-monosubstituted derivatives. Addition of oxygen-containing nucleophiles has

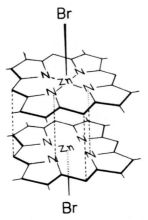

Fig. 2. Typical structure of a metalloporphyrin radical dimer. The distance between planes is 4.2 Å. Such dimer produces strong absorption bands around 900 nm and is quite stable ($\Delta H \simeq 70$ kJ·mol^{-1}).

only been reported with porphyrin di-cations, which are formed at potentials approximately 500 mV higher than those for radical formation [20]. The primary product of these reactions is a substituted phlorin, either with a positive charge or an unpaired electron:

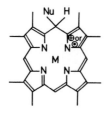

phlorin cation or
phlorin radical

If one removes the hydrogen on C-15' by strong bases and oxidizes the carbanion with oxygen, very short-living radicals are formed which immediately add nucleophiles at C-15'. This "allomerisation" [25,26] proceeds further towards complete destruction of the chlorin macrocycle if the base is not removed. The vinyl groups [6,8,9] and probably the phytyl ester chain and the 13'-carbonyl group [7] in decarboxylated chlorophylls, can be hydrated in the presence of catalytic amounts of protons. Finally water adds as a fifth ligand to the central magnesium ions and forms a hydrogen bridge to a neighbouring chlorophyll molecule. Such magnesium hydrates form π-radicals more easily than water-free pigment molecules. Fig. 3 summarizes the points of attachment of water to the functional groups in chlorophyll a as determined in chlorophyll itself and related porphyrins.

Reversibility of most of these hydrations has been demonstrated under a variety of conditions (heat, reductants, acid etc.). No photochemical reaction of chlorophyll-bound water, however, is known. A possible exception is water bound to magnesium in chlorophyll-dimers. These dimers form radicals on irradiation which are not seen in the water-free monomers [22,27].

Photo-oxygenations

With chlorophyll-containing biological preparations and model systems oxygen is often consumed instead of

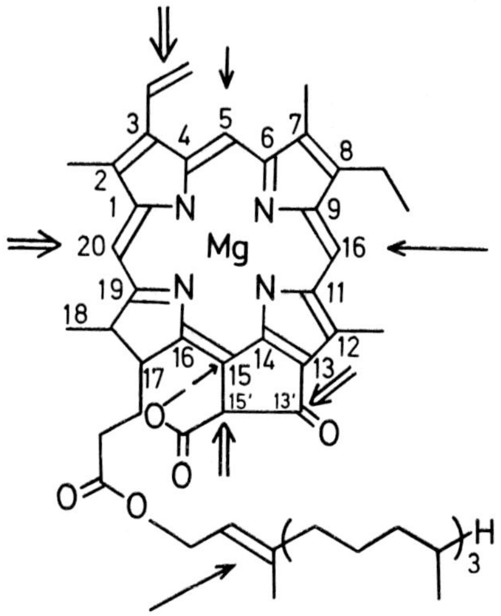

Fig. 3. Structure and reactivity of the chlorophyll α molecule. Double arrows indicate most reactive sites for an electrophilic attack or for a nucleophilic reaction of the π-cation radical or chlorophyll. Single arrows indicate less susceptible sites to such reactions.

being produced [21,31]. Many organic compounds can, of course, be photo-oxygenated in the presence of chlorophyll as a sensitizer, but chlorophyll itself is probably the best candidate. However, because of its structural complexity, a variety of reactions takes place at the same time and well-defined products have never been isolated. Simple porphyrins or their magnesium complexes have been used to elucidate the reactions of the porphyrin macrocycle with molecular oxygen. If one irradiates magnesium octaethylporphyrin with visible light the open chain formylbiliverdinate is formed in quantitative yield [18]. Magnesium and zinc octaethylchlorin give the same

formyl bilatriene

reaction and the macrocycle is regio-selectively cleaved at the methine bridges next to the hydrogenated pyrrole ring [32]. The zinc formylbilatrienes are stable and can be crystallized from aprotic, dry solvents, but water often leads to rapid further degradation. A metal-free acetyl-bilatriene has been obtained from chlorophyll a which bears a methyl group on C-20 [23]. The X-ray analysis of zinc octaethyl-formylbiliverdinate revealed either a square planar or a bis-helical structure depending on the solvent [29]. The most probable site of cleavage in chlorophyll a would be C-20, and the planar conformation should be stabilized by the rigid isocyclic ring E.

Chemical oxidation of magnesium and zinc porphyrinates leads to *meso*-oxyporphyrins. Subsequent photo-oxygenation cleaves off the methine bridge as carbon monoxide and oxoniaporphyrins are formed [12]. These are cleaved to bilatrienes by water [14]. This is also a chemically feasible pathway for the phycocyanobilin formation in algae from chlorophyll precursors.

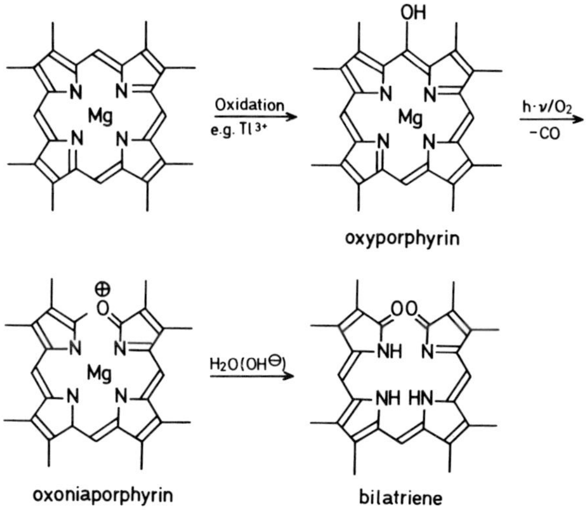

oxyporphyrin

oxoniaporphyrin

bilatriene

The vinyl group of porphyrins can also be photo-oxygenized in good yield. Formylethylidene hydroxy chlorins are the main products.

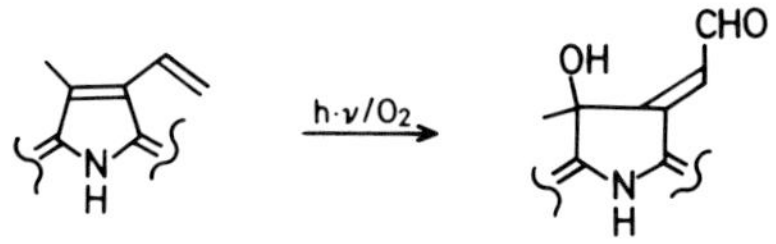

If one considers all the possible irreversible addition reactions to chlorophyll, it is understandable that possible reversible photoreactions such as hydration or carboxylation on C-13' have not been detected so far. Clearly chemical investigations on simple porphyrins are needed to evaluate these possibilities.

Protection of porphyrins against irreversible nucleophilic attack and autoxidation

An often cited and obvious protection factor of chlorophyll in living organisms are the carotenes [24]. Photo-oxygenation of magnesium porphyrinates in benzene solution is also slowed down in the presence of β-carotene [16]. Whether or not the quenching of metalloporphyrin triplet state is the only mechanism of this autoxidation inhibition is not clear. Chemical reactions of carotenes in such

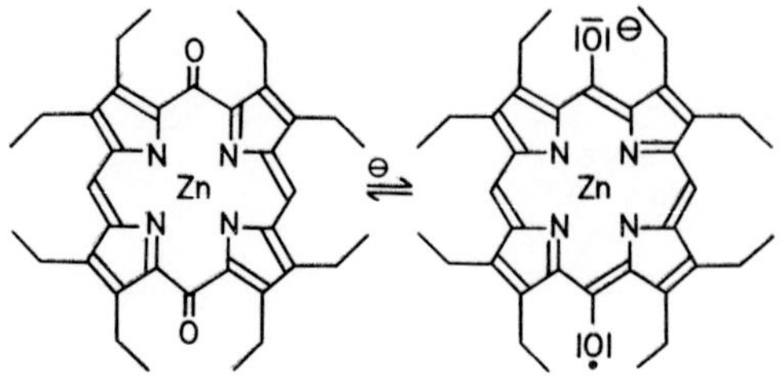

Scheme 6a

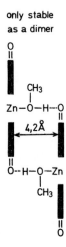

only stable as a dimer

Scheme 6b

systems should also be investigated.

Another stabilizing factor would be the distribution of the radical's single electron over two molecules. An example for such an effect is the methanol-bridged dimer of zinc octaethyl-5,15-dioxoporphodimethene [11]. This dimer produces on reduction a rather stable anion radical, whereas a similar reaction mixture containing the monomer decolorizes rapidly.

The partial neutralization of the negative charge by hydrogen bonding may also be important. Water-bridged chlorophyll dimers are more complicated than this symmetrical example [22], but the chemical effect could be similar. Iron(III)- and cobalt(III)-porphyrins catalyze effectively the peroxidation of olefins and their own destruction by the peroxides formed. Both reactions can be completely inhibited by fixation of a copper ion close to the porphyrin centre. Under favourable conditions the second metal needs not be fixed to the porphyrin, but simple addition of a second metal complex in a 1:1 ratio to the solution is sufficient to block catalysis completely [1]. Paramagnetic metal complexes close to the chlorophyll could provide comparable protection of chlorophyll against photo-oxygenation.

Finally the rigidity of the chemical environment of

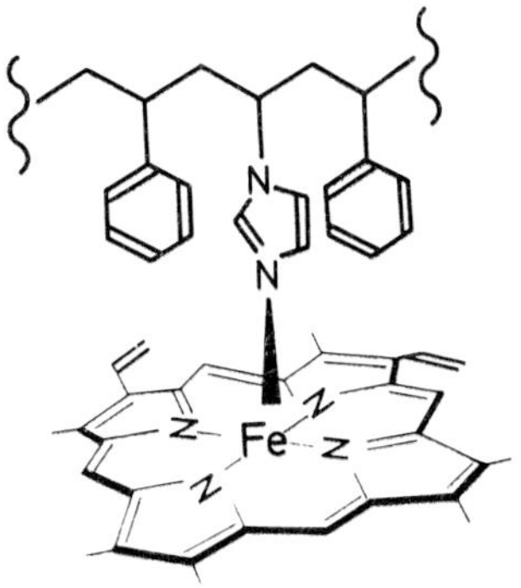

Fig. 4a. A copolymer of 1-vinylimidazole and styrene to which a heme molecule is attached. Such polymer adsorbs oxygen very slowly.

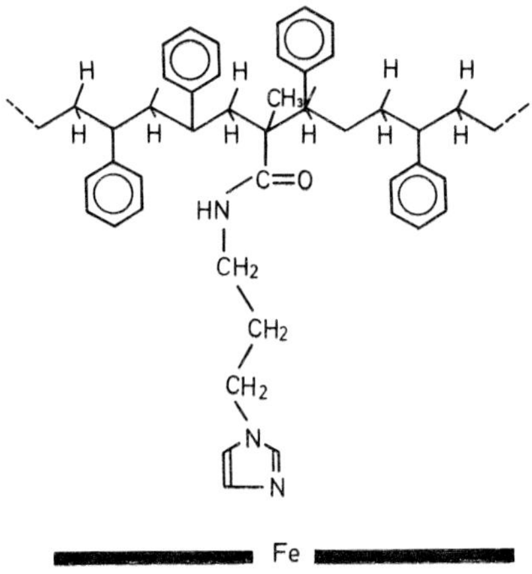

Fig. 4b. The imidazole has been rendered mobile, by attaching it to a flexible chain. This heme-containing polymer adsorbs oxygen rapidly.

porphyrins in solid matrices may play a role. The reversible addition of molecular oxygen to the iron(II) ion of heme in a polymer containing imidazole for example is greatly accelerated, when the axial base is not a part of the polymer backbone, but is attached to a polymethylene chain. The iron ion in the imidazole adduct is supposedly out of plane by 0.8 Å, whereas it moves into the plane when oxygen is added. The rigid polymer ligand makes such movement difficult, whereas the flexible side chain allows it [11].

The attachment of the chlorophyll molecule by the phytol side chain in a well organized matrix could certainly also have the effect that chemical reactions, other than electron transport, are retarded.

Conclusions

Irreversible destruction of chlorophyll *a* by oxygen and light makes *in vitro* experiments of the photo-oxidation of water difficult. Simple, mono-functional metalloporphyrins in an optimized environment are more promising. Obvious candidates are dimers of porphyrins hydrated on the metal or on carbonyl groups, but many other possibilities exist.

References

1. Bacchouche, M., Furhop, J.-H., Schlözer, R. and Arzoumanian, H. (1977). *Chem. Comm.* in press.
2. Barnett, G.H., Evans, B., Smith, K.M., Besecke, S. and Fuhrhop, J.-H. (1976). *Tetr. Lett.* 4009-4012.
3. Barnett, G.H. and Smith, K.M. (1974). *J. Chem. Soc., Chem. Comm.* 772-773.
4. Besecke, S., Evans, B., Barnett, G.H., Smith, K.M. and Fuhrhop, J.-H. (1976). *Angew. Chem.* **88**, 616.
5. Clayton, R.K. (1965). "Molecular Physics in Photosynthesis", Blaisdell, New York.
6. Clezy, P.S. and Barrett, J. (1961). *Biochem. J.* **78**, 798-806.
7. Fischer, H., Moldenhauer, O. and Sus, O. (1931). *Liebigs Ann. Chem.* **485**, 1-25.
8. Fischer, H. and Müller, R. (1925). *Hoppe-Seyler's Z. physiol. Chem.* **142**, 120-140.
9. Fischer, H. and Müller, R. (1925). *Hopper-Seyler's Z. physiol.*

Chem. **142**, 155-174.
10. Fuhrhop, J.-H. (1970). *Z. Naturforschg.* **25b**, 225-265.
11. Fuhrhop, J.-H., unpublished.
12. Fuhrhop, J.-H., Besecke, S., Subramanian, J., Mengersen, Ch. and Riesner, D. (1975). *J. Am. Chem. Soc.* **97**, 7141-7152.
13. Fuhrhop, J.-H., Kadish, K.M. and Davis, D.G. (1973). *J. Am. Chem. Soc.* **95**, 5140-5147.
14. Fuhrhop, J.-H. and Kruger, P. (1977). *Liebigs Ann. Chem.* 360-370.
15. Fuhrhop, J.-H. and Mauzerall, D. (1969). *J. Am. Chem. Soc.* **91**, 4174-4181.
16. Fuhrhop, J.-H. and Mauzerall, D. (1971).
17. Fuhrhop, J.-H. and Wasser, P., Riesner, D. and Mauzerall, D. (1972). *J. Amer. Chem. Soc.* **94**, 7996-8001.
18. Fuhrhop, J.-H., Wasser, P.K.W., Subramanian, J. and Schrader, U. (1974). *Liebigs Ann. Chem.* 1450-1466.
19. Inhoffen, H.H., Brockmann, H. jr. and Bliesener, K.-M. (1969). *Liebigs Ann. Chem.* **730**, 173-185.
20. Johnson, E.C. and Dolphin, D. (1976). *Tetr. Lett.* 2197-2200.
21. Kameke, E.V. and Wegmann, K. (1977), this volume.
22. Katz, J.J., Oettmeier, W. and Norris, J.R. (1976). *Phil. Trans. Roy. Soc. London B* **273**, 227-253.
23. Kenner, G.W., Rimmer, J., Smith, K.M. and Unsworth, J.F. (1976). *Phil. Trans. Roy. Soc. London B.* **273**, 255-276.
24. Krinsky, N.I. (1971). *In* "Carotenoids" (O. Isler, ed.). p. 669. Birkhäuser-Verlag, Basel.
25. Pennington, F.C., Strain, H.H., Svec, W.A. and Katz, J.J. (1964). *J. Am. Chem. Soc.* **86**, 1418-1426.
26. Seely, G.R. (1966). *In* "The Chlorophylls" (L.P. Vernon and G.R. Seely, eds), pp. 67-109, Academic Press, New York.
27. Sherman, G. and Fujimori, E. (1969). *In* "Progress in Photosynthesis Research" (H. Metzner, ed.), Vol. II, pp. 763-770, International Union of Biological Sciences, Tübingen.
28. Smith, K.M., Barnett, G.H., Hudson, M.F. and Combie, S.W. (1972). *J. Chem. Soc., Perkin Trans. I*, 1471-1475.
29. Struckmeier, G., Thewalt, U. and Fuhrhop, J.-H. (1976). *J. Am. Chem. Soc.* **98**, 278-279.
30. Subramanian, J., Mengersen, C. and Fuhrhop, J.-H. (1976). *Mol. Phys.* **32**, 893-897.
31. Tien, T.H. (1977), this volume).
32. Wasser, P.K.W. and Fuhrhop, J.-H. (1973). *Ann. N.Y. Acad. Sci.* **206**, 533-548.

PHOTOREDOX REACTIONS OF MANGANESE

A. HARRIMAN, G. PORTER and I. DUNCAN

The Royal Institution, London UK

Over the last few years or so, it has become accepted that manganese plays a vital role in PS II of the natural photosynthetic process. Little is known about the actual manganese complex - neither the oxidation state, the type of binding atoms, or the environment of the manganese ion are understood as yet. However, it seems highly probable that the principal role played by manganese in the PS II reaction scheme involves electron transfer. In fact, manganese forms a wide range of complexes with oxidation states of +2, +3, and +4. Usually, the stable state is +2 so that, in principle, a complex can undergo a two-electron oxidation to form a +4 complex. Most manganese (IV) complexes adopt oxygen-bridged binuclear structures, hence an overall four-electron oxidation can be achieved. This provides a means of charge accumulation which is particularly important considering that the oxidation of water to oxygen requires a four-electron change. However, despite the importance and interest in the general redox properties of manganese complexes, there have been very few investigations into the photoredox reactions of these compounds. We have studied in detail several different types of manganese complexes in an attempt to develop a model system capable of undergoing a four-electron oxidation.

Manganese (II) forms many simply octahedral complexes with chelating ligands such as 2,2'-bipyridyl. The corresponding manganese (IV) complexes have two oxygen atoms bridging two manganese nuclei:

$$\left[(bipy)_2 Mn^{IV} \underset{O}{\overset{O}{\diamondsuit}} Mn^{IV} (bipy)_2 \right]^{4+}$$

Attempts were made to photooxidise the manganese (II) complexes, firstly to manganese (III) and then to the binuclear manganese (IV) complex. Since octahedral manganese (II) compounds have negligible absorption in the visible region, the experiments were conducted with a dye acting as both light absorber and electron acceptor. The dye used for most experiments was chlorophyll and ground state manganese (II) complexes were found to quench both singlet and triplet excited states of chlorophyll [1]. The quenching rate constants are collected in Table 1, where k_S

TABLE 1

Quenching of photo-excited chlorophyll a by a series of manganese compounds

Quencher	$10^{-8} k_S$ ($M^{-1}s^{-1}$)	$10^{-6} k_T$ ($M^{-1}s^{-1}$)	$E_{\frac{1}{2}}$ (V)
$Mn(H_2O)_6^{II}$	1.5	1.6	-1.51
$Mn(en)_3^{II}$	4.0	4.2	-1.44
$Mn(acac)_2^{II}$	3.7	6.5	-1.40
$Mn(bipy)_3^{II}$	7.2	16.0	-1.32
$Mn(phen)_3^{II}$	9.2	60.0	-1.25
$Mn(acac)_2^{III}$	49.	140.	+0.45
$[Mn(bipy)_2O]_2^{III/IV}$	-	230.	+0.90
$[Mn(phen)_2O]_2^{IV/IV}$	-	400.	+1.10

en: 1,2-diaminoethane; acac: acetylacetone; bipy: 2,2'-bipyridyl; phen: 1,10-phenanthroline;

$[Mn(bipy)_2O]_2^{III/IV} \equiv [bipy_2Mn^{III} \underset{O}{\overset{O}{\diamondsuit}} Mn^{IV} bipy_2]^{3+}$

$[Mn(phen)_2O]_2^{IV/IV} \equiv [phen_2Mn^{IV} \underset{O}{\overset{O}{\diamondsuit}} Mn^{IV} phen_2]^{4+}$

refers to quenching of the singlet state as measured by steady state fluorescence and single photon counting techniques. The corresponding triplet quenching rate constants, k_T, were determined by conventional flash photolysis. The quenching efficiency is seen to depend upon the structure of the actual manganese chelate used as quencher but, in all cases, it is low. In fact, the quenching rate constants for both singlet and triplet reactions show a good correlation with the redox potential of the quencher. This situation is consistent with a charge transfer quenching mechanism. However, using flash spectroscopy, we were not able to detect the transient formation of either reduced chlorophyll or manganese (III) so that reaction does not lead to full electron transfer. Energy transfer is unimportant since manganese (II) does not possess low energy excited states and the quenching effects can be attributed to enhanced nonradiative decay of the excited state induced by the paramagnetic metal ion.

$$Chl_T + Mn^{2+} \rightleftharpoons (Chl_T \ldots Mn^{2+}) \longrightarrow Chl + Mn^{2+}$$

Other dyes [3] gave identical results and in no case could manganese (III) be detected as a transient photoproduct. During these experiments, it was noticed that when the manganese complex was excited directly then full electron transfer could take place. However, this requires irradiation with UV light and does not represent a useful system.

Also shown in Table 1 are quenching rate constants for the corresponding manganese (III) and (IV) complexes. These compounds are strongly oxidising and react rapidly with photoexcited chlorophyll. Here, full electron transfer does take place but, for our purposes, this occurs in the wrong direction. That is, chlorophyll is photooxidised and the manganese complex is reduced.

Obviously, the redox potential of the manganese complex

needs to be lowered whilst still retaining the mononuclear manganese (II)- binuclear manganese (IV) structure relationship. This requirement is fulfilled by manganese gluconate complexes [4]. Again, manganese (II) has little absorption in the visible region and the purple dye thionine was used as light absorber and electron acceptor. Thionine is weakly fluorescent and has a singlet state lifetime of 350 ps. Addition of manganese (II) gluconate results in diffusional controlled quenching of the thionine singlet state but since the lifetime is so short, it is only at very high quencher concentrations that singlet quenching adopts any real importance. On the other hand, triplet thionine is formed in high yield and has a much longer lifetime, 65 µs in outgassed solution. Manganese (II) gluconate reacts quite efficiently with triplet thionine forming an ion-pair.

$$Th_T + Mn^{2+} \longrightarrow Th^- + Mn^{3+}$$

Flash spectroscopy shows the formation of both redox products which can be detected by their characteristic absorption spectra. Under the experimental conditions, manganese (III) is stable but semithionine (Th$^-$) undergoes disproportionation to give the fully reduced leucothionine. The rate of this disproportionation reaction is diffusional controlled and this helps to minimise the importance of back electron transfer. Leucothionine is colourless, it cannot be monitored by flash spectroscopy but by observing any return to ground state thionine at 600 nm, the concentration of the fully reduced dye can be followed indirectly. Addition of oxygen to the system, oxidises leucothionine back to the starting compound and there is no loss of material.

Similarly, manganese (III) reacts with triplet thionine to give manganese (IV) and semithionine. Manganese (IV) can be detected by its characteristic absorption spectrum

$$Th_T + Mn^{3+} \longrightarrow Th^- + Mn^{4+}$$

Although some leucothionine is formed by disproportionation of the semithionine, most of the semithionine decays by simple back electron transfer with manganese (IV). Manganese (IV) is also reduced back to manganese (III) by leucothionine. This result highlights one of the most persistant problems encountered with photoredox reactions in fluid solution, that is, some means of stabilising the reaction products to back electron transfer must be found.

Thus, in this system, manganese (II) can be photooxidised with visible light to give a good yield of manganese (IV) but this is unstable with respect to the reduced electron acceptor and is converted into manganese (III). This back reaction is so efficient that any alternative reaction between manganese (IV) and the aqueous substrate cannot be observed.

A further class of compounds that has been studied consists of manganese phthalocyanine and porphyrin complexes. There are several important differences between these compounds and the simple chelates. From an experimental viewpoint, the most obvious difference is that the porphyrin compounds show intense absorption in the visible region so that they can be irradiated directly. A second point (2) is that in aerated solution the stable oxidation state of a manganese porphyrin is +3 and not +2 as normally found with manganese complexes. Although there are differences between phthalocyanine and porphyrin compounds, their photochemical properties are very similar and the work described here will be restricted to phthalocyanines (MnPc).

The absorption spectra of the various oxidation states of MnPc are given in Figs. 1-3. Fig. 1 shows the spectrum obtained by dissolving MnPc in air-equilibrated solvent. In all solvents, there is an intense absorption band at about 720 nm. However, if the solution is prepared by condensing solid MnPc into the cell and adding thoroughly outgassed solvent, then a different absorption spectrum is

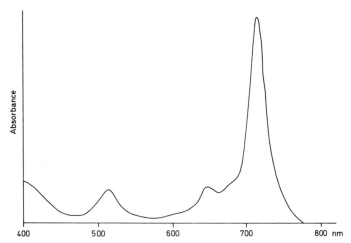

Fig. 1. Absorption spectrum of manganese (III) phthalocyanine in chloronaphthalene.

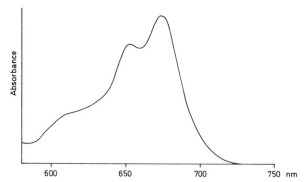

Fig. 2. Absorption spectrum of manganese (II) phthalocyanine in thoroughly outgassed ethanol.

obtained. This is shown in Fig. 2 and consists of a strong absorption band at 660 nm. Addition of air or any other mild oxidising agent results in complete conversion to the 720 nm absorbing species. Similarly, addition of a reducing agent to the air-equilibrated solution gives complete conversion to the 660 nm band. These results can be interpreted to show that the 720 nm band is characteristic of Mn(III)Pc whilst the 660 nm band is due to the reduced Mn(II)Pc. Oxidation of Mn(III)Pc is more difficult than reduction but can be achieved to generate the corresponding

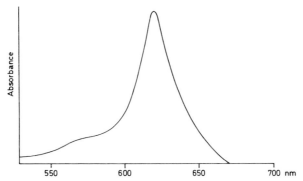

Fig. 3. Absorption spectrum of manganese (IV) phthalocyanine in chloronaphthalene.

Mn(IV)Pc. The absorption spectrum of this species is shown in Fig. 3 and has a strong band at 620 nm. Therefore, the absorption bands of the three oxidation states of MnPc are well resolved. This makes flash spectroscopy a particularly suitable technique to study the photochemistry of these compounds.

Irradiation of Mn(III)Pc in outgassed solution leads to inefficient photoreduction to the corresponding manganese (II) compound. As might be expected, reduction is only important in basic solvents and reaches a quantum yield value of about 0.1 in dilute sodium hydroxide solution. If air is added to the photolysis solution then manganese (III) is regenerated with very little overall loss of chromophore. Since we are concerned with oxidation rather than reduction of the compounds, we choose to start reaction with Mn(II)Pc. Irradiation of the manganese (II) compound in the presence of an electron acceptor (Q) leads to efficient photooxidation to Mn(III)Pc. This reaction sequence can be represented as;

$$\underset{H_2O}{\overset{H_2O}{\boxed{II}}} + Q \xrightarrow{h\nu} \underset{H_2O}{\overset{OH}{\boxed{III}}} + QH^\bullet$$

Flash spectroscopy shows the appearance of Mn(III)Pc and the formation and decay of the semireduced electron acceptor.

Subsequently, irradiation of Mn(III)Pc in the presence of excess electron acceptor results in formation of Mn(IV)Pc.

$$\underset{H_2O}{\overset{OH}{\boxed{III}}} + Q \xrightarrow{h\nu} \underset{H_2O}{\overset{O}{\boxed{IV}}} + QH^\bullet$$

This step is much less efficient than oxidation of Mn(II)Pc but, overall, the process allows photooxidation of manganese (II) to manganese (IV) with corresponding reduction of the electron acceptor. At this stage, it is necessary that the reduced electron acceptor does not reduce either manganese (IV) or (III). If this problem is overcome then the manganese (IV) compound is available for reaction with the substrate.

Now, the stability of manganese (IV) depends upon the nature of the solvent. In toluene, it is fairly stable but is reduced on addition of water or ethanol. In fact in water, the stability of manganese (IV) depends upon pH.

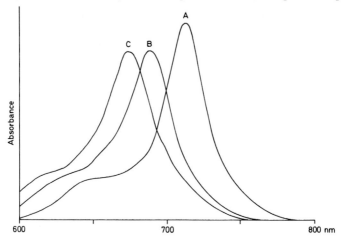

Fig. 4. The effect of pH on the absorption spectrum of manganese (III) phthalocyanine in water; A) pH 6; B) pH 9; and C) pH 11.

At high pH the manganese (IV) complex is stable over several hours but lowering the pH reduces the stability until at pH about 6, we would describe manganese (IV) as being unstable with respect to water. This pH effect is important for all aspects of MnPc chemistry. Fig. 4 shows the pH dependence of Mn(III)Pc. Here, the normal 720 nm absorption band undergoes successive acid/base transitions as the pH is increased. These acid/base forms are associated with deprotonation of axial water molecules co-ordinated to the manganese complex.

$$\begin{array}{c}\text{OH}\\|\\\boxed{\text{III}}\\|\\\text{H}_2\text{O}\end{array} \rightleftharpoons \text{H}^+ + \begin{array}{c}\text{O}\\|\\\boxed{\text{III}}\\|\\\text{H}_2\text{O}\end{array}^- \rightleftharpoons 2\text{H}^+ + \begin{array}{c}\text{O}\\|\\\boxed{\text{III}}\\|\\\text{OH}\end{array}^{2-}$$

Each acid/base form has its own characteristic rate of oxidation. We find that both manganese (II) and (III) can be oxidised readily at high pH whilst oxidation is much less efficient at neutral pH.

Reduction of Mn(IV)Pc results in formation of the corresponding manganese (II) compound. During this reduction step, hydroxyl radicals are also formed. Normally, hydroxyl radicals rapidly attack any available aromatic ring, forming an adduct which leads to destruction of the chromophore. However, this product formation process can be minimised by heavily substituting the aromatic rings with sulphonic acid groups. Under such conditions, hydroxyl radicals probably combine to form hydrogen peroxide. Control experiments have shown that manganese porphyrins catalytically react with hydrogen peroxide to evolve molecular oxygen.

Thus, hydrogen peroxide is a good oxidant for both the lower valent oxidation states of MnPc and, in aqueous solution, results in efficient oxidation of manganese (II) to the corresponding manganese (III) complex.

$$\underset{H_2O}{\overset{H_2O}{\boxed{II}}} + H_2O_2 \longrightarrow \underset{H_2O}{\overset{OH}{\boxed{III}}} + H_2O + OH^{\bullet}$$

Similarly, addition of hydrogen peroxide to manganese (III) results in further oxidation to manganese (IV), again with the subsequent formation of hydroxyl radicals.

$$\underset{H_2O}{\overset{OH}{\boxed{III}}} + H_2O_2 \longrightarrow \underset{H_2O_2}{\overset{O}{\boxed{IV}}} + H_2O + OH^{\bullet}$$

In turn, manganese (IV) reacts with hydrogen peroxide but this reaction leads to reduction of manganese.

$$\underset{H_2O}{\overset{O}{\boxed{IV}}} + H_2O_2 \longrightarrow \underset{H_2O}{\overset{H_2O}{\boxed{II}}} + O_2$$

Hence, hydrogen peroxide can function as both oxidant and reductant in its reactions with manganese porphyrins. This leads to a catalytic evolution of oxygen. These reactions occur in the dark and do not need irradiation.

Thus, there are two possible sources of oxygen during the reduction of manganese (IV). That is, oxygen can be formed by secondary reactions of hydroxyl radicals or else it may be formed as an integral part of the reduction process. The relevance of these alternative processes can be shown by consideration of possible structures for the manganese (IV) complex. In neutral solution, manganese (III) can be represented as the neutral complex so that its subsequent oxidation may be drawn;

$$\underset{H_2O}{\overset{OH}{\boxed{III}}} \overset{Q}{\longrightarrow} \underset{H_2O}{\overset{OH}{\boxed{IV}}}{}^{+}$$

Now, reduction of this monomeric manganese (IV) species

will give hydroxyl radicals and manganese (II). Here, oxygen could only be formed as a result of secondary reactions of the radicals.

However, in alkaline solution the manganese (III) complex will be present in a basic form so that oxidation will will become;

$$\begin{array}{c} O \\ | \\ \boxed{III}^{-} \\ | \\ H_2O \end{array} \xrightarrow{Q} \begin{array}{c} O \\ | \\ \boxed{IV} \\ | \\ H_2O \end{array}$$

Under alkaline conditions, we might expect that oxidation results in the formation of a peroxo-bridged binuclear manganese (IV) complex. For example, this could have the structure shown below.

$$\left[H_2O - \boxed{IV} - O - O - \boxed{IV} - H_2O \right]^{2+}$$

Now, the acid catalysed reduction of this binuclear complex would lead to formation of manganese (II), hydroxyl radicals, and molecular oxygen. Our present research is aimed at distinguishing between these possible alternative reaction mechanisms.

References

1. Brown, R.G., Harriman, A. and Porter, G. (1977). *J. Chem. Soc. Faraday II* **73**, 113-119.
2. Engelsman, G., Yamamoto, A., Makrham, E. and Calvin, M. (1962). *J. Phys. Chem.* **66**, 2517-2531.
3. Ferreira, M.I.C. and Harriman, A. (1977). *J. Chem. Soc. Faraday I* **73**, 1085-1092.
4. Sawyer, D.T. and Bodini, M.E. (1975). *J. Am. Chem. Soc.* **97**, 6588-6590.

PHOTOSENSITIZATION BY TITANIUM DIOXIDE AND ZINC OXIDE: OXYGEN AND HYDROGEN EVOLUTION

A.A. KRASNOVSKY and G.P. BRIN

*A.A. Bakh Institute of Biochemistry,
USSR Academy of Sciences, Moscow (USSR)*

It is generally assumed, that the primary reaction in photosynthesis is a charge separation. An electron passes through a series of intermediates to $NADP^+$, whereas a hole finally reaches water molecules, and oxygen is released. We wished to have a simple system to study: how under the action of light electrons and holes can be generated. Such a system we can find out in the case of inorganic photocatalysts-semiconductors like titanium dioxide, zinc oxide, etc. In this case under the action of light electrons come to the conduction band leaving behind holes in the valence band. It encouraged us to use these systems for our model experiments.

About 50 years ago Emil Baur in Switzerland discovered that under the action of light zinc oxide and some other inorganic compounds gained the property to cause oxidoreductions. Baur came to photochemistry from electrochemistry. He described the phenomenon of charge separation as "molecular electrolysis" [1]. Considering an excited sensitizer there were two sites - anodic and cathodic- and then oxidoreduction proceeded at the phase boundary. According to contemporary knowledge this may be regarded as a primitive point of view, but his basic idea looks quite plausible. We tried to construct models, which in their action resembled oxidoreduction taking place in chloroplasts by using titanium dioxide, zinc oxide and tungsten trioxide (see review [9]). Absorption bands of titanium and zinc oxides are situated near the border between visible light and UV. Titanium and zinc oxides are white powders, a sharp absorption begins at ∼400 nm, so they are "black" in ultraviolet. In most cases we used excitation at 365 nm, but even if you choose other UV frequencies, you will see the effects described. I selected these experiments here to compare them with Metzner's silver chloride models [13], which should be mentioned in this context.

In our experiments we used fine powders of photocatalysts, having specific surfaces up to ∼20 m^2g^1. We suspended them in water and studied the system in our first experiments inside quartz Warburg vessels. If we illuminate these TiO_2, ZnO or WO_3 suspensions in the presence of electron acceptors, we observe an evolution of gas. In our first experiments we observed sensitised photoevolution of oxygen in solutions of some oxidants including potassium permanganate [6].

Then we tried to simulate Hill reaction [7]. So we added ferricyanide or ferric ions. Then electrons on the phase boundary come to these acceptors, and so we measure oxygen evolution coupled to the reduction of an electron donor. The stoichiometry of the reaction strictly corresponds to the equation

$$2\ Fe^{3+} + H_2O \xrightarrow{h\nu} 2\ Fe^{2+} + 2H^+ + 1/2\ O_2$$

Then we performed experiments of this kind with water labelled with heavy oxygen. It was demonstrated, that in this case the oxygen isotope ratio of the released O_2 was the same as in the suspension water used [3]. So it was confirmed, that oxygen has really originated from water.

It is possible to reduce not only ferric compounds and ferricyanides but also quinones, like *p*-benzoquinone [10].

The quantum yield of the reaction studied did not exceed 1%. So these models show the reduction of an electron acceptor and the simultaneous evolution of oxygen molecules, which we primarily described in 1961 [6].

Summarizing these series of experiments we can say, that under the excitation of photocatalysts charge separation proceeds; if electrons can be taken into the hole from OH$^-$ anions or H_2O molecules, radicals may be produced, which can recombine on the boundary to give molecular oxygen. The formation of OH· radicals under the described conditions had been really found earlier [5]. The electrons are accepted at the phase boundary by oxidants (electron accepting molecules).

A second type of reaction to simulate was the Mehler reaction of chloroplasts. In our model system oxygen may act as terminal electron acceptor, water molecules (or OH$^-$ ions) as donor. The first observation of hydrogen peroxide formation on ZnO was made by Baur and Neuweiler [2]. We used a sensitive measurement of the chemiluminescence of luminol in the presence of peroxidase, so we could quantitatively determine the hydrogen peroxide produced, not only in the case of ZnO, but also in aqueous suspension of TiO_2. If we added methylviologen to this system, the reaction was greatly enhanced [11]. Here the methylviologen probably functioned as intermediate electron carrier. So we could simulate a Mehler reaction in the photocatalysts suspension in the presence of oxygen.

There is a third type of reaction. If we excluded oxygen from the suspension of photocatalysts to work under anaerobic conditions, we observed the reduction of methylviologen [8], which in this case acts as a terminal electron acceptor. The only possible electron donor in this system was H_2O or OH$^-$ anions, but we could not observe oxygen evolution in this case. Maybe the OH· radicals formed react with viologen, which may function as a scavenger

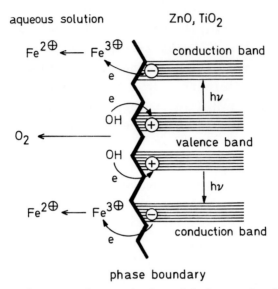

Fig. 1. Scheme of oxygen photoevolution with inorganic photocatalysts.

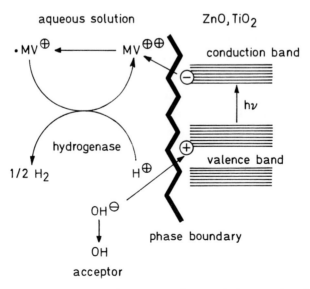

Fig. 2. Scheme of hydrogen photoevolution with inorganic photocatalysts.

of these radicals. This, however, is not clear yet.

It is well known that the redox potential of methylviologen slightly exceeds that of the hydrogen electrode. So by introduction of suitable catalysts we may expect to observe hydrogen gas evolution. Using gas chromatography we observed a hydrogen release indeed, if we supplied our system with a hydrogenase from photosynthetic bacteria [12]. So we found a possibility to simulate this chloroplast reaction, too.

One side of our system works like an oxygen electrode, releasing molecular oxygen; the other side may be compared to a hydrogen electrode.

In this connection I would like to mention the very interesting studies of Fujishima and Honda [4]. They are an electrochemical version of our experiments. In their essence they demonstrate a charge separation on TiO_2 "membranes". Making these layers part of electrochemical cells it could be demonstrated, that these cells produce both hydrogen and oxygen. It is remarkable, that Honda obtained a great enhancement of oxygen evolution, if he added ferric ions to his system. This makes his set-up a remarkable modification of the experiments, which we described 16 years ago.

It is important to emphasize, that the semiconductors studied function not only as photosensitizers but also as catalysts. On the interfaces there may be a recombination of the primary produced intermediates, which probably makes these experiments successful. I suppose, that this simple model may be of interest for those, who are working on the important problem of conversion of light energy into chemical energy.

References

1. Baur, E. (1928). *Z. phys. Chem.* **131**, 143-148.
2. Baur, E. and Neuweiler, C. (1927). *Helv. Chim. Acta* **10**, 901-906.
3. Fomin, G.V., Brin, G.P., Genkin, M.V., Liubimova, A.K., Blumenfeld, L.A. and Krasnovsky, A.A. (1973). *Dokl. Akad. Nauk*

SSSR **212**, 424-427.
4. Fujishima, A. and Honda, K. (1972). *Nature* **238**, 37-38.
5. Korsunovsky, G.A. (1960). *J. Fiz. Chem.* **34**, 510-517.
6. Krasnovsky, A.A. and Brin, G.P. (1961). *Dokl. Akad. Nauk SSSR* **139**, 142-145.
7. Krasnovsky, A.A. and Brin, G.P. (1962). *Dokl. Akad. Nauk SSSR* **147**, 656-659.
8. Krasnovsky, A.A. and Brin, G.P. (1965). *Dokl. Akad. Nauk SSSR* **163**, 761-764.
9. Krasnovsky, A.A. and Brin, G.P. (1970). *In* "Molecular Photonics" ("Nauka" Publ. House, ed.), 161-178, Leningrad.
10. Krasnovsky, A.A., Brin, G.P. and Aliev, Z.Sh. (1971). *Dokl. Akad. Nauk SSSR* **199**, 952-955.
11. Krasnovsky, A.A., Brin, G.P. and Nikandrov, V.V. (1976). *Dokl. Akad. Nauk SSSR* **229**, 990-993.
12. Krasnovsky, A.A., Nikandrov, V.V., Brin, G.P., Gogotov, I.N. and Ochshepkov, V.P. (1975). *Dokl. Akad. Nauk SSSR* **225**, 711-713.
13. Metzner, H. (1968). *Z. physiol. Chem.* **349**, 1586-1588.

BILAYER LIPID MEMBRANES IN AQUEOUS MEDIA: INCORPORATION OF
PHOTOSYNTHETIC MATERIAL FROM BROKEN THYLAKOIDS

H. TI TIEN

*Department of Biophysics, Michigan State University,
East Lansing, Michigan (U.S.A.)*

Introduction

The thylakoid membrane of chloroplasts as seen from electron micrographs is interpreted to consist of an ultrathin (<100 Å) highly organized assembly of lipids, pigments, proteins, etc. [5,9]. Structurally, among the various models proposed for the thylakoid membrane, consistent with the current biomembrane thinking, is the one based on the lipid bilayer organization [15,25]. Functionally, the three most vital processes - the transduction of light, the evolution of oxygen, and the phosphorylation of ADP - are believed to be carried out by the membranes of thylakoids [2,18,19,31]. Concerning the transduction of light (the primary event), a number of hypotheses have been suggested in which the mechanism of light conversion by the photosynthetic apparatus in chloroplasts might be similar to that taking place in semiconductors [1]. If the thylakoid membrane, in which the photosynthetic apparatus is believed to be located, functioned as a semiconductor photovoltaic device (e.g., the silicon solar cell), the most direct kind of experiment would be to place electrodes transversely across it and monitor its voltage when it is excited by light. This hypothetical experiment is shown in Fig. 1. However, the minuteness and complexity of the thylakoids preclude such a straightforward experiment from being carried out with available methods, unless many technical difficulties can be overcome [6]. Thus it seems unlikely at present that detailed physical chemical processes of the thylakoid membrane will be elucidated without using model membrane systems. This paper describes the results of our experiments using artificial bilayer lipid membranes for the understanding of certain aspects of quantum conversion and oxygen evolution.

In our experiments two types of photosynthetic material-

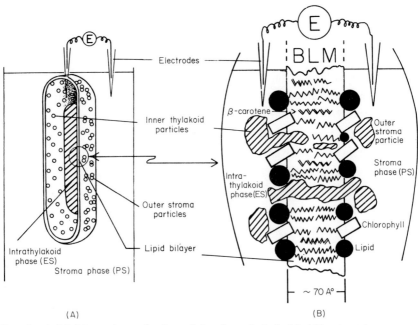

Fig. 1. A highly schematical model of a thylakoid illustrating a portion of the lipid bilayer of lamellar membrane structure (patterned after Mühlethaler [15] with the notations suggested by Branton *et al.* [5]). (A) Showing a hypothetical situation for measurements of membrane electrical properties with one microelectrode inserted transversely through the thylakoid membrane and the other microelectrode in the stroma phase. (B) An enlarged view of the hypothetical experiment indicated in (A). Also shown in (B) is a model of thylakoid membrane depicting an arrangement of amphipathic lipids with their hydrophobic chains inside and hydrophilic portions in contact with aqueous phases forming the lipid bilayer. This lipid bilayer is visualized as a two-dimensional liquid crystal matrix, onto which pigments, proteins and the like are interpolated (see Fig. 2C).

containing bilayer lipid membranes have been used: (i) bilayer lipid membranes of planar configuration, commonly known as BLM, and (ii) bilayer lipid membranes of spherical configuration, generally referred to as liposomes. BLM are formed by spreading a lipid solution on a hydrophobic aperture separating two aqueous solutions (see Fig. 2A). Bilayer lipid membranes of this type are best suited for measurements of transmembrane electrical properties induced by light and/or chemical asymmetry. Liposomes are formed by mechanical agitation of a dried lipid layer in aqueous solution; the lipid molecules acquire the bilayer structure and are formed in a series of concentric spheres, which can be disrupted by sonication into single-shelled bilayer vesicles separating two aqueous phases (Fig. 2B). Liposomes are best suited for studies of permeability, conventional spectroscopy, chemical reactions, and oxygen evolution. The growing interest in these model membranes is principally due to the fact that they appear to exhibit a close resemblance to the biological membranes of cells and their organelles. A number of comprehensive reviews giving up-to-date accounts of these experimental lipid bilayers, as models of biological membranes, are available [13,14,17,20,28]. For more complete technical details, two pertinent publications may be consulted [3,28].

Planar bilayer lipid membranes (BLM)

The preparation of planar BLM has been described in detail elsewhere [28]. Briefly, the apparatus consists essentially of a Teflon beaker (10 ml) with a tiny aperture (<2 mm) in the side, which sits in a glass container (~20 ml). The inner beaker and outer container are filled with a salt solution above the aperture. A small quantity (0.2 µl) of a lipid solution is applied over the aperture to form a thin lipid membrane which thins spontaneously to a "black" or bilayer lipid membrane (BLM) separating two aqueous solutions (Fig. 2A). By placing non-polarizable

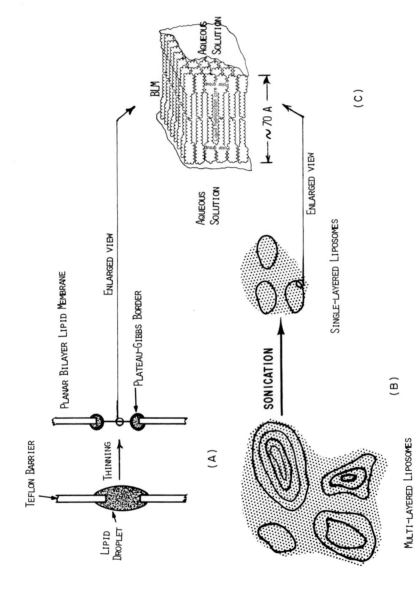

Fig. 2. (For details see opposite page.)

electrodes across the BLM, transmembrane electrical properties of the system can be measured. Photosynthetic material of interest may be incorporated into the BLM either by dissolving it in the lipid solution or by adding it to the aqueous solution after the BLM has formed. In the latter case, one speaks of sensitization. Obviously a combination of these two procedures is also feasible and has been used. The results of some of the relevant experiments are summarized in the following paragraphs.

Typical photovoltage responses for a pigmented BLM and a pigmented sensitized BLM are shown in Fig. 3. The pigmented BLM were formed from spinach chloroplast extracts in 0.1M sodium acetate at pH 5. The BLM of this type have been referred to as Chl-BLM; their physical properties in the dark are as follows: the permeability to water is 50 μM/s, electrical resistance $10^6 \Omega \cdot cm^2$. capacitance 0.5 μF/cm^2, and the dielectric breakdown strength is about 200,000 volts per centimetre using an estimated membrane thickness of 70 Å. It should be mentioned that Chl-BLM were sensitive to H$^+$ ions. In the early studies the magnitude of the observed photoresponse was very small, in the order of a few mV. It was soon discovered that the photoresponse of Chl-BLM could be greatly enhanced by adding redox compounds to the bathing solutions. Fig. 3(A) shows the photoelectric response of a Chl-BLM with 10^{-3}M FeCl$_3$ present in one side of the bathing solution. The Fe^{3+}-free side was positive. The light-induced photovoltage was more than doubled on addition of an equal molar cysteine/cystine to the opposite side plus 10^{-3}M FeCl$_2$ added to the Fe^{3+}

Fig. 2. Schematic illustration of experimental processes for forming bilayer lipid membranes. (A) Two stages of thinning of a planar lipid membrane: A lipid droplet is placed on aperture and thins spontaneously to a bilayer lipid membrane (BLM) supported by Plateau-Gibbs border [28]. (B) Formation of multi- and uni-layer bilayer lipid membranes (liposomes) by mechanical agitation and sonication [3]. (C) A three-dimensional view of a bilayer lipid membrane illustrating the fluid state of the lipid bilayer. The lipid chosen is that of phosphatidyl choline. (Reproduced with permission from J.Theoret.Biol., 16, 107, 1967).

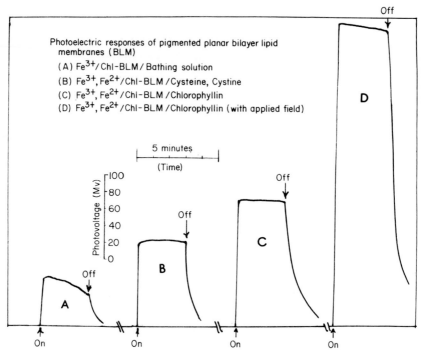

Fig. 3. Photo-induced redox reactions in pigmented bilayer lipid membranes (BLM). (A) The Chl-BLM was formed from chloroplast extract. The bathing solution was 0.1 M sodium acetate at pH 5. The outer bathing solution also contained 0.001 M $FeCl_3$. (B) The same as (A) except the outer solution contained 10^{-3} M $FeCl_3$ and $FeCl_2$ and equal molar (10^{-3}) of cystine and cysteine were added to the inner solution. (C) The Chl-BLM was sensitized by adding water-soluble chlorophyllin to the inner solution. In this case the bathing solution was 0.01 M KCl with equal molar (10^{-3}) $FeCl_3/FeCl_2$ added to the outer solution. (D) Same membrane as in (C) except the membrane was under an external potential of 180 mV (applied through a $10^9 \Omega$ shunt resistor with the chlorophyllin side negative [21]).

side (Fig. 3B). Note that the waveforms are different. The photovoltage attained a steady value when different redox couples were present in opposite sides of the Chl-BLM. A variety of redox compounds affecting the photoresponse of Chl-BLM have been investigated [21,28].

Non-pigmented BLMs, which are normally insensitive to light, can be sensitized by dyes, water soluble or dispersible pigments (e.g., bacteriorhodopsin), and pigmented

organelle extracts (e.g., broken thylakoids, see below).
We have observed photoelectric effects in phospholipid/
oxidized cholesterol BLM sensitized by a number of dyes
such as methylene blue, methyl red, thionine, rhodamine B,
methylviologen, and water-soluble chlorophyll (chlorophyllin). The photoresponse of pigmented BLM can also be sensitized by adding, for example chlorophyllin to one side of
the bathing solution [11]. The result of a typical experiment is shown in Fig. 3(C), where it is seen that light
elicited a voltage of 150 mV with the chlorophyllin side
positive. It should be noted that a dark voltage of 180-
200 mV was generated upon addition of chlorophyllin to the
bathing solution (unbuffered 0.01 M KCl). This dark voltage
was owing to the pH gradient (0.01 M KCl, pH 5.8; chlorophyllin, pH 8.8) across the membrane. The photoresponse of
this system could be further enhanced through application
of external electrical potential. For example, when dark
membrane voltage of 200 mV was imposed across the BLM
through a $10^9 \Omega$ resistor with the chlorophyll side negative,
the photovoltage observed was nearly 360 mV (Fig. 3D). The
response was completely reversible and could be repeated
numerous times without any deleterious effects. This system, a sensitized pigmented BLM, has certain merits in that
two kinds of photosynthetic pigments are involved with the
lipid bilayer: one is hydrophobic chlorophylls submerged
in the lipid bilayer of the membrane and the other (chlorophyllin) is hydrophilic and located at the membrane/solu-
tion interface. A naive suggestion to put forth is that the
reconstituted membrane has two photosystems connected in
series effectively transferring electrons from one side to
the other side of the membrane. The driving force is, of
course, provided by the quanta of light. A more detailed
explanation will be given in the discussion section. It is
worth mentioning that the photoresponse of the system was
greatly diminished or totally abolished by compounds such
as DCMU and 2,4-dinitrophenol.

Since extremely rapid electrical transients are seen to originate across Chl-BLM in response to light excitation under a variety of asymmetrical conditions discussed above, it is necessary to study their kinetics in a special manner. The flash excitation technique offers an approach that gives the needed insight into the individual processes that under continuous illumination constitute usually a single

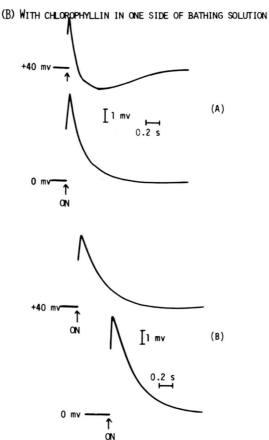

Fig. 4. Photoelectric responses of pigmented BLM excited by 3 μs flashes. The traces illustrate the photovoltages which resulted from (A) a 3 mM Ce^{4+} present in one side of the bathing solution at 0 mV and 40 mV (with the Ce^{4+} side negative). (B) 3 mM $(NH_4)_2 Ce(NO_3)_6$ and chlorophyllin were present in opposite solutions [11].

steady-state response. Experimentally, a different set-up was used [11]. The non-polarizable electrodes across the BLM were connected to a high input impedance buffer amplifier mounted as close as possible to the electrodes to minimize input cable length. This was necessary because the high impedance BLM circuit was quite sensitive to spurious external signals. This amplifier also provided satisfactorily low output impedance that guaranteed minimal sensitivity to noise and rapid signal rise time. The amplifier output was displayed and recorded on an oscilloscope screen. Excitation of the membrane was provided by a xenon flash tube that produced a peak value of 10^9 lux at the membrane with 1/3 peak times separated three microseconds. For resolving microsecond transients, an optional transistor triggering device synchronized stroboscopic flash and oscilloscope sweep. We shall again use the Chl-BLM and the sensitized Chl-BLM for illustration, although a different redox compound was used. Fig. 4(A) shows the photovoltage transient which resulted from the flash excitation in the presence of a 3 mM $(NH_4)_2Ce(NO_3)_6$ in one side at 0 mV and 40 mV applied through a $10^8 \Omega$ shunt resistor. As expected, the electron acceptor side of the Chl-BLM became negative. The initial rapid rise, which has been called component A occurred during the irradiation period. A much slower component (Component B) was seen to occur reaching a peak value about 10-20 ms after illumination. From this observation, it has been concluded that some sort of exciton mechanism must be involved [26]. A still slower trace shows the transmembrane voltage decaying to the pre-illuminated voltage. This has been known as Component D and was the discharge of the membrane capacitance through the membrane and shunt resistors. If an external potential was applied, a new component (Component C) was observed. This component could be also generated by a pH gradient, and since the Chl-BLM was sensitive to H^+, it responded to the net transmembrane electrochemical gradient for H$^+$ ions. Thus it seems probable that Component C results from a light-activated transport of hydrogen ions. Similar to the results obtained under continuous illumination, the largest photovoltage transients of Chl-BLM were obtained when chlorophyllin was added to one side of the bathing solution. The photoresponse often exceeded 200 mV with one 3 μs flash. Moreover, two other effects were noted. First, under applied voltage, Component C did not appear under these conditions, which is shown in Fig. 4(B). Second, the Chl-BLM became very fragile when near trace quantities of chlorophyllin were present on both sides. This has been interpreted to mean that the chlorophyllin interacts with the adjacent monolayer of the BLM in such a manner as to disrupt its structure and cohesion. The fact that BLM may survive several hours with high concentration of chlorophyllin in one side only (with no

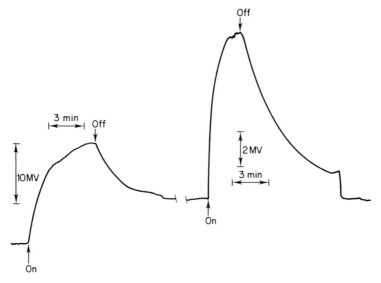

Fig. 5. Photoresponses of a planar BLM formed from a mixture of phosphatidyl choline (1%), cholesterol (0.8%) and dioctadecyl phosphite (0.08%) in n-dodecane. Bathing solution was 0.1 M KCl plus 0.1 M potassium acetate at pH 5. The inner bathing solution also contained 0.5 ml broken thylakoid preparation and 5 x 10^{-3} KI saturated with I$_2$; ascorbic acid (10^{-2}M) was added to the outer bathing solution. The BLM was illuminated by 500 W tungsten projector lamp filtered through 5 cm CuSO$_4$ solution (5%).

electron acceptor present in the other side) suggests that the chlorophyllin may not penetrate the lipid bilayer at an appreciable rate. This supposition will be used in our later discussion on the mechanisms of quantum conversion by sensitized pigmented BLM.

In a new series of experiments, attempts were made to incorporate oxygen evolving complexes into planar BLM since it has been observed that, upon incorporation of this material, isolated from broken thylakoids, into phosphatidyl choline liposomes, the stability and oxygen evolving ability are greatly increased [24]. The results of oxygen-evolving liposomes will be presented in the next section. The association of broken thylakoids containing "intact" oxygen evolving complexes (OEC) with planar non-pigmented BLM is described here. Basically, measurements of light-induced transmembrane electric potential differences were carried out as described earlier. The result of a preliminary experiment is shown in Fig. 5. Compared with the results of the chlorophyllin experiments, the speed of photoresponses appeared sluggish and the magnitude mediocre. Several explanations can be offered; one of the most plausible is that we have not yet mastered the incorporation technique. Nevertheless, the initial trial seems promising and will be prosecuted with vigour in future experiments.

Spherical bilayer lipid membranes (liposomes)

The second type of artificial bilayer lipid membranes, known as liposomes, used in our experiments are illustrated in Fig. 2(B). Techniques of liposome formation and their studies have been authoritatively reviewed by Bangham et al. [3]. Since the liposomes are very stable and can be easily made in quantity, the tremendous interfacial areas afforded by liposomes are therefore ideally suited for binding studies and electrochemical measurements. Much earlier, we attempted to test the idea of water splitting

by light by incorporating photosynthetic pigments into
liposomes, very much as has been done for the planar BLM.
Preliminary experiments showed O_2 evolution from Chl-
containing liposomes, as detected by a Clark-type electrode (Sahu and Tien, unpublished results, 1971-1972; see
ref. [27]). However, later experiments with better temperature control indicated otherwise. The apparent oxygen
evolution as sensed by the electrode was owing to heating
effects on the electrode itself. Our failure notwithstand-
ing, Toyoshima et al. [30], using a similar system,
recently reported light-assisted oxidation of water in
chlorophyll-containing liposomes. The rate of oxygen production, as calculated from the initial slope, was 4.2 x
10^{-4} per mole at 10^5 lux. Under similar conditions to
those reported by Toyoshima et al. but employing rigid
temperature controls (within ± 0.2C), we observed oxygen
consumption instead, as the pigments in the liposomes became irreversibly photooxidized [24]. On the basis of
these findings, evidently a different approach is needed.
What follows is a description of the background and the
results of our current approach to the problem of oxygen
evolution using liposomes.

Initially, we were also interested in finding out the
reason(s) for the differences in chlorophyll absorption
spectra and photochemical activities *in vivo* and in organic solvents. For example, the absorption maximum of Chl a
in acetone is at 663 nm compared with 678 nm *in vivo*. This
difference in absorption maxima has been attributed to
solvent effects, state of aggregation of the tetrapyrrole
moieties or specific interactions with lipids and proteins,
as pointed out by Boardman [4]. It was reasoned that the
liposomes made from plant lipids and chlorophyll could be
used in absorption study to give an absorption spectrum
closely similar to that *in vivo*, if the postulated structure of the thylakoid membrane were correct [15]. The
reasoning was that the lipid bilayer of the liposomes

would provide a favourable site so that the phytol chain of the chlorophyll would anchor while the hydrophilic porphyrin ring would orient at the membrane/solution interface [25,28,32]. With these thoughts in mind, the following experiments were performed. The Chl-containing liposomes were prepared by sonication of a lipid-chlorophyll mixture (obtained commercially) for 45 min in ice cold 0.1 M phosphate buffer at pH 7.5. The sonicated mixture was passed through a Sephadex G-50 column previously equilibrated with the same buffer. The liposomes collected were those contained in the first two ml of the fastest green band to pass down the column. The absorption spectrum was taken of the Chl-liposomes and the peak was at 673 nm (Beall and Tien, unpublished study, 1969-1970). The result strongly suggested that the Chl in liposomes was in an environment more similar to that *in vivo* than when it was simply dissolved in organic solvents.

The Chl-containing liposomes were also applied to a possible O_2 evolution by a simple Hill reaction using one of the known electron acceptors. The oxygen evolution experiment was not tried, however, with planar BLM for the simple reason of insufficient available area (a simple calculation indicates that, when 0.1 g of lecithin is being formed into spherical bilayer lipid membranes, the result is a membrane area of some 60 m^2 as compared with 1 mm^2 area of a typical BLM). As already mentioned, Chl-liposomes under the condition tested did not evolve oxygen but rather consumed oxygen at a rapid rate [23]. The results of some carefully executed experiments are shown in Fig. 6. Chl-containing liposomes used in these studies were prepared by drying a chloroform solution of 300 mg phosphatidyl choline and the 10 mg of Chl extract (by the method given in ref. [28] in the dark under N_2 in a 50 ml round-bottom flask. Thirty ml of the test solution (0.1 M KCl, 0.01 M of either TRIS or potassium acetate at pH 7.5 or 5.0, respectively, to which redox agents were added) were vigorously mixed in the dark for 15 min. The pigmented liposomes thus formed were sonicated for 30 s. 10 ml samples were removed and added to 140 ml of the test solution, 60 ml of which was used in the irradiation experiments. The reaction box consisted of a YSI Model 5302 Macro Bath (Yellow Springs Instrument Co., Ohio) and a bank of 5 lamps with an intensity at the reaction chamber of 1.7×10^5 erg.cm^{-2}s^{-1}. The temperature of the reaction chamber was controlled at 25°C±0.2 by a constant temperature circulating bath. The pH was determined to within an accuracy of 0.002 pH units by a Beckman pH meter (Expandomatic Model). A YSI Oxygen Monitor (Model 53) was used for O_2 measurements. Other experimental details are given elsewhere [4]. Clearly as shown in Fig. 6, when Chl-liposomes under aerobic conditions were illuminated, the media became more acidic and oxygen was taken up from the bathing solution. Concurrently, the liposomes were bleached from green to yellow. The bleaching which was followed spectrophotometrically, resulted in a decrease in the

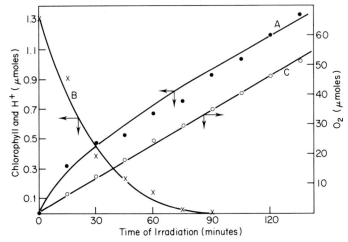

Fig. 6. Light-induced effects in chlorophyll-containing liposomes [23]. Curve A- acidification, Curve B- chlorophyll bleaching, Curve C- oxygen uptake. The liposomes were made from egg lecithin and the chlorophyll extract by the method previously described [28].

total Chl content as well as the Chl a/b ratio.

Further comments should be made concerning light-induced pH decrease. A very large acidification of the bathing solution was observed at pH 7.5 in the presence of ascorbic acid. This large pH decrease was not influenced by KI. No light-induced pH change could be measured when the Chl-liposomes were kept in the dark (for hours) or sealed under N_2. Also, the plain liposomes without Chl resulted in no acidification upon irradiation. In an earlier series of experiments, the acidification reaction was examined as a function of pH gradient across the membrane, the pH outside being higher in all cases studied. The results are presented in Fig. 7. The maximum light-induced pH change occurred between 8.2 and 8.5 depending on the pH gradient across the membrane (Sahu and Tien, unpublished studies, 1971-1972). The preceding experiments

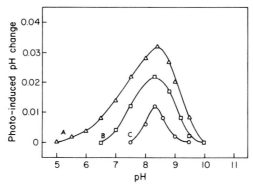

Fig. 7. Light-induced pH change in chlorophyll-containing liposomes as a function of pH [27]. The liposomes were prepared by coating a flat-bottomed flask with the chloroform solution containing 1 µM chlorophyll and 9 µM egg lecithin. After evaporating the solvent under nitrogen in the dark, 35 ml of unbuffered 0.1 M NaCl of known pH (previously saturated with N_2) was added. The mixture was allowed to swell under N_2 for 3-4 hours and then gently shaken to form chlorophyll-containing liposomes. The results show that the higher was the pH gradient, the larger was ΔpH. (a) Inside pH 5.0 (b) Inside pH 6.5 (c) Inside pH 7.5

although unsuccessful insofar as O_2 evolution is concerned, have provided us with experience and impetus for the next series of experiments described below.

In view of the fact that simple Chl-liposomes did not generate oxygen, we decided to incorporate the oxygen evolving complexes associated with broken thylakoids into bilayer lipid membranes since we have observed that: (i) phospholipids protect Chl from photooxidation [22], (ii) the absorption peak of Chl in liposomes corresponds to that of Chl *in vivo*, and (iii) an intact membrane system (i.e., separating two aqueous phases) is a pre-requisite for the generation and separation of charges by light in pigmented BLM. Thus it seems that a cogent experiment is

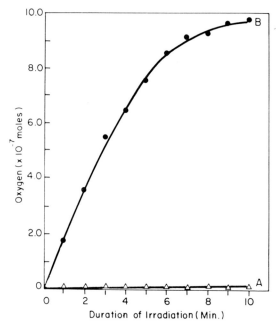

Fig. 8. Oxygen evolution by broken thylakoids [24]. Chloroplasts were isolated from fresh spinach by blending at top speed in an ice cold blender for 5 minutes followed by sonication in ice for 10 minutes with a Branson Model 140D Cell Disruptor. The material was centrifuged at 8300 RPM in a Servall centrifuge for 5 minutes to remove unbroken chloroplasts. The green supernatant was then made to 10 mM tricine, 5 mM $K_3Fe(CN)_6$ and adjusted to pH 7.5. (A) After blending in distilled water. (B) After sonication.

to demonstrate that a sealed vesicle containing oxygen evolving complexes should generate O_2 by light under appropriate conditions.

There are strong beliefs and some evidence to support the idea that the thylakoid membrane, in the form of a sealed vesicle (i.e., a thylakoid), is an indispensable structure for O_2 evolution and photophosphorylation [9]. Indeed, we have found that osmotically shocked or broken thylakoids did not readily generate O_2. If, however, these broken thylakoids were sonicated, some of the O_2 evolving ability was revived. The results of these experiments are shown in Fig. 8. The large increase in O_2 evolving capacity upon sonication is attributed to the formation of resealed vesicles from broken thylakoids, even though the reconstituted vesicles were not stable and their ability to evolve O_2 diminished within about two hours. In contrast to the above findings, broken thylakoids when fused with phosphatidyl choline liposomes generated not only a much larger quantity of O_2 but also sustained the ability of oxygen evolution for 7 hours or more. Two methods of incorporation of oxygen evolving complexes (OEC) from broken thylakoids were used. In the first method, the OEC were added to preformed liposomes by the method described [24]. Briefly, 50 ml of the broken thylakoids (containing OEC) were mixed with 10 ml liposomes and the mixture was tested for O_2. The second method consisted of adding the OEC to a flask containing 0.3 g dried phospholipids and the complexes were directly incorporated into the lipid bilayer in the process of liposome formation. Fig. 9 shows the ability of O_2 evolution by the broken thylakoids fused with liposomes as a function of time. The "hump" in the curve apparently signifies the rate of incorporation of OEC into liposomes. As the fusion progressed, the ability of the system to evolve O_2 increased for the first hour. After that, the "normal" decrease of O_2 evolution, presumably due to the degradation of the OEC, became predominant.

If the incorporation of broken thylakoids containing OEC into liposomes initiates a forward step in the understanding of photooxidation of water, the reconstituted system should then behave similarly under the experimental conditions which govern the Hill reaction. For examples, the Hill reaction is inhibited by DCMU, hydroxylamine, chaotropic agents, heat treatment, etc. and stimulated by manganese, chloride, electron acceptors (benzoquinone methylviologen, ferricyanide), bicarbonate, etc. We have decided therefore to test some of these parameters on the OEC-liposomes so that some information about the nature and behaviour of the reconstituted system might be obtained [24].

In the present series of experiments, the OEC-containing

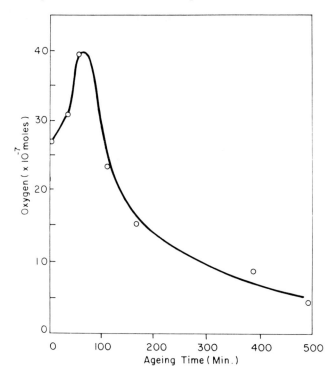

Fig. 9. Time-dependent oxygen evolution by broken thylakoids fused with liposomes [24]. The broken thylakoids were added to preformed liposomes. The hump in the curve suggests that the rate of fusion is greater than that of the rate degradation of OEC in the liposomes.

liposomes were prepared and tested as follows. Chloroplasts from 240 g of fresh spinach were mixed in ice-water for 5 min (~230 ml), sonicated for 10 min at 0°C, and then centrifuged at 8300 RPM in a Servall centrifuge (Model SS-1). From this, two 10 ml samples were taken for Chl determination. 200 ml of the remainder was added to the preweighed dry mixture to give the desired salt/buffer concentrations. The pH of the solution was adjusted to 7.5. Aliquots of this solution (containing OEC associated with broken thylakoids) were then mixed with preformed liposomes and the compound to be tested. Sixty ml of this final mixture was loaded into the reaction cell for illumination and O_2 measurements. To date, six kinds of compounds have been tested; these include electron acceptors (benzoquinone, ferricyanide, methylviologen, DCPIP), stimulators (Mn^{2+}, Cl^-, HCO_3^-), inhibitor (DCMU), electron donor (hydroxylamine), uncoupler (2,4-dinitrophenol), chaotropic agents (KSCN), and detergents (Triton X-100, sodium cholate). The results are summarized below: We have found that

(a) At 1 mM concentration, benzoquinone (BQ) was more effective than ferricyanide in effecting O_2 evolution in OEC-containing liposomes (80 µM vs. 49 µM after 5 min illumination), whereas methylviologen and DCPIP did not work,

(b) Chloride ions at 5-25 mM definitely stimulated O_2 production in the presence of either benzoquinone or ferricyanide; again it was more effective in the presence of BQ; however at 100 mM KCl in the presence of ferricyanide both the rate and quantity of O_2 decreased sharply, only a slight decrease was observed with BQ;

(c) Mn^{2+} ions at 1 mM stimulated the O_2 production by about 1.4 times, however at 5 mM, liposomes precipitated out of the solution,

(d) HCO_3^- at 5 mM did not seem to have much effect on O_2 evolution; at 25 mM no O_2 was evolved,

(e) DCMU was not effective at 10^{-8} M but almost completely stopped O_2 evolution at 10^{-6} M,

(f) 2,4-DNP had no effect at $10^{-7} - 10^{-5}$ M, some decrease was observed at 10^{-3} M,

(g) Chaotropic agents such as KSCN were very effective at large concentration (100 mM), only moderately effective at 25 mM, hardly effective at all at 5 mM,

(h) Hydroxylamine even at 5 mM completely destroyed O_2 evolving complexes,

(i) Detergents such as Triton X-100 and Na cholate totally inhibited O_2 evolution, and

(j) Heat treatment of sonicated broken thylakoids at 50°C for 3 min completely stopped O_2 evolution, whereas some O_2 was still observed for similarly treated OEC-containing liposomes.

The results presented above strongly indicate that the OEC-containing liposomes resemble the intact thylakoids in behaviour insofar as O_2 evolution is concerned. Further, as already noted, upon incorporation of OEC into liposomes O_2 evolving ability is greatly prolonged. In the past the lability of OEC has been a major obstacle in its isolation and study. We suggest that the technique described may be used as a novel method for isolation and purification of the oxygen evolving complexes [24]. Such studies are in progress in our laboratory.

Discussion

In literature on photosynthesis suggestions have been repeatedly made that solid-state mechanisms for energy conversion might be relevant [1], for reviews see refs [28,29]. The results obtained as photoelectric effects seem to be best explained by a solid-state mechanism. If we assume that the BLM possesses a liquid-crystalline structure (Fig. 2C) and the chlorophyll molecule can function either as an electron acceptor or a donor, then the task of interpreting and explaining the observed light-induced events in bilayer lipid membranes is greatly simplified. Fig. 10 shows the mechanism by which quantum conversion is accomplished in pigmented BLM. The interface between the bathing solution and the pigmented BLM is likened to that of a semiconductor in contact with an electrolyte [16]. The energy levels of a chlorophyll-containing BLM separating redox solutions of different composition (or concentration) is shown in Fig. 10(A). At equilibrium in the dark, the two interfacial regions are considered to be isolated from each other owing to the insulating lipid bilayer. Across each interface (the inner side is designated by primes) the electrochemical potential is therefore equal but the electrochemical potentials of the two interfaces may not be the same as depicted in Fig. 10(A) for the reasons just given. In drawing the energy levels both the redox potential of the bathing solution and the pigmented BLM are referred to the energy level of an electron *in vacuo* at infinity (i.e., E_{redox} in volts = E_{redox} in eV + 4.5 eV). Upon illumination by light with energy greater than the optical band gap (i.e., $E_c - V_v$), charge separation may occur leading to a potential difference across the interface. Since there are two interfaces (or a biface, a term introduced to stress the two coexisting solution/membrane interfaces, through which material, energy, and charge transfer are possible), the overall maximum photovoltage observable would be the sum of $(E'_{f(hv)} - E'_{redox}) + (E_{redox} - E_{f(hv)})$ as shown in Fig. 10(B). In this situation it is assumed that the redox potential of the bathing solution is no longer equal to the Fermi energy level in the pigmented membrane. In this scheme, it is evident that the magnitude of photovoltage (V + V') is dependent upon the standard redox potential of the compounds in the bathing solutions. One should be aware however, that standard potentials are derived from equilibrium measurements in the dark; they might not be directly applicable to electrodes excited by light. For electron transfer processes in BLM, the half-wave potentials derived from polarographic studies are perhaps more appropriate and should be used, if available.

As shown in Fig. 3(A,B) and reported previously [21,28], the magnitude of observed photovoltage is related to the redox potential couples in the bathing solution. For the system comprising Fe^{3+}, Fe^{2+}/Chl-BLM/cysteine, cystine in 0.1 M acetate at pH 5, the estimated maximum value from half-wave potential data is 180 mV, whereas the observed

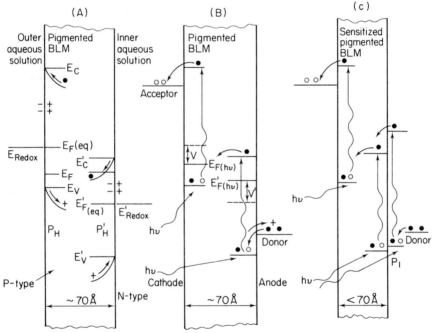

Fig. 10. Mechanism of charge generation, separation, and energy storage in pigmented bilayer lipid membranes (BLM). The pigmented BLM/redox solution interface is treated as a simple Schottky-like barrier [16]. (A) A pigmented BLM in the dark at equilibrium with the redox solutions. E_c=conduction band, E_v=valence band, E_f=Fermi level, E_{redox}= redox potential of the outer aqueous solution; those with primes are the corresponding energy levels for the inner side. (B) The same system showing in (A) under illumination. P_h and P'_h refer to the pigments in the hydrophobic phase (i.e., lipid bilayer). (C) A pigmented BLM is sensitized by the presence of an aqueous soluble pigment (P_i) or dye. Note a change in membrane thickness is depicted. (See text for further details.)

value was 110 mV. The large discrepancy might be easily due to the insufficient data on which the estimate was based. As to the electron transfer from the conduction band of the pigmented membrane to the acceptor in the bathing solution, the process is also illustrated in Fig. 10(B), where the energy levels of the electron acceptor and donor are assumed to lie, respectively, below the conduction state and above the valence state of the photoactive species in the BLM. Upon the absorption of a quantum by the photoactive pigment, an electron is promoted to the energy level above the acceptor level, thereby permitting a thermodynamically favoured electron transfer. Meanwhile since the level of the electron donor is higher than that of the vacant site, a transfer of an electron is also possible from the donor. The end result is, of course, a reduced acceptor and an oxidized donor. How does an electron move from one side of the biface to the other side? The most plausible mechanism is the quantum-mechanical tunnelling [7,10] assuming the height of the insulating lipid bilayer to be 1 eV. For practical considerations at room temperature, such a high energy barrier reduces the probability of the electron jumping over to a negligible value. Thus, the photoactive pigment in the BLM serves merely as a light-driven "pump" (a transducer) transferring electrons from one side to the other side of the membrane. To prevent any immediate back reactions, the photogenerated redox products are separated by the lipid bilayer [12].

To split water a minimum energy of 1.23 eV is necessary, which in theory can be easily supplied by a quantum of red light. However, in the scheme shown in Fig. 10(A,B), various energy losses have to be considered, which are inherent in the proposed process of quantum conversion. Some of these are the internal resistance of the lipid bilayer, and the overpotential associated with each interface. Here we have specifically assumed that the outer and the inner membrane interface function, respectively, as a photocathode and photoanode [28]. These electrode overpotentials are unavoidable and subject to the usual considerations of electrochemistry. A more urgent question that so far has not been raised is "How can one ever hope to have BLM or for that matter thylakoid membranes about 70 Å thickness of sufficient dielectric strength to withstand 1230 mV, a minimum potential needed to electrolyze water, assuming such a large photovoltage could be generated?" (One is reminded that air breaks down at about 10^6 V/cm.) It seems unlikely that the thylakoid membranes of the chloroplast could withstand even 500 mV, let alone more than twice the value that is believed to be required for water decomposition, unless the thylakoid membrane were much thicker than that revealed by the electron microscope!

Faced with this dilemma, I suggest that, for the photo-oxidation of water in thylakoid membranes and in pigmented BLM, a light-induced potential difference of much less than 1.23 V across the membrane should be sufficient, if the membrane behaves as a semi-conductor electrode as postulated earlier [26,28]. The reasoning is as follows: First, let us take a closer look at the left side of the pigmented membrane/redox solution interface (Fig. 10B). Upon illumination with photons of energy equal to or greater than ($E_c - E_v$), excitons (electron-hole pairs) are generated at the interface, which under the influence of the electric field are separated into electrons and holes. Electrons, on the one hand, move to the front of the pigmented membrane where they are captured by electron acceptors to produce a reduction. On the other hand, holes drift towards the interior of the lipid bilayer. In a similar manner, an oxidation reaction takes place at the inner (or right side) membrane/redox solution interface where electron donors are located. In this case, electrons move to the left; holes move to the front of the pigmented membrane and are injected into the bathing solution. The interesting result is a sum of the two photoevents operating in series (see Fig. 10B). Moreover, the photo-induced chemical reactions are carried out by one electron-hole pair for each two photons absorbed. It can be readily seen that these processes can create electron-hole pairs that are potentially more energetic than the energy of exciting photons. Accordingly, the decomposition of water by photo-generated holes, mediated by the pigmented membrane process, has been proposed [26]. This has been stated succinctly in eq. (1):

$$\oplus + \tfrac{1}{2} H_2O \longrightarrow H^+ + \tfrac{1}{4} O_2 \qquad (1)$$

At present, the most unequivocal experimental evidence in support of the proposed scheme is the observation by Fujishima and Honda [8], who have reported O_2 evolution by light at a semiconductor TiO_2 electrode at potentials less than 500 mV. They have also used eq. (1) to describe their most exciting finding.

To carry the above scheme one step further, pigments with different spectral characteristics should be associated with each interface (or environment) for more efficient light utilization. This cannot be easily done with Chl-BLM, since the pigments were dissolved in the membrane-forming solution. To overcome this technical problem, a dye or a water-soluble pigment (e.g. chlorophyllin) has been added to one side of the bathing solution with the result shown in Fig. 3(C). Clearly the sensitivity of the photoelectric BLM has been extended. In order for spectral sensitization to be effective, the water-soluble pigment molecules (or dye) must interact intimately with the lipid bilayer, which can be accomplished by having a portion of the pigment molecule hydrophobic. A partially submerged pigment molecule in the hydrophobic lipid bilayer seems to facilitate the electron transfer process, both across the interface and through the lipid bilayer phase.

Consistent with the idea of strong interaction between the lipid bilayer and the water-soluble pigment, the action spectrum of dye-sensitized Chl-BLM, as expected, was modified by the dye used [28]. Finally to account for the marked increase in efficiency of the electron transfer, the decrease in distance that the charge carriers have to move is suggested. This is shown in Fig. 10(C) in the form of reduced membrane thickness. As a result of sensitization, both the height and the width of the energy barriers are believed to be lowered. This explanation can be supported by the further enhancement of the photovoltage when the sensitized pigmented BLM was under anodic bias as shown in Fig. 3(D).

Summary

1. Planar bilayer lipid membranes (BLM) and spherical bilayer lipid membranes (liposomes) containing material from broken thylakoids are used as models for the thylakoid membrane for the understanding of the quantum conversion and oxygen evolution.

2. Photoelectric effects are seen in pigmented BLM whose effects can be increased by the process of dye or water-soluble pigment (chlorophyllin) sensitization. The photoresponse of sensitized pigmented BLM can be further enhanced by electrical polarization.

3. Incorporation of broken thylakoids containing oxygen evolving complexes into non-pigmented BLM can also generate a potential difference upon illumination.

4. Simple chlorophyll-containing liposomes evolve no O_2 when irradiated by light; oxygen is consumed instead. This is accompanied by a rise of H^+ concentration in the bathing medium while the chlorophyll pigments become irreversibly photooxidized.

5. The maximum light-induced acidification of Chl-containing liposomes occurs between pH 8.2 and pH 8.5 when a pH gradient is imposed across the membrane.

6. Broken thylakoids containing oxygen evolving complexes (OEC) when fused with phosphatidyl choline liposomes evolve O_2 profusely as compared with broken thylakoids alone. Also, incorporation of the O_2 evolving complexes into liposomes apparently stabilizes the OEC and prolongs the system to generate O_2.

7. A number of factors which affect photosystem II are also seen to be operative in liposomes containing the OEC.

8. All light-induced phenomena in experimental bilayer lipid membranes containing material from broken thylakoids can be best explained in terms of the band theory of solids. The pigmented BLM is considered as a semiconductor. Thus the operational mechanism of the system (redox solution/pigmented BLM/redox solution) is most easily described in terms similar to those that explain a semiconductor in contact with a redox solution. In essence, the pigmented membrane/redox solution interface acts like a Schottky-type barrier when illuminated.

9. Since the pigmented BLM system possesses two solution/membrane interfaces, the absorption of photons at each interface can be considered as two separate events operated in concert and in series across the lipid bilayer. The highly energetic species (such as positive "holes"), thus generated, may be of sufficient energy to effect water decomposition.

References

1. Arnold, W. and Sherwood, H. (1957). *Proc. Natl. Acad. Sci. Washington* **43**, 105-114.
2. Arntzen, C.J., Armond, P.A., Briantais, J.M., Burke, J.J. and Novitzky, W.P. (1977). *Broohaven Symp. Biol.* **28**, 316-336.
3. Bangham, A.D., Hill, M.W. and Miller, N.G.A. (1974). *In* "Methods in Membrane Biology" (E.D. Korn, ed.), Vol. 1, 1-68, Plenum Press, New York.
4. Boardman, N.K. (1968). *In* "Advances in Enzymology and Related Areas of Molecular Biology" (F.F. Nord, ed.), **30**, 1-79.
5. Branton, D., Bullivant, S., Gilula, N.B., Karnovsky, M., Moor, H., Mühlethaler, K., Northcote, D.H., Packer, L., Satir, B., Satir, P., Speth, V., Staehlin, L.A., Steere, R.L. and Weinstein, R.S. (1975). *Science* **190**, 54-56.

6. Bulychev, A.A., Andrianov, V.K., Kurella, G.A. and Litvin, F.F. (1976). *Biochim. Biophys. Acta* **430**, 336-351.
7. Devault, D., Parkes, J.H. and Chance, B. (1967). *Nature* **215**, 642-644.
8. Fujishima, A. and Honda, K. (1972). *Nature* **238**, 37-38.
9. Gregory, R.P.F. (1977). "Biochemistry of Photosynthesis", Wiley & Sons, Inc., New York.
10. Gutmann, F. (1968). *Nature* **219**, 1359.
11. Huebner, J.S. and Tien, H.T. (1973). *Bioenergetics* **4**, 469-478.
12. Kobamoto, N. and Tien, H.T. (1971). *Biochim. Biophys. Acta* **241**, 129-146.
13. Lahshminarayaniah, N. (1975). *In* "Electrochemistry" (H.R. Thirsk, ed.), Vol. 5, 184-219, The Chemical Society, London.
14. McLaughlin, S. and Eisenberg, M. (1975). *Ann. Rev. Biophys. Bioeng.* **4**, 335-366.
15. Mühlethaler, K. (1966). *In* "Biochemistry of Chloroplasts" (T.W. Goodwin, ed.), 49-64. Academic Press, New York.
16. Myamlin, V.A. and Pleskov, Y.V. (1967). "Electrochemistry of Semiconductors", Plenum Press, New York.
17. Ohki, S. (1976). *Progr. Surface and Membrane Sci.* **10**, 117-252.
18. Olson, J.M. and Hind, G. (eds) (1977). "Chlorophyll-Proteins, Reaction Centers, and Photosynthetic Membranes", Brookhaven Symp. Biol., Vol. 28, National Technical Information Service, Springfield.
19. Sauer, K. (1975). *In* "Bioenergetics of Photosynthesis" (Govindjee, ed.), 115-181, Academic Press, New York.
20. Shamoo, A.E. and Goldstein, D.A. (1977). *Biochim. Biophys. Acta* **472**, 13-53.
21. Shieh, P.K. and Tien, H.T. (1974). *J. Bioenerg.* **6**, 45-55.
22. Stillwell, W. and Tien, H.T. (1977). *Biochem. Biophys. Res. Comm.* **76**, 232-238.
23. Stillwell, W. and Tien, H.T. (1977). *Biochim. Biophys. Acta*, in press.
24. Stillwell, W. and Tien, H.T. (1977), in preparation.
25. Tien, H.T. (1967). *J. Theor. Biol.* **16**, 97-110.
26. Tien, H.T. (1968). *J. Phys. Chem.* **72**, 4512-4519.
27. Tien, H.T. (1972). *In* "MTP International Review of Science" (M. Kerker, ed.), vol. 7, 27-81, University Press, Baltimore.
28. Tien, H.T. (1974). "Bilayer Lipid Membranes (BLM): Theory and Practice", Marcel Dekker, Inc., New York.
29. Tien, H.T. and Karvaly, B. (1977). *In* "Photochemical Conversion and Storage of Solar Energy" (J.R. Bolton, ed.), Academic Press, New York.
30. Toyoshima, Y., Morino, M., Motoki, H. and Sukigara, M. (1977). *Nature* **265**, 187-189.
31. Trebst, A. (1974). *Ann. Rev. Plant Physiol.* **25**, 423-458.
32. Weller, H.G. and Tien, H.T. (1973). *Biochim. Biophys. Acta* **325**, 433-440.

CATHODIC REDUCTION OF OXYGEN ON CHELATES

H. JAHNKE, M. SCHÖNBORN and G. ZIMMERMANN

*Robert Bosch GmbH, Research Centre,
Gerlingen (F.R.G.)*

Introduction

In contrast to the theme of this Symposium, the present work treats of not an oxidative but a reducing process, namely the electrochemical cathodic reduction of oxygen. The oxygen is reduced by transition-metal chelates which show a certain structural similarity to several biochemically active compounds such as, for example, chlorophyll or cytochromes. Results have been obtained on the phenomenology and the mechanism of the cathodic reduction of oxygen on chelates, and these results can perhaps lead to new insight into the mechanism of photosynthetic oxygen evolution.

The starting point of our work in this field was a technical problem. For fuel cells and for metal/air batteries, electrode materials are needed which can reduce the oxygen involved. This reaction proceeds at the cathode (right-hand electrode in Fig. 1) with the following basic equation:

$$O_2 + 4e^- + 4H^+ \rightleftharpoons 2 H_2O \qquad (1)$$

and with the standard potential $\varphi_o = 1.229$ V. As the oxygen is very inactive, the cathode must contain a catalyst which accelerates its reduction.

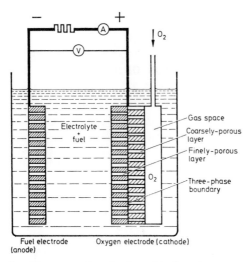

Fig. 1. Schematic diagram of a fuel cell for fuel soluble in electrolyte.

Up to the middle sixties, the only catalysts known for the reduction of oxygen in the particularly advantageous acid media were the platinum metals. These are, however, too costly for general use, and the search for cheaper catalysts has been actively pursued in many research laboratories. Important results have been obtained in the field of chelates, and these are summarized here (for details see [12]).

Transition-metal chelates as catalysts in acid media

Our work on chelates began with the consideration that N_4-chelates can both transport and reduce the oxygen in the human or the animal body. As the compounds involved, namely cytochrome and haemoglobin, and their prosthetic groups are not stable in sulphuric acid, we investigated the phthalocyanines, in which, as also for example in haem (Fig. 2), the central atom in the plane of the molecule is connected to the four central nitrogen atoms of the porphyrine ring (N_4-chelates). The other two bonds from the sixfold co-ordination of the iron are perpendicular to

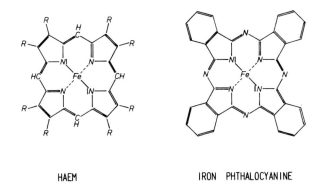

Fig. 2. Structures of haem and iron phthalocyanine

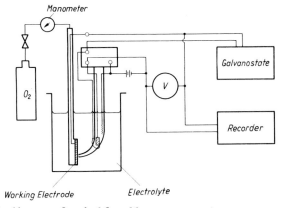

Fig. 3. Block diagram for half cell measurements

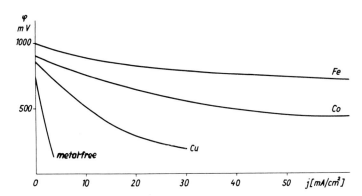

Fig. 4. Comparison of activity of phthalocyanines with different central atoms.

the plane of the molecule and can form links with other groups [15,21].

Monomeric and polymeric phthalocyanines were precipitated on carbon from concentrated sulphuric acid by addition of water and were pressed to form electrodes with a polyethylene binder. Such electrodes were tested in a normal electrochemical half-cell (Fig. 3) under galvanostatic conditions. The most important results are as follows [9,10].

a) The central atom of the phthalocyanine has a strong influence on the activity, the latter increasing markedly in the order

$$\text{metal-free} < Cu < Co < Fe$$

The ability of the metal to attain a higher valency state increases in the same order. Polymeric iron phthalocyanine has the highest activity, that of the monomeric material is somewhat lower.

It is notable that the most highly developed living beings have iron in the porphyrine nucleus as oxygen carrier, while the lower beings (molluscs, crustaceae) have copper in the same position. Cobalt as the central atom of a porphyrine-like nucleus plays an important part in vitamin B_{12}, which is involved in blood formation in humans.

b) The nature of the carbon substrate has a considerable influence on the activity of the phthalocyanic carbon electrode (Table 1). The specific surface area and the electrical conductivity are relatively unimportant. The critical characteristic is the nature of the oxygen-containing groups on the carbon surface: alkaline groups must be present if the phthalocyanine is to exert a catalytic effect.

c) The decisive influence of the carbon substrate is shown by the following experiment. The catalyst was not

TABLE 1
Physical properties and catalytic activities of some carbon supports

Substance	Current density mA·cm^{-2} at 700 mV with FePc	Specific resistance [Ωcm]	Specific surface [m^2·g^{-1}]	Type of surface groups
Acetylene black	51	0.069	57.5	alkaline
Norit BRX	31	0.116	1740	alkaline
Vulkan 3H	24	0.063	89.9	alkaline
Corax 6	21	0.066	105	alkaline
Corax A	-	0.053	44.7	alkaline
Thermax	-	0.26	8.1	alkaline
CK3	-	0.905	76.2	acid
Durex 0	-	0.040	19.0	acid
E$_o$ (Graphite)	-	0.012	4.3	acid
S + E (Graphite)	-	0.013	17.5	acid

pressed to form an electrode, but was suspended in a solution in sulphuric acid, as suggested by Gerischer [7,8] and Schlygin [17,20]. Intensive stirring projects the catalyst particles against an inert gold network, where charge transfer is effected. The oxygen is blown into the solution through sintered glass. This process has the advantages that the catalytic function is locally separated from the current-carrying function and that in addition the influence of the electrode structure is entirely eliminated.

Measurements by this method show that pure iron phthalocyanine is completely inactive, and that acetylene black also catalyses the reduction of oxygen only very slightly. If the two substances are present together in the suspension, the catalytic properties are clearly improved. The rate of reaction is, however, increased nearly tenfold if the iron phthalocyanine is precipitated on acetylene carbon.

Obviously, intimate contact between the phthalocyanine and the carbon substrate is necessary to attain high activity. The effect is probably due to interaction between the central atom of the phthalocyanine and the alkaline surface groups of the carbon.

Further very active chelates for the reduction of oxygen in acids, discovered in the early seventies, are the complexes tetraphenylporphyrine (TPP) and its derivatives and dihydro-dibenzo-tetraazaannulene (TAA) and anthraquincyanine (ACC) [5]. Fig. 5 shows the structure of these compounds, and Fig. 6 typical current/voltage-curves for oxygen reduction on tetra-(p-methoxy-phenyl)-porphyrines

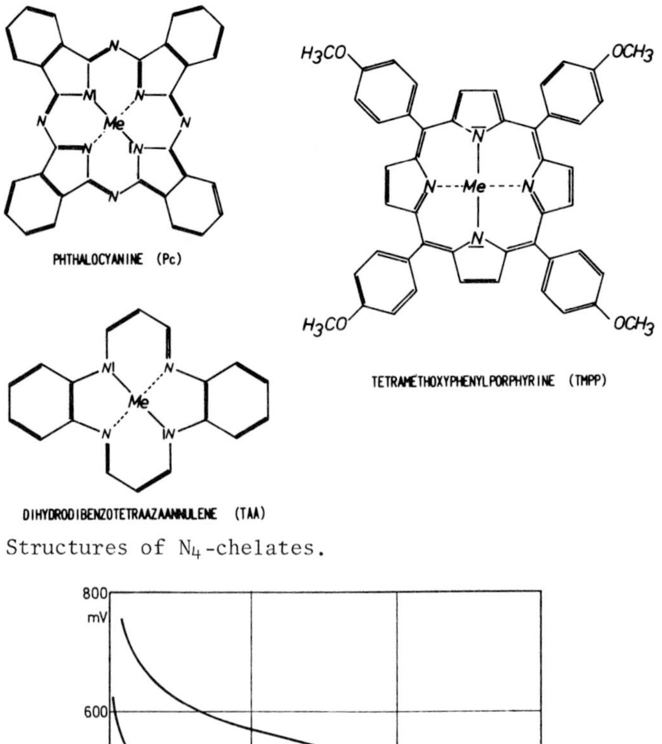

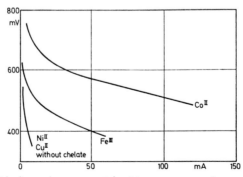

Fig. 5. Structures of N_4-chelates.

Fig. 6. Potentiodynamic current/voltage curves for oxygen reduction at tetra-(p-methoxyphenyl)-porphyrine catalysts in 3 N H_2SO_4 at 30°C. Sweep rate: 40 mV/min.

with various central atoms.

It must be emphasised that all the compounds named above are too inactive and too unstable for convenient application.

Comparison of the activities of various co-ordinated chelates in buffer solution

All the chelates hitherto mentioned are N_4 complexes. This raises the question, whether chelates with other than N_4 co-ordination are also catalytically active. In this connection, 67 compounds were studied [11,12]. Description of the structures of all 67 compounds is outside the scope of this work, but Fig. 7 gives a general survey. In the

Coordination	Number of complexes investigated	Central atoms	Example		
			Structure	Name	Abbreviation
N_4	12	Fe Co Ni Cu Mn		Phthalocyanine	Pc
				Dihydro-dibenzo-tetraaza-annulene	TAA
N_2O_2	4	Fe Co N Cu Mn		"Pfeiffer -complexes"	
N_2S_2	2	Fe Co Cu Ni		Complexes of o-aminothiophenol	
O_4	4	Co Ni Cu		Complex of salicyl-aldehyde	
S_4	2	Fe Co Ni Cu			

Fig. 7. Survey of the substances investigated. Total number of complexes investigated: 67.

left-hand column is the type of co-ordination, and adjoining is the number of compounds of this type. As central transition-metal atom, Fe, Co, Cu, Ni and Mn were used, but it was not possible to synthesise all chelates with all central atoms. The chelates whose structural formulae and names are given in the right-hand portion of Fig. 7 are to be regarded as only examples of the type of compound

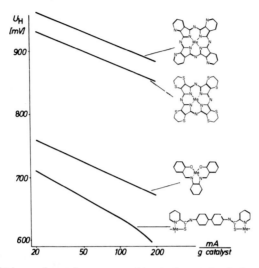

Fig. 8. Activities of various co-ordinated metal chelates with Fe as central atom in the reduction of oxygen (electrolyte: CO_3^{--}/HCO_3^-, pH = 9.3).

in question. As examples of N_4 chelates we have chosen the particularly important phthalocyanines (Pc) and dihydro-dibenzo-tetraazaannulenes (TAA). All the N_2O_2 chelates are the so-called "Pfeiffer complexes" with ethylene (as in

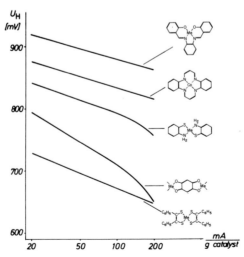

Fig. 9. Activities of various co-ordinated metal chelates with Co as central atom in the reduction of oxygen (electrolyte: CO_3^{--}/HCO_3^-; pH = 9.3).

Fig. 7), propylene or phenylene bridges. Similarly the N_2S_2, O_4 and S_4 chelates have been modified, as shown in the corresponding examples.

Despite these considerable differences in the structure of the organic skeletons, general rules for the influence of the ligand closest to the central atom on the catalytic activity can be deduced. An order of activity can be established which is approximately the same for all central atoms. These activity values are shown in the form of current/voltage curves in Fig. 8 for Fe and in Fig. 9 for Co. For comparison, we have given the current in proportion to the mass of catalyst used.

With iron as central atom, the activity decreases in the order:

$$N_4 > N_2O_2 \gg N_2S_2 > O_4 \approx S_4$$

Chelates with O_4 and S_4 are so inactive that their characteristic curves cannot be shown in Fig. 8.

In the case of the cobalt complexes (Fig. 9) the most active two ligands change places:

$$N_2O_2 > N_4 > N_2S_2 > O_4 \approx S_4$$

It may be supposed that the similar order in all these chelates has its basic cause in the electronic structure at the central atom.

The influence of the central atom can be shown by comparison of N_4 and N_2O_2 complexes. For N_4 complexes (Fig. 10) the order is:

$$Fe > Co > Cu > Ni \approx Mn$$

while in N_2O_2 complexes (Fig. 11) the activity decreases in the order

$$Co > Mn > Fe > Cu > Ni$$

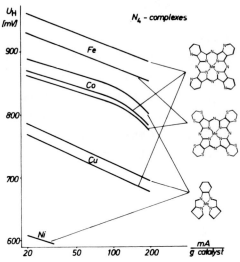

Fig. 10. Influence of the central atom on the oxygen activity of N_4-chelates (electrolyte: CO_3^{--}/HCO_3^-, pH = 9.3).

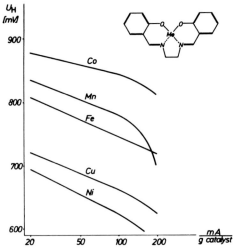

Fig. 11. Influence of the central atom on the oxygen activity of N_2O_2-chelates (electrolyte: CO_3^{--}/HCO_3^-, pH = 9.3).

In this case the Co chelates have the highest activity and Mn is even more active than Fe.

With the same N_4 co-ordination and the same central atom (Co), the influence of the external organic skeleton is relatively slight (Fig. 12). The reason is presumably

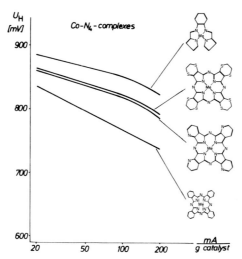

Fig. 12. Influence of the organic skeleton on the oxygen activity of various Co-N_4-complexes (electrolyte: CO_3^{--}/HCO_3^-, pH = 9.3).

the slight influence of the external skeleton on the electron density at the central atom.

As several of the substances here investigated are extremely unstable in sulphuric acid, which is the medium of interest in practical applications, we have carried out all investigations in a carbonate/bicarbonate equilibrium electrolyte at pH 9.3. For comparison, we have not precipitated the chelates on a substrate, as this treatment can produce an increase of activity, as shown in the case of the phthalocyanines. Instead, in the present work the catalyst was mechanically mixed with powdered carbon (acetylene black), whose activity in the reduction of oxygen is negligibly small. The 1:1 mixture of catalyst and carbon is mixed to a paste with methanol and spread on a porous PTFE sheet. The coverage density of the chelate is about 3.5 mg/cm². The PTFE foil, about 2 cm², was built into a plexiglass holder. The catalyst layer was covered with a graphite felt and a perforated metal sheet to improve the electrical conductivity. Oxygen was then passed through this device from the PTFE side. An autogenous hydrogen electrode was used as reference electrode, and all potentials are given against hydrogen in the same solution. When not otherwise stated, all measurements were made under galvanostatic conditions.

On the mechanism of oxygen reduction by chelates

The results given above on the order of activity of the complexes with various co-ordinations can be qualitatively interpreted by MO observations [12], although quantitative interpretation is not possible.

The question of the reaction mechanism of an electrochemical reaction is, however, only completely answered when a sequence of reaction steps can be drawn up on the basis of the experimental results and when it can be shown which electron transfer determines the velocity. Many investigations of the mechanism of the cathodic reduction of oxygen by chelates have been published, but the results are partly contradictory. The oxygen reduction obviously proceeds by different mechanisms on different chelates. Even a single chelate can show different mechanisms at different pH values. Thus Sandstede and co-workers [6] have shown, by investigations with the rotating ring-disc electrode, that with CoTAA in a strongly alkaline medium the transfer of the second electron is rate-determining. The O_2^- ion, formed primarily by the reaction:

$$O_2 + e^- \rightleftharpoons O_2^-$$

is further reduced, the proton donor being probably water:

$$O_2^- + H_2O + e^- \longrightarrow O_2H^- + OH^-$$

At low pH values, however, the first electron transfer is the slowest:

$$O_2 + H^+ + e^- \longrightarrow O_2H$$

Further reduction, with formation of H_2O_2, follows:

$$O_2H + H^+ + e^- \longrightarrow H_2O_2$$

Although some authors postulate the formation of H_2O_2 in the reduction on phthalocyanines, with subsequent decomposition by virtue of the catalase activity of the phthalocyanine [13,14,16], nevertheless here too, at least under suitable conditions, the reduction seems to proceed exclusively to water [4,13,19], *i.e.* to its final stage, without formation of the intermediate product H_2O_2.

References

1. Alt, H., Binder, H., Klempert, G., Köhling, A., Lindner, W. and Sandstede, G. (1971). *Battelle Frankfurt, Informationsheft* **11**, 64-68.
2. Alt, H., Binder, H., Lindner, W. and Sandstede, G. (1971). *In* "From Electrocatalysis to Fuel Cells" (G. Sandstede ed.), pp. 113-125, University of Washington Press, Seattle.
3. Alt, H., Binder, H., Lindner, W. and Sandstede, G. (1971). *J. Electroanal. Chem. Interfacial Electrochem.* **31**, 19-22.
4. Appleby, A.J., Fleisch, J. and Savy, M. (1976). *J. Catal.* **44**, 281-292.
5. Beck, F., Dammert, W., Heiss, J., Hiller, H. and Polster, R. (1973). *Z. Naturforschg.* **28a**, 1009-1021.
6. Behret, M., Binder, H., Clauberg, W. and Sandstede, G. (1977). Unpublished.
7. Gerischer, H. (1963). *Ber. Bunsenges. Chem.* **67**, 164-167.
8. Held, J. and Gerischer, H. (1963). *Ber. Bunsenges. physik. Chem.* **67**, 921-929.
9. Jahnke, H. (1968). *Ber. Bunsenges. physik. Chem.* **72**, 1053.
10. Jahnke, H. and Schönborn, M. (1969). *In* "Comptes rendus d'Etudes des Piles à Combustible, pp. 60-65, Press Acad. Européennes, Brussels.
11. Jahnke, H., Schönborn, M. and Zimmermann, G. (1974). *In* "Electrocatalysis" (M.W. Breiter, ed.), pp. 303-318, Electrochem. Soc., Princeton.
12. Jahnke, J., Schönborn, M. and Zimmermann, G. (1976). *Topics in Current Chemistry* **61**, 133-181.
13. Kozawa, A., Zilionis, W.E. and Brodd, R.J. (1970). *J. Electrochem. Soc.* **117**, 1470-1474.
14. Kretzschmar, Ch. and Wiesener, K. (1976). *Z. phys. Chem. (Leipzig)* **257**, 39-48.
15. Lemberg, R. and Legge, J.W. (1949). "Hematin Compounds and Bile Pigments", pp. 230-245, Wiley, New York.
16. Meier, H., Albrecht, W., Tschirwitz, U. and Zimmerhackl, E. (1973). *Ber. Bunsenges. physik. Chem.* **77**, 843-849.

17. Podwjaskin, J.A. and Schlygin, A.J. (1957). *J. phys. Chem. (russ.)* **31**, 1305-1308.
18. Sandstede, G. (1971). *Chem. Ing. Techn.* **43**, 495.
19. Tarasevich, M.R. and Bogdanovskaya, V.A. (1975). *Bioelectrochem. Bioenerget.* **2**, 69-74.
20. Tjurin, J.M. and Schlygin, A.J. (1958). *J. phys. Chem. (russ.)* **32**, 2487-2491.
21. Williams, R.P.J. (1961). *In* "Haematin Enzymes" (J.E. Falk, R. Lemberg and R.K. Morton, eds), pp. 41-55, Pergamon Press, New York.

PHOTOSYSTEM II UNIT: ASSEMBLY, GROWTH AND ITS INTERACTIONS DURING DEVELOPMENT OF HIGHER PLANT THYLAKOIDS

G. AKOYUNOGLOU and J.H. ARGYROUDI-AKOYUNOGLOU

Biology Department, Nuclear Research Centre, "Democritos", Athens (Greece)

Introduction

The etioplast to chloroplast differentiation in angiosperms depends on light. The differentiation consists essentially in the transformation of the prolamellar body to primary thylakoids and their subsequent growth and development into grana and stroma thylakoids. The growth and development of the chloroplast membrane is a multistep process. Under continuous illumination the differentiation of etioplasts to chloroplasts proceeds normally, i.e., after a lag-phase which depends on the age of the etiolated tissue [3,7,8], the different components are synthesized in a parallel way and integrated into the developing membrane, and in a short time the characteristic features of the mature chloroplasts are observed [7,26]. It has been found that it is possible to distinguish the different steps in thylakoid development, by changing the mode of illumination [6,15]. For example, in periodic light (2 min light-98 min dark) the etioplasts are transformed to protochloroplasts. The protochloroplasts can be transformed to chloroplasts only after transfer of the plants to continuous light [15]. Contrary to chloroplasts, the protochloroplasts have no grana stacks but unstacked ("primary") thylakoids [18]. They synthesize selectively Chl a, are devoid of Chl b, and deficient in the Chl b rich Chl-protein Complex II [15,16,18,19]. In spite of the incomplete development of the protochloroplast thylakoid they are photosynthetically very active with high PS I and PS II activity, cyclic and non-cyclic photophosphorylation [7], and CO_2 fixation of a rate 4 to 5 times higher on a Chl basis than that of mature chloroplasts [4].

Making use of this system some aspects of chloroplast membrane biogenesis and development have been studied. In this paper emphasis will be put on the development of the photosystem II unit, and the establishment of the regulation of the excitation energy distribution among the two photosystems.

Materials and Methods

Etiolated bean leaves (*Phaseolus vulgaris*, red kidney var.) and etiolated pea leaves (*Pisum sativum*, Cephalinia var.) were used for the study. The growth and handling of the bean and pea seedlings were done as previously described [15]. The etiolated leaves were exposed to periodic light-dark cycles (2 min light-98 min dark) or to continuous light [15]. For activity determination the plastids were isolated in dim green light from the developing bean leaves as previously described [1]. The isolation medium consisted of a 10 mM Tricine buffer, pH 7.8, containing 0.4 M sucrose, 10 mM NaCl, 5 mM $MgCl_2$, 1 mM $MnCl_2$ and 1 mg/ml bovine serum albumin.

PS II activity was measured as previously described [1,7], by determining the rate of DCPIP reduction in the absence or presence of DPC. Fluorescence measurements were done with a setup described by Argyroudi-Akoyunoglou and Akoyunoglou [17]. Broad band blue actinic light, filtered from the light of a projection lamp, was admitted to the sample chamber through a photographic shutter (opening time about 1 ms). The kinetics of the rise of the fluorescence at 685 nm was recorded either on a Tektronix 549 storage oscilloscope or on a recorder.

For digitonin disruption experiments the plastids were isolated from spinach (*Spinacia oleracea*), bean or pea leaves. The plastids were isolated as pellets after centrifugation for 10 min at 6,000 g (periodic light plastids), or 3,000 g (continuous light plastids), or 1,000 g (mature chloroplasts) as previously described [5,10,13], and they were washed once with the homogenization buffer (0.3 M sucrose-0.01 M NaCl-0.05M phosphate buffer, pH 7.2).

The grana were disorganized by treatment with 0.05 M Tricine-NaOH buffer, pH 7.2, and they were reconstituted by addition of cations [5,13,17,21]. Digitonin disruption of the plastids and the isolation of the subchloroplast fractions were done as previously described [5,10,13,17,21]. Gel electrophoreses of lipid free thylakoids were performed according to Hoober [36].

Results and Discussion

Onset of photosystem II activity

Reaction center Young etiolated bean leaves exposed to continuous light show PS II activity shortly after exposure. The activity on a Chl basis increases with the time of illumination and reaches a plateau after 4 hours of illumination in 5-day old leaves. The value reached at the plateau is approximately that found in mature chloroplasts [7]. In continuous illumination, therefore, the complete PS II unit is formed from the beginning of greening, and prolonged illumination results only in an increase in the number of the complete units.

When etiolated bean leaves are exposed to periodic light, however, the rate of the PS II activity, on a Chl basis, increases with time and after 2-3 days it reaches a plateau, with values at the plateau almost 5-times higher than those of mature chloroplasts (Table 1). The increase in the rate observed early during greening reflects the presence of unorganized Chl in a high

TABLE 1

Onset of photosystem II activity in developing protochloroplasts from 5-day old etiolated bean leaves exposed to 2 min light - 98 min dark cycles (LDC) and then transferred to continuous light (CL).

Sample	Chl a µg/g fresh wt	Chl b	µmoles DCPIP/mg.Chl h $H_2O \rightarrow$ DCPIP	DPC $\rightarrow$ DCPIP	µmoles DCPIP/g fresh wt.h $H_2O \rightarrow$ DCPIP	DPC $\rightarrow$ DCPIP
6 LDC	45	0	150	350	7	16
14 LDC	110	1	500	750	55	83
42 LDC	313	12	950	1050	309	341
28 LDC	230	7	750	1150	178	272
28 LDC+2h CL	452	78	480	540	254	286
28 LDC+4h CL	638	156	320	350	254	278
Chloroplasts	1980	640	175	190	67	73

The reaction mixture contained in a final volume of 1 ml: DCPIP, 0.1 mM; 40 mM; NaCl, 5 mM; and plastids containing 2-3 µg Chl. DPC whenever used was present at 0.5 mM. The reaction mixture was illuminated by two projector lamps providing, after passing through a 1% w/v aqueous $CuSO_4$ solution 5 cm thick, a light intensity of 100,000 lux at the level of the cuvette. The DCPIP reduction was calculated from the absorption change during the first 10 s of illumination after subtraction of the dark control.

proportion, which decreases as the time of illumination increases. This conclusion is supported by the results of dark incubation experiments where etiolated leaves were exposed first to 14 or 28 LDC and then left in the dark for 24 hours. In this experiment we expected the concentration of unorganized Chl to decrease during the dark period, and this would affect the rate of DCPP reduction. The results are shown in Table 2. Dark incubation either after 14 or 28 LDC increases the rate of DPC $\rightarrow$ DCPIP reduction on a Chl basis, and the values reached are much higher than those obtained from the same leaves left under periodic light. This reflects the existence of unorganized Chl, which in the dark is used for the formation of new small PS units. During the remainder of the greening process under periodic light the activity per gram fresh weight increases, but on a Chl basis remains almost constant, indicating that PS II units of the same size are formed.

Transfer of the periodic-light leaves to continuous light induces an increase in the rate of Chl a and Chl b synthesis with the concomitant gradual decrease, on a Chl basis, of the PS II activity to the value of mature chloroplasts.

TABLE 2

Effect of dark incubation on the rate of DCPIP reduction by developing protochloroplasts of bean leaves

Sample	µmoles DCPIP reduced/mg Chl.h		F_{max}/F_o
	$H_2O \rightarrow$ DCPIP	DPC $\rightarrow$ DCPIP	
5-day + 14 LDC	500	750	3.15
" + 14 LDC+24 h dark	480	1450	5.30
" + 28 LDC	750	1150	4.38
" + 28 LDC+24 h dark	700	1400	5.60
" + 42 LDC	950	1050	5.20

Experimental conditions as in Table 1.

Taking into account that high light intensities are needed for maximum photosynthetic activity [4,7], the results shown in Table 1 suggest that under periodic light the Chl a formed is used for the formation of small PS units. The excess Chl a and Chl b formed after transfer of the plants to continuous light is used as antenna and light-harvesting pigments of the existing units, increasing thus their size. Moreover, the results suggest that the PS II unit is formed stepwise.

Water-splitting capacity In the beginning of greening under periodic light the rate of $H_2O \rightarrow DCPIP$ reduction is lower than that of $DPC \rightarrow DCPIP$, but the difference decreases as the number of LDC increases (Table 1). Moreover, dark incubation after a number of LDC increases the rate of $DPC \rightarrow DCPIP$ but not the rate of $H_2O \rightarrow DCPIP$ (see Table 2). These results indicate that the reaction centre of PS II is assembled first and then the H_2O-splitting enzymes are added; and that the H_2O-splitting enzymes, or their organization into active PS II units are light-induced or activated. This is supported also from the results of Sironval and coworkers [32,50], who found no oxygen evolution nor variable fluorescence in leaves greened under ms flashes, but their isolated plastids could reduce DCPIP with DPC as electron donor [37,58]. However, the rate on a Chl basis was lower than that of mature chloroplasts [37,58]. Transfer of these plants to continuous light resulted in the gradual appearance of the H_2O-splitting capacity [37,58,61,62]. These result indicate the presence of PS II in an inactive form in such leaves, which becomes active after exposing them to continuous light. It should be mentioned that the plastids formed under the ms flashes are also agranal [59], and they contain mainly Chl a [6].

As we found recently, plastids with high rates, on a Chl basis, of $DPC \rightarrow DCPIP$ reduction, but with no $H_2O \rightarrow DCPIP$, are formed when etiolated bean leaves are exposed

TABLE 3

Onset of photosystem II activity in developing plastids from 5-day old etiolated bean leaves exposed to 1 s flashes separated by 80 min dark period and then transferred to continuous light (CL)

Sample	Chl a	Chl b	µmoles DCPIP/g fresh wt.h	
	µg/g fresh wt		$H_2O \longrightarrow$ DCPIP	DPC $\longrightarrow$ DCPIP
65 (1 s) Flashes	197	5.2	60	1050
" + 40 min CL	203	7.1	680	1000

Experimental conditions as in Table 1.

to intermittent light (1 s light-80 min dark). The intensity of the light used was adequate to reduce most of the protochlorophyllide present in the etiolated leaves during the 1 s illumination period. Such leaves show also no variable fluorescence. Transfer of the plants to continuous light results in the gradual appearance of variable fluorescence and the $H_2O \longrightarrow$ DCPIP reduction. Results of a

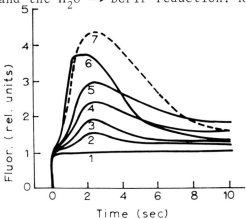

Fig. 1. Time course of the *in vivo* Chl a fluorescence emission at 685 nm, of 5-day-old etiolated bean leaves. A: exposed first to 72 1 s LDC (1 s light-80 min dark) (curve 1), and then transferred to continuous light for 20 min (2), 45 min (3), 140 min (4), and 24 h (6); or to 72 LDC, then to 140 min CL and then allowed to remain in the dark for 20 h (5). B: exposed to 58 LDC (2 min light-98 min dark) (7). The leaves were kept in darkness for 8 min before measurement. Measurements at room temperature; exciting light intensity; 575 µW/cm^2.

representative experiment are shown in Fig. 1 and Table 3. As it is evident, in these plastids the PS II units are small in size, and the H_2O-splitting capacity is missing, but it is induced by continuous illumination. It is not known if other components, in addition to the H_2O-splitting enzymes, are also missing in these plastids. This is quite possible, as judged from the results of Fig. 1, curves 7 and 3. As shown, the variable fluorescence of leaves exposed to 2 min LDC (curve 7) is higher than that of leaves exposed to 1 s LDC and then transferred to continuous light (curve 3), even though these two samples have the same PS II activity. Experiments are under way in order to clarify this point.

Photosystem II unit size

The development of the PS II unit size and their interaction have been studied by fluorescence emission spectroscopy [1]. Figs 2 and 3 show the normalized at F_o fluorescence rise kinetics curves of developing protochloroplasts in the presence or absence of DCMU, and Table 4 some of the fluorescence kinetics values. The curves have

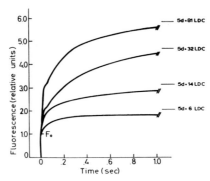

Fig. 2. Fluorescence induction of developing protochloroplasts from etiolated 5-day-old bean leaves exposed to LDC (2 min light-98 min dark). Room temperature; 5 µg Chl/ml; exciting light intensity: 2610 µW/cm^2. The curves were obtained from actual oscilloscope traces and normalized to the same F_o.

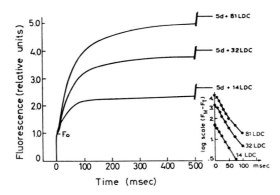

Fig. 3. Fluorescence induction in the presence of 12 μM DCMU of developing protochloroplasts. Exciting light intensity; 575 μW/cm². Inset: log plot of $F_M - F_T$ for the same curves. Other experimental conditions as in Fig. 2.

the typical characteristics of the fluorescence induction, i.e. the initial fast rise to the F_o level, and the slower rise to a final maximum yield F_{max}. The F_{max} level increases as the number of light-dark cycles increases, and values twice as high as those of mature chloroplasts are reached. Addition of NH_2OH, which is believed to feed electrons directly to the reaction center of PS II, increases the F_{max} level even higher. Moreover, addition of dithionite, which reduces directly the electron acceptor Q or PS II, increases the F_{max} to a much higher level. The half-rise time in the presence or absence of DCMU is larger (at least 7-fold) than the corresponding of mature chloroplasts. In the presence of DCMU the slow rise kinetics is exponential in the beginning of greening, but later becomes sigmoidal (see inset of Fig. 3). The sigmoidal shape becomes more pronounced as the number of LDC increases. Transfer of the periodic-light leaves to continuous light results in a dramatic decrease of the half-rise time and of the F_{max}/F_o ratio (see Table 4).

According to the Duysens and Sweers hypothesis [33], the fluorescence rise reflects the photoreduction of Q, a

TABLE 4

Fluorescence induction of protochloroplasts and chloroplasts from 5-day old etiolated bean leaves exposed to 2 min light–98 min dark cycles (LDC) and then transferred to continuous light (CL)

Sample	Chl a	Chl b	F_{max}/F_o	F_{NH_2OH}/F_o	Half-rise time of fluorescence	
	µg/g fresh wt				$I=296\ \mu W/cm^2$ +DCMU	$I=2610\ \mu W/cm^2$ −DCMU
					(ms)	
14 LDC	105	0	3.15	3.80	110	120
41 LDC	315	12	5.30	6.20	95	105
81 LDC	495	42	5.90	6.45	70	80
28 LDC	230	7	4.30	4.95	98	110
28 LDC + 3.5 h CL	594	124	3.30	3.30	20	21
28 LDC + 14 h CL	1210	378	3.30	3.30	13	14
Chloroplasts	1980	640	3.30	3.30	13	14

fluorescence quencher, to the non-quenching form Q^-. If the electron flow from Q^- is not inhibited, then Q^- can reduce the secondary electron carriers and finally the reaction center of PS I [47]. In the presence of an inhibitor (e.g. DCMU) the rate of the fluorescence rise represents the rate by which Q is reduced. The rate of Q reduction depends on the rate by which the absorbed photons reach the reaction center. Thus, it depends on the quality and intensity of the actinic light, the number of antenna Chl per PS II unit, and the efficiency of the units to utilize the absorbed photons [31]. With the same actinic light, and assuming that the efficiency remains constant, the difference in half-rise time in different samples reflects the difference in the size of the PS II unit. In protochloroplasts the half-rise time of the fluorescence rise is 7-times larger than that of mature chloroplasts, indicating that the size of their PS II unit is several times smaller than that of the mature chloroplasts; moreover, PS II units of similar size and organization are formed during the intermittent illumination, since the half-rise time remains constant throughout exposure.

The 5-fold increase in the rate of the fluorescence rise in only 3.5 h of continuous illumination after the 28 LDC (see Table 4) indicates a fast increase in the size of the PS II unit. This suggests that the Chl formed during the first hours of continuous light is inserted into preexisting units, increasing thus their size, and making them more efficient in absorbing the incident light.

The sigmoidal shape of the fluorescence induction in the presence of DCMU has been attributed to the intersystem II unit-unit interaction [39]. The change, therefore, from exponential to the sigmoidal shape can be considered as an indication of the development of energy transfer and connection between the PS II units. It seems, therefore, that at the beginning of greening under periodic light the system II units are structurally isolated but later they come into contact. This unit interaction may result either from unit aggregation or from insertion of newly synthesized membrane components in between the units rather than from the increase in unit size. Since the aggregation takes place even when the units are small in size, the reaction centers of the connected units must be close to each other, like multicenter units, so that a small number of Chl molecules, already packed around each unit, is adequate for the onset of the excitation energy transfer.

It is accepted that at room temperature only Chl a of PS II fluoresces. According to Kitajima and Butler [42], in isolated chloroplasts the non-variable (F_0) and the variable ($F_{max}-F_0=F_v$) fluorescence emission are both emitted from the bulk Chl a of PS II. Structural changes of the thylakoids induce changes in the F_v level. For example, suspension of granal chloroplasts in low salt medium dissociates the grana, and decreases the F_v level. Addition of salts reverses the process. Protochloroplasts have single thylakoids, and the presence or absence of salts has no effect on the F_{max} level in the presence of DCMU [17,29]. They have, however, very high F_v, and the F_{max}/F_0 is almost twice as high as that of mature chloroplasts. This implies that either in mature chloroplasts all F_0 does not originate from the bulk Chl of PS II [42], or that the intrinsic F_0 is lower in protochloroplasts. In the latter case it indicates that in protochloroplasts the bulk Chl is organized in such a way so that their PS II units are more efficient in utilizing the absorbed light energy.

The change observed in the F_{max}/F_0 ratio with time in periodic light reflects the decrease in the relative concentration of the unorganized Chl with time, which contributes only to the F_0 level. This is supported also from the results of the dark incubation experiments (see Table 2).

The decrease in the F_{max}/F_0 observed after transfer of the periodic light leaves to continuous light reflects the different changes that take place during the continuous illumination; i.e., formation of new Chl which increases the relative concentration of unorganized Chl (effect on the F_o); increase in size of the PS II unit, which has an effect on both F_o and F_{max} level; and finally, formation of grana with the concomitant appearance of the Mg^{++} effect on the spillover phenomena, which affect the F_V level.

In vivo *fluorescence measurements of periodic-light leaves*

That the periodic-light leaves (2 min LDC) are indeed active in PS II is also obvious from the *in vivo* fluorescence transient kinetics of intact leaves. Fig. 4 shows the normalized at F_o fluorescence induction kinetics of etiolated bean leaves exposed to different number of LDC.

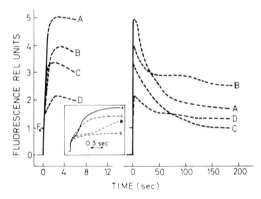

Fig. 4. Time course of the Chl a fluorescence emission at 685 nm *in vivo* of 5-day-old etiolated bean leaves exposed to periodic light (2 min light-98 min dark). Room temperature; exciting light intensity: 575 μW/cm². Curves: (A): 87 LCD; (B): 27 LDC; (D): LDC; (C): mature green bean leaf (control). The leaves were kept in darkness for 8 min before fluorescence measurements were taken.

In the same figure the fluorescence induction kinetics of a green bean leaf is also shown, for comparison (curve C). It is interesting to note that the transient variations in fluorescence of the periodic-light leaves are similar to those of intact green leaves and on intact algal cells,

[22,40,44,56], i.e., a fast increase to the F_o level, and then a slower rise to the steady state level through transient variations resulting in three peaks. The first and second peaks appear in a few ms and in 2-3 s after the onset of illumination, respectively; the third peak appears after 2-3 min. Moreover, the height of the second peak of the periodic-light leaves reaches values much higher than the corresponding one of the green control. This is similar to the results observed with the isolated protochloroplasts where their F_{max} level reaches values much higher than the F_{max} of the control chloroplasts (see Fig. 2).

The fluorescence variations observed in intact algal cells, green leaves and isolated intact chloroplasts have been attributed to the redistribution of absorbed light energy among the two photosystems [27,52] to the cation movement in chloroplasts and through the thylakoid membrane [23,45,55], to conformational changes occurring within the thylakoid membrane [63], etc. It is interesting to note here that the lack of the regulation of the energy distribution among the two photosystems observed in protochloroplasts (see Fig. 12) has no effect on the *in vivo* transient variations in fluorescence of the corresponding periodic-light leaves.

Development of the heterogeneous PS II units

It has been shown [54], that the area over the fluorescence induction curve of isolated chloroplasts is a measure of the total number of charge separations that occur at PS II during the fluorescence induction period. A first order kinetic analysis of the area growth in the presence of DCMU revealed that it occurred in two distinct linear phases, a fast and a slow phase. This suggested a functional differentiation in PS II, and the existence of two types of PS II centers, the efficient α-centers and the relatively less efficient β-centers [49]. The kinetics of the area accumulation over the fluorescence induction curve of DCMU-poisoned bean protochloroplasts which received 23 LDC, and a first order analysis of the area as a function of time are shown in Fig. 5. As observed, this

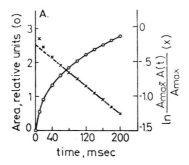

Fig. 5. Kinetics of the area accumulation over the fluorescence induction curve of protochloroplasts (o) from 5-day-old etiolated bean leaves exposed to 23 LDC (2 min light-98 min dark), and a first order analysis of the area as a function of time (x); 12 μM DCMU; exciting light intensity; 2610 μW/cm^2; 4 μg Chl/ml. The total area (A_{max}) was equal to 3.8 r.u.

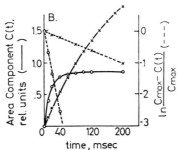

Fig. 6. Kinetics of the fast and the slow component (o and, x, respectively) of the area growth of isolated protochloroplasts (23 LDC, 2 min light-98 min dark), and a first order analysis of their time course. The area covered by the α component (o) was 0.8 r.u., that covered by the β component (x) was 3.0 r.u. $K'_\alpha = 62$ s^{-1}, $K_\beta = 5$ s^{-1}. For other experimental conditions see Fig. 5.

accumulation is biphasic. The slope of the line corresponding to the second phase of the area growth gives the value of the rate constant K_β of the β-centers. The intercept of this line with the logarithmic ordinate at zero time gives the log β_{max}, the maximum value of the area accumulation due to β-centers. The value of β_{max} permits the calculation of the relative concentration of the α and β-centers in the plastids [49]. The kinetics of the fast (α-centers) and the slow (β-centers) component of the area growth of the same protochloroplasts are shown in Fig. 6. As we see, the α and the β component of the area are exponential functions of time, and the slope of the line of the log plot of the α component of the area gives the value of the rate constant K'_α of the α-centers. In mature chloroplasts the log plot of the α component is not exponential [49], and a comparable value of K'_α is taken from the half-time of the growth of the α component.

The differential effect of Mg^{++} ions on the function of the two types of PS II centers led to the hypothesis that the chloroplast grana are the loci of the efficient α-centers while the stroma thylakoids are the loci of the β-centers [49].

The development of the two types of units (α and β) has been studied [48] in an effort to understand how they develop, what's their interrelationship, and, moreover, if the heterogeneity of the photochemical centers of the PS II units is an endogenous property of the chloroplast lamellae or a consequence of their differentiation into grana and stroma thylakoids.

It was found that in continuous illumination the first photosynthetic membranes formed contain both the α and the β-centers. During the first hours of continuous light the relative concentration of β-centers is high, but gradually decreases, and in 6-8 hours it reaches the steady-state value of mature chloroplasts (about 65%). The magnitude of the photochemical rate constants K'_α and K_β increases with time in continuous light and it reaches the steady-state value of the mature chloroplast. Most of the increase is accomplished within 4 hours of continuous light. The K'_α and K_β depends among other things on the size of the PS II unit. Under our experimental conditions the change in the value of K'_α and K_β during greening gives an idea of the change in the size of the PS II unit. Obviously, under continuous light there is a fast growth of the units around the α- and β-centers.

Since protochloroplasts do not contain any grana structures they can be used to ascertain whether the formation of the α-centers is a phenomenon related to grana formation. Figs. 7 and 8 show the change in the relative

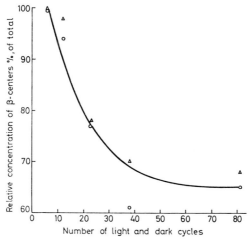

Fig. 7. Intermittent illumination-induced change in the relative concentration of β-centres in developing protochloroplasts. For experimental conditions see Fig. 5.

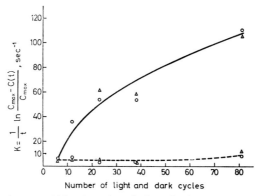

Fig. 8. Intermittent illumination-induced change in the photochemical rate constants K'_α (—) and K_β (- - -) in developing protochloroplasts. For experimental conditions see Fig. 5. For mature chloroplasts, $K'_\alpha = 480$ s^{-1}, $K_\beta = 44$ s^{-1}.

concentration of the α and β-centers, and their respective rate constants K'_α and K_β during greening under periodic light. It is seen that the originally high relative concentration of the β-centers decreases, and finally it reaches the steady-state value of the green control (65%). The results indicate that the kinetic expression of the α-centers does exist in protochloroplasts and, therefore, that the formation of the α-centers is a phenomenon that precedes the biosynthesis of Chl b, the CP II, and the formation of grana (see also Fig. 5 and 6). In the beginning the values of both K'_α and K_β are almost equal. Subsequently, however, there is a dissimilar dependence of K'_α and K_β on the number of LDC. The value of K'_α remains constant, while the value of K'_α increases steadily with time in periodic light, but none ever reaches the value of the green control.

Transfer of the periodic-light leaves to continuous light induces a fast and parallel increase of K'_α and K_β up to the value of the green control. The results of a representative experiment, where etiolated leaves were exposed first to 14 LDC and then transferred to continuous light, showed that in 1 hour the K'_α and K_β increases almost 5-times, and in 3 hours they both reach the value of mature chloroplasts.

These results indicate that the α and β-centers are both present in the photosynthetic membranes from the beginning of the light-induced greening; the α-centers and the photosynthetic units associated with them develop independently of the β-centers; the α-centers do not share the same pigment bed with the β-centers, otherwise, there should have been a parallel change in K'_α and K_β under periodic light; the presence of Chl b, CP II and grana is not a prerequisite for the formation of the α-centers; and the heterogeneity of the photochemical centers in PS II is an endogenous property of the chloroplast lamellae.

Onset of the control of excitation energy distribution Correlation between grana and Chl a fluorescence rise The most characteristic change occurring during the phototransformation of the etioplast or protochloroplast to the mature chloroplast stage is the differentiation of the prolamellar body or of the primary thylakoid, respectively, to the stroma and grana thylakoids. The high PS II activity found in primary thylakoids is gradually decreased during this differentiation to the value of the mature stage, and finally, it is found associated mainly with the grana stacks.

Grana stacking, apart from being responsible for more efficient harvesting of the light energy, appears also to be necessary for controlling the excitation energy distribution among the two photosystems. Recent studies have shown [17], that the structural changes leading to grana stacking are most probably involved in the control of the excitation energy distribution between the two photosystems. This conclusion was suggested from the following observations: As it is known, the degree of grana stacking, apart from being dependent on light conditions, greatly depends on the ionic environment. Thus grana dissociate in low-salt media giving rise to single lamellae, and reassociate upon addition of cations [28,38,60]. This dissociation-reconstitution process is correlated with the very efficient formation of light or heavy subchloroplast fractions, respectively, produced after digitonin disruption of chloroplasts [5,13]. Digitonin disruption separates stacked (grana) from unstacked (stroma) lamellae [5]. Fig. 9 shows the effect of cation addition to low-salt agranal spinach chloroplasts on thylakoid aggregation. As shown, addition of cations to Tricine-treated, "low-salt", disorganized plastids induces the formation of heavy subchloroplast fractions (grana), the yield of which depends on the concentration of cations. Based on yield, composition and photochemical activities of the disorganized and

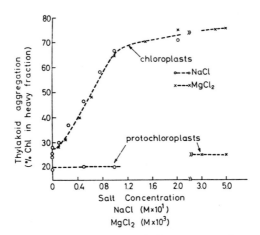

Fig. 9. The effect of monovalent and divalent cations on the yield of heavy subchloroplast fractions in spinach chloroplasts or bean protochloroplasts washed and suspended in 0.05 M Tricine-NaOH, pH 7.2. Heavy fractions, obtained by differential centrifugation of digitonin disrupted plastids (disruption at 0° for 30 min with 0.5% final digitonin concentration) represent 1 K (1,000 g, 10 min) + 10 K (10,000 g, 30 min) for spinach chloroplasts, or 5 K (5,000 g, 10 min) + 10 K for protochloroplasts.

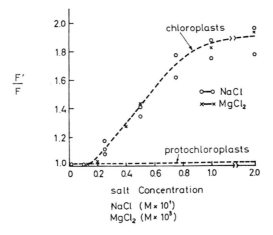

cation reconstituted fractions [5,13,17], as well as on electron microscopy evidence [20], one can conclude that the cation-induced aggregation of the thylakoids is a specific binding leading to grana reconstitution. Adopting the low-salt disorganization-cation reconstitution approach, we further tested whether the cation-induced reconstitution of grana is in any way correlated with the cation-induced increase in the Chl a fluorescence yield, considered to monitor the control of excitation energy distribution between the two photosystems [52]. Fig. 10 shows the effect of cation addition to low-salt agranal spinach chloroplasts on the Chl a fluorescence yield. Comparison of Figs 9 and 10 shows, that both parameters increase drastically upon addition of monovalent or divalent cations to low-salt spinach chloroplasts, and that divalent cations are more effective than monovalent in both cases. The close parallelism of the processes is shown in Fig. 11, where a plot of the percent change of both, the Chl a fluorescence yield increase and the degree of thylakoid aggregation, as a function of the concentration of monovalent or divalent cations is presented. It is clear that a very good correlation exists between the two parameters. Such a good correlation was also found to exist when cations were added to light subchloroplast fractions obtained after sonication of low-salt, Tricine-suspended, pea chloroplasts [17].

These results, therefore, show that the structural changes leading to grana stacking are implicated in the

---

Fig. 10. The effect of cation concentration on the Chl a fluorescence yield of Tricine-washed spinach chloroplasts or bean protochloroplasts. The results are presented as the ratio of the fluorescence yield in the presence (F') and in the absence (F) of cations. Exciting light intensity for spinach chloroplasts: 574 $\mu W/cm^2$, for protochloroplasts 2610 $\mu W/cm^2$. Fluorescence was measured at 680 nm. The suspension contained 10 μg Chl/ml (spinach chloroplasts) or 5 μg Chl/ml (protochloroplasts) in 50 mM Tricine-NaOH, pH 7.2, and the concentrations of cations shown. The plastids were incubated with the cation at least for 5 min before measurement. DCMU at 10 μM was also present.

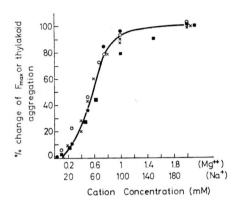

Fig. 11. Effect of cation concentration on the percent increase in F_{max} and in the yield of heavy subchloroplast fractions in Tricine-washed spinach chloroplasts. The values obtained at 2 mM $MgCl_2$ or 100 mM NaCl represent the 100%; those in Tricine, before addition of cations, represent the 0 value. F_{max}: ● (Na^+), ■ (Mg^{2+}); thylakoid aggregation: o (Na^+), x (Mg^{2+}). For experimental conditions see Figs 9 and 10.

control of the excitation energy distribution among the two photosystems.

As shown in Figs. 9 and 12, cation addition to low-salt, Tricine-suspended, pea protochloroplasts does not affect the degree of thylakoid aggregation nor their Chl a fluorescence yield. This suggests that components necessary for

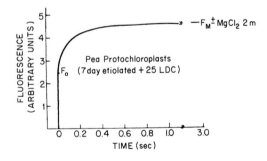

Fig. 12. The Chl a fluorescence rise on illumination of dark-adapted Tricine-washed pea protochloroplasts. Exciting light intensity; 2610 µW/cm²; 5 µg Chl/ml and 10 µM DCMU in 50 mM Tricine. Other conditions as in Fig. 10.

stacking and for the control of spillover are missing, and that these components are formed gradually during thylakoid maturation in continuous light.

Development of grana stacking The good correlation found to exist between the extent of grana reconstitution and the increase in the Chl a fluorescence yield caused by addition of cations to low-salt chloroplasts, point out to a parallel development of grana and of the capacity for the regulation of the excitation energy distribution among the two photosystems. We studied, therefore, the development of the cation-induced stacking capacity (i.e., grana formation) during the transformation of protochloroplasts to chloroplasts, and we tried to correlate it with the various thylakoid components synthesized during this transformation. Fig. 13 shows the development of the cation-induced stacking capacity in etiolated leaves which received 42 LDC and then were transferred to continuous light. Only 10 to 30% of the Chl is pelletable in the heavy fractions in the Tricine-suspended samples, depending on the

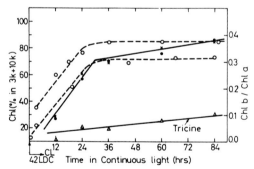

Fig. 13. Development of the cation-induced stacking capacity of thylakoids of pea leaves exposed to 42 LDC (2 min light-98 min dark) and then transferred to continuous light. The stacking capacity is monitored by the yield of heavy fractions (Chl % of recovered) after digitonin disruption of plastids suspended in Tricine prior to or after monovalent or divalent cation addition. Chlorophyll in heavy fraction prior to cation addition, △—△; after addition of Na^+ (100 mM), ●—●; after addition of Mg^{2+} (5 mM), x--x; Chl b to Chl a ratio in plastids (upper curve) or leaves (lower curve), o--o. The Chl concentration in plastids prior to digitonin addition was 100 μg/ml.

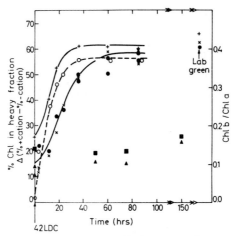

Fig. 14. Effect of illumination conditions on the capacity of pea thylakoids to stack after addition of cations to plastids suspended in Tricine-NaOH (50 mM, pH 7.2). This capacity is monitored by the difference in the amount of Chl pelletable in the heavy fractions after cation addition minus that pelletable prior to cation addition. The subchloroplast fractions were isolated after digitonin disruption. Periodic light leaves: 5 mM Mg^{2+}, (▲); 100 mM Na^+, (■); leaves transferred to continuous light: 5 mM Mg^{2+}, (x); 100 mM Mg^{2+}, (+), or 100 mM Na^+ (o). Chl b/Chl a ratio: o-o. Lab green: 19 day plants, night-day conditions. The Chl concentration in plastids, prior to digitonin disruption was 100 μg/ml or 400 μg/ml.

time of exposure to continuous light. As exposure is prolonged the effectiveness of Tricine washing to disorganize the grana formed is diminished. The cation-induced stacking capacity increases with time in continuous light, and appears in parallel with the synthesis of Chl b. In Fig. 14 comparison is made of the development of the cation effect between plastids of leaves transferred from periodic to continuous light with those that remained in periodic light. In this figure the net value of the cation effect is shown, i.e., after subtracting the respective values found in Tricine. The biphasic type of curve found in Fig. 13 is no longer noticed. Contrary to the strong cation effect found in continuous light, only a small effect begins to be noticed in periodic light, and this only

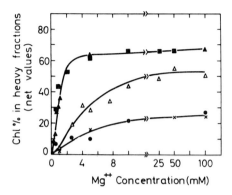

Fig. 15. Effect of the thylakoid developmental stage on the Mg^{2+} concentration dependence. The stacking capacity determined as in Fig. 14. Plastids were obtained from leaves exposed to 40 (o) or 72 (x) LDC; transferred to CL for 18 (Δ) or 144 (■) hours after 43 LDC; grown in laboratory day-night conditions for 19 days (▲). The Chl concentration prior to digitonin disruption was in the order above and in μg/ml: 70, 155, 300, 277, and 400.

after prolonged exposure.

The gradual appearance of the cation effect with greening brought up the question whether these results reflect also a different saturation pattern between the mature and the developing plastids. As shown in Fig. 15, in plastids of pea leaves grown in the laboratory (day-night conditions) for 19 days, as well as in plastids of pea leaves that received 43 LDC and then were transferred to continuous light for 144 hours, the effect saturates at low cation concentrations, 1-2 Mg^{++}, as in spinach chloroplasts. In plastids, however, of pea leaves that received 43 LDC and then were transferred to continuous light for 18 hours, the effect does not saturate up to about 50 mM Mg^{++}. Two possibilities can be considered to account for this difference: First, the thylakoids at the early stages of development may lose easily their ionic content, and thus have lower internal cation concentration. Higher concentrations of cations are, therefore, needed to overcome this loss. This possibility, however, does not seem to be substantiated, since, as shown in Table 5, the difference in the cation content of thylakoids isolated at different developmental stages is not such to account for the 10-fold difference in the cation requirement for thylakoid aggregation. Another possibility is that grana formation occurs gradually, so that grana thylakoids at various stages of development, incompletely to fully developed ones, are present, and they may aggregate at different cation concentrations, depending on the developmental stage.

The gradual appearance of the stacking capacity after transfer of the periodic-light leaves to continuous light strongly suggested the parallel formation of thylakoid components which are either directly or indirectly involved in this process. A first indication of such involvement was offered by the parallel appearance of Chl b. The appearance of Chl b in a way can be considered as

TABLE 5

Mg^{2+} content of pea thylakoids during development

Light conditions	Mg^{2+} (μmoles/mg dry wt)	Chl (μg/mg dry wt)	Mg^{2+} (μmoles/mg Chl)
A. Peas			
40 LDC[a]	0.007	3.2	
40 LDC[b]	0.007	5.0	
43 LDC + 24 h CL[a]	0.040	62.2	0.64
43 LDC + 73 h CL[a]	0.147	96.8	1.52
43 LDC + 73 h CL[b]	0.070	91.7	0.76
green[a]	0.112	83.7	1.34
green[b]	0.072	86.8	0.83
B. Spinach[b]	0.180	93.6	1.92

Determinations were done by atomic absorption. Plastids were isolated from 4-5 g fresh weight leaves as pellets at 3,000 g (intermittent light) or 1,000 g (continuous light and mature green leaves). They were suspended in 20 ml 0.3 M sucrose-0.01 M KCl-0.05 M phosphate buffer, pH 7.2, and were collected in two parts. One part was washed with 0.05 M phosphate-0.01 M KCl, pH 7.2, (a), the other with 0.05 M Tricine-NaOH, pH 7.2 (b), 20 ml each. The thylakoids were recovered by centrifugation and were taken to dryness.

monitoring the formation of the CP II, since it is needed for the stabilization of the pigments on the protein of CP II [16,18,19,34]. The lack of this complex from primary thylakoids, and its formation after transfer of the periodic-light leaves to continuous light [16,18,19], along with grana formation [20], as well as the fact that isolated grana fractions or granal chloroplasts are enriched in this complex, while stroma lamellae or agranal chloroplasts are enriched in CP I [14,16], further supported the involvement of CP II in the stacking process. In addition, it was shown lately that the onset of the cation-induced control on spillover parallels the appearance of CP II in pea plants transferred to continuous light after exposure to periodic light [29]. However, since the presence or absence of CP II, as monitored by pigment binding on the CP II protein, depends highly on the presence of Chl b, the stabilizing pigment, the crucial question to be answered is whether the polypeptides derived from CP II are also gradually synthesized in continuous light. Early results have shown that in mature wild type or mutant plants or algae having stacked or unstacked membranes, the polypeptides of so-called group II (having M.W. between 25,000 and 30,000), and believed to be derived from CP II, are indeed present in higher amounts in the granal plastids [11,41,46]. Our studies have shown that this is true also for plastids transformed from the agranal to the granal stage, i.e., during protochloroplast to chloroplast differentiation. This is shown clearly in Figs 16 and 17. Fig. 17 shows the electrophoretic analyses of lipid free thylakoids at various stages of development; Fig. 16 shows the electrophoretic pattern of the mature pea thylakoids along with that of simultaneously run molecular weight markers. The pattern obtained from the mature thylakoids resembles that obtained from lipid free lamellae of a variety of sources [11,12,35,41,43,46,57]. A marked difference is evident in the popypeptide region marked 8-10, corresponding to

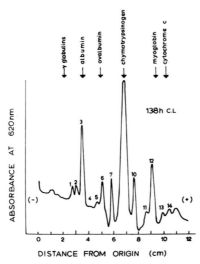

Fig. 16. SDS-gel electrophoresis run of lipid-free thylakoids of pea leaves exposed to CL for 138 hours. Marker protein molecular weights are from left to right: 160,000, 67,000, 45,000, 25,000, 17,800, and 12,400.

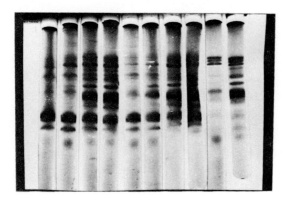

Fig. 17A. Electrophoretic patterns of SDS-solubilized lipid-free thylakoids obtained from leaves exposed from left to right to: 28 LDS, 70 LDC, 44 LDC+70 h CL, 18 h CL, 42 h CL, 114 h CL and 138 h CL. The last two gels at the right show the pattern of the 144 K fraction, left, and 10 K fraction, right, obtained after digitonin disruption of 138 h CL plastids.

DEVELOPMENT OF PS II UNIT

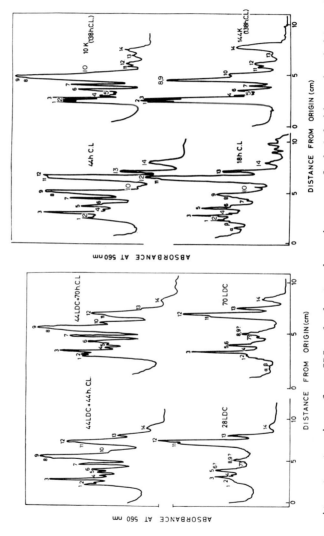

Fig. 17B. Densitometer tracings from SDS gel electrophoresis runs of thylakoids and subchloroplast fractions of plastids at various stages of development.

molecular weights between 25,000 and 30,000, between the primary thylakoids and the mature green ones. This group of peptides increases gradually with greening in continuous light, and predominates also in the 10 K fraction (grana) obtained from mature chloroplasts, in accordance to earlier results [35].

These results, therefore, strongly suggest that these polypeptides are required for the cation-induced stacking and for the control of the spillover phenomena. Cations are generally believed to act by binding on the fixed negative charges of the thylakoids, the isoelectric point of mature chloroplasts being very low, vis. 4.7 [30]. Stacking of adjacent thylakoids may be facilitated by either divalent salt-bridge formation or by monovalent cation suppression of the electrostatic repulsion, through neutralization of some charges. Electrolytes are also expected to lower the water solubility of the membranes, and enhance the hydrophobic forces which prevail between adjacent membranes [51]. Although both lipids and proteins may contribute to the negative charges, proteins are more strongly implicated in the cation binding [24]. If the parallel increase in the cation stacking capacity and the synthesis of polypeptides with molecular weights between 15,000 and 30,000 does indeed reflect the involvment of these peptides in the cation binding, one has to visualize a process by which these peptides, along with Chl a and Chl b, are incorporated in the membrane as it grows in continuous light; this incorporation results in the increase in the PS unit size, and offers the necessary organization and charge distribution to induce grana stacking. The structural changes involved in this stacking set up the control of the excitation energy distribution between the two photosystems. In thylakoids, therefore, in which some of these polypeptides are missing or present in minor amounts, one might expect that cations would not induce thylakoid appression, nor would they set up the control of the spillover. Similarly, in thylakoids which lack some of these peptides, it might be possible to find, that higher concentrations of cations may be effective, as in the case of the early stages of thylakoid development (see Fig. 15), where saturation of the cation effect occurs at high cation concentration (25 mM Mg^{++}). It should be mentioned here, that preliminary experiments showed that the 10 K fraction pelletable at low cation concentration (Fig. 15, 43 LDC + 18 h CL) is enriched in the mentioned polypeptides, contrary to that pelletable at high cation concentration. In the latter case peptides resolved at the positions of peptides numbered 6 and 7 (Fig. 16) seem to be in excess. It should be of interest, therefore, to see whether the thylakoids of the Chl b - less barley mutant, which lack the CP II, are deficient in some of the group II peptides (II_b and II_c of ref. [52]), and show only a few appressed lamellae and a small cation effect on the Chl a fluorescence yield at 0.67 mM Ca^{++}, [25], have the capacity for a higher cation effect at higher cation concentrations.

Photocontrol of PS II development

It was mentioned, that the differentiation of etioplasts to chloroplasts depends on light, and that the reduction of protochlorophyllide to Chl(ide) is a photochemical reaction. Light is not only necessary for the reduction of protochlorophyllide, but it also induces or activates some of the steps of chloroplast development. For example, the H_2O-splitting enzymes are light-induced or light-activated. The results of the dark incubation experiments support this conclusion. Moreover, greening in 1 s LDC or in ms flashes induces the formation of PS II units in etiolated leaves which shows DCPIP reduction with DPC, but not with H_2O, as electron donor.

The synthesis of Chl b, the polypeptides with molecular weights between 25,000 and 30,000, the CP II and grana, and also the development of the ability for cation-induced stacking and cation-induced Chl a fluorescence yield, are induced by continuous illumination [16,18,21]. Since there is an interrelationship between these membrane components, the results indicate that either the synthesis of all or some of them, or the rate of their integration into the developing membrane, is controlled by continuous illumination.

Recent studies in our laboratory have shown [2], that if the etiolated leaves are exposed to red light LDC (2 min red light-98 min dark) instead of to white light LDC, then the PS II activity (DCPIP reduction) as well as the rate of $^{14}CO_2$-fixation, on a Chl basis, reach the plateau in less than 14 LDC. Since the rise in the activity in the beginning of greening reflects the rate by which the unorganized Chl is integrated into the developing membrane, the results indicate that the organization of Chl into active units is faster in red light LDC than in white light LDC. The integration of Chl into the developing membrane takes place together with the integration of the necessary polypeptides which are synthesized during the light and dark period. So, the rate of Chl organization will depend on the rate by which the polypeptides are synthesized, and on the rate by which all components are integrated into the developing membrane. It seems, therefore, that red light enhances the rate of either of these two processes.

Conclusions and Summary

The development of the PS II unit is a stepwise process. During the biogenesis and growth of the PS II unit in periodic light first the reaction center is formed with a few antenna Chl closely packed around. The unit is thus very efficient for photochemistry. This structure represents the core of the PS II unit, and does not require the formation of CP II. The orientation of the pigments which are close to the reaction center seems to be different from that of the rest of the pigments, as judged by the *in vivo* 518 nm absorbance changes of such leaves [9]. The H_2O-splitting enzymes are then formed, which are light-activated or induced. As the number of small units increases they form clusters which have the reaction centers very close to each other. The aggregation of the units is controlled by the structural development and organization of the membrane and not by the concentration and type of Chl. The heterogeneity of the PS II units is an intrinsic property of the membrane. The α-units and β-units develop independently of each other from the beginning of the light-induced greening. They do not share the same pigment bed. The presence of Chl b, CP II and grana is not a prerequisite for the formation or development of the α-centers.

Further greening in continuous light results in the formation of Chl b and the polypeptides with molecular weights between 25,000 and 30,000 and more Chl a and carotenoids. The Chl, which is formed with a high rate, is inserted into preexisting units increasing their size, and making them more efficient for absorbing the incident light. At the same time the stacking component, CP II, is formed, which has as a consequence the formation of grana, and the development of the ability for cation-induced stacking and cation-induced Chl a fluorescence rise.

Acknowledgements

The experiments on the development of the heterogeneous PS II units have been conducted in collaboration with Dr. A. Melis. We wish to thank Mr. S. Daoussis for his excellent technical assistance, and Dr. A. Souliotis and Mrs. V. Dimitrelli of the Analytical Laboratory, Chemistry Department, N.R.C. "Democritos", for the determination of the Mg^{++} content in the developing thylakoids.

References

1. Akoyunoglou, G. (1977). *Arch. Biochem. Biophys.*, in press.
2. Akoyunoglou, G. (1977). *Photochem. Photobiol.*, in press.
3. Akoyunoglou, G. and Argyroudi-Akoyunoglou, J.H. (1969). *Physiol. Plantarum* **22**, 288-295.
4. Akoyunoglou, G. and Argyroudi-Akoyunoglou, J.H. (1972). *In* "Proceedings of the IInd International Congress on Photosynthesis Research" (G. Forti, M. Avron and A. Melandri, eds), Vol. III, 2427-2436, Dr. W. Junk N.V. Publishers, The Hague.
5. Akoyunoglou, G. and Argyroudi-Akoyunoglou, J.H. (1974). *FEBS Lett.* **42**, 135-140.
6. Akoyunoglou, G., Argyroudi-Akoyunoglou, J.H., Michel-Wolwertz, M.R. and Sironval, C. (1966). *Physiol. Plantarum* **19**, 1101-1104.
7. Akoyunoglou, G. and Michelinaki-Maneta, M. (1975). *In* "Proceedings of the Third International Congress on Photosynthesis" (M. Avron, ed.), Vol. III, 1885-1896, Elsevier Scientific Publishing Company, Amsterdam, Oxford, New York.
8. Akoyunoglou, G. and Siegelman, H.W. (1968). *Plant Physiol.* **43**, 66-68.
9. Akoyunoglou, G. and Tsimilli-Michael, M. (1977). *Plant Sci. Lett.*, in press.
10. Anderson, J.M. and Boardman, N.K. (1966). *Biochim. Biophys. Acta* **112**, 403-421.
11. Anderson, J.M. and Levine, R.P. (1974). *Biochim. Biophys. Acta* **333**, 378-387.
12. Anderson, J.M. and Levine, R.P. (1974). *Biochim. Biophys. Acta* **357**, 118-126.
13. Argyroudi-Akoyunoglou, J.H. (1976). *Arch. Biochem. Biophys.* **176**, 267-274.
14. Argyroudi-Akoyunoglou, J.H. (1976). Abstracts of the 7th International Congress on Photobiology, p. 87, Rome.
15. Argyroudi-Akoyunoglou, J.H. and Akoyunoglou, G. (1970). *Plant Physiol.* **46**, 247-249.
16. Argyroudi-Akoyunoglou, J.H. and Akoyunoglou, G. (1973). *Photochem. Photobiol.* **18**, 219-228.
17. Argyroudi-Akoyunoglou, J.H. and Akoyunoglou, G. (1977). *Arch. Biochem. Biophys.* **179**, 370-377.
18. Argyroudi-Akoyunoglou, J.H., Feleki, Z. and Akoyunoglou, G. (1971). *Biochem. Biophys. Res. Comm.* **45**, 606-614.
19. Argyroudi-Akoyunoglou, J.H., Feleki, Z. and Akoyunoglou, G. (1972). *In* "Proceedings of the IInd International Congress on Photosynthesis Research" (G. Forti, M. Avron and A. Melandri, eds), Vol. III, 2417-2426, Dr. W. Junk N.V. Publishers, The Hague.
20. Argyroudi-Akoyunoglou, J.H., Kondylaki, S. and Akoyunoglou, G. (1976). *Plant Cell Physiol.* **17**, 939-954.
21. Argyroudi-Akoyunoglou, J.H. and Tsakiris, S. (1977). *Arch. Biochem. Biophys.*, in press.
22. Bannister, T.T. and Rice, G. (1968). *Biochim. Biophys. Acta* **162**, 555-580.
23. Barber, J., Telfer, A. and Nicolson, J. (1974). *Biochim. Biophys. Acta* **357**, 161-165.

24. Berg, S., Dodge, S., Krogmann, D.W. and Dilley, R.A. (1974). *Plant Physiol.* **53**, 619-627.
25. Boardman, N.K. and Thorne, S.W. (1976). *Plant Sci. Lett.* **7**, 219-224.
26. Boardman, N.K., Anderson, J.M., Hiller, R.G., Kahn, A., Roughan, P.G., Treffy, T.E., and Thorne, S.W. (1972). *In* "Proceedings of the IInd International Congress on Photosynthesis Research" (G. Forti, M. Avron and A. Melandri, eds), Vol. III, 2265-2287, Dr. W. Junk N.V. Publishers, The Hague.
27. Bonaventura, C. and Myers, J. (1969). *Biochim. Biophys. Acta* **189**, 366-383.
28. Brangeon, J. (1974). *J. Microscopy* **21**, 75-84.
29. Davis, D.J., Armond, P.A., Gross, E.L. and Arntzen, C.J. (1976). *Arch. Biochem. Biophys.* **175**, 64-70.
30. Dilley, R.A. and Rothstein, A. (1967). *Biochim. Biophys. Acta* **135**, 427-443.
31. Dubertret, G. and Joliot, P. (1974). *Biochim. Biophys. Acta* **357**, 399-411.
32. Dujardin, E., De Kouchkovsky, Y. and Sironval, C. (1970). *Photosynthetica* **4**, 223-227.
33. Duysens, L.N.M. and Sweers, H.E. (1963). *In* "Studies on Microalgae and Photosynthetic Bacteria" (Japanese Society of Plant Physiologists, ed), 353-372, The University of Tokyo Press, Tokyo.
34. Genge, S., Pilger, D. and Hiller, R.G. (1974). *Biochim. Biophys. Acta* **347**, 22-30.
35. Henriques, F. and Park, R.B. (1974). *Plant Physiol.* **54**, 386-391.
36. Hoober, D.K. (1970). *J. Biol. Chem.* **245**, 4327-4334.
37. Inoue, Y., Kobayashi, Y., Sakamoto, E. and Shibata, K. (1974). *Physiol. Plantarum* **32**, 228-232.
38. Izawa, S. and Good, N.E. (1966). *Plant Physiol.* **41**, 544-552.
39. Joliot, A. and Joliot, P. (1964). *C.r. Acad. Sci. Paris* **258**, 4622-4625.
40. Kautsky, H., Appel, W. and Amann, H. (1960). *Biochem. Z.* **332**, 277-292.
41. Kirchanski, S.J. and Park, R.B. (1976). *Plant Physiol.* **58**, 345-349.
42. Kitajima, M. and Butler, W.L. (1975). *Biochim. Biophys. Acta* **376**, 105-115.
43. Klein, S.M. and Vernon, L.P. (1974). *Photochem. Photobiol.* **19**, 43-49.
44. Kobayashi, Y., Inoue, Y. and Shibata, K. (1975). *Plant Cell Physiol.* **16**, 767-776.
45. Krause, G.H. (1974). *Biochim. Biophys. Acta* **333**, 301-313.
46. Levine, R.P. and Duram, H.A. (1973). *Biochim. Biophys. Acta* **325**, 565-572.
47. Malkin, S. and Kok B. (1966). *Biochim. Biophys. Acta* **126**, 413-432.
48. Melis, A. and Akoyunoglou, G. (1977). *Plant Physiol.* **59**, 1156-1160.
49. Melis, A. and Homann, P.H. (1976). *Photochem. Photobiol.* **23**, 343-350.
50. Michel, J.M. and Sironval, C. (1972). *FEBS Lett.* **27**, 231-234.
51. Murakami, S. and Packer, L. (1971). *Arch. Biochem. Biophys.* **146**, 337-347.
52. Murata, N. (1969). *Biochim. Biophys. Acta* **172**, 242-251.

53. Murata, N. (1969). *Biochim. Biophys. Acta* **189**, 171-181.
54. Murata, N. Nishimura, M. and Takamiya, A. (1966). *Biochim. Biophys. Acta* **120**, 23-33.
55. Murata, N., Tashiro, H. and Takamiya, A. (1970). *Biochim. Biophys. Acta* **197**, 250-256.
56. Papageorgiou, G. and Govindjee (1968). *Biophys. J.* **8**, 1316-1328.
57. Remy, R. (1971). *FEBS Lett.* **13**, 313-317.
58. Remy, R. (1973). *Photochem. Photobiol.* **18**, 409-416.
59. Sironval, C., Bronchart, R., Michel, J.M., Brouers, M. and Kuyper, Y. (1968). *Bull. Soc. Franc. Physiol. Vég.* **14**, 195-225.
60. Smillie, R.M., Henningsen, K.W., Nielsen, N.C. and von Wettstein, D. (1976). *Carlsberg Res. Comm.* **41**, 27-32.
61. Strasser, R.J. and Sironval, C. (1972). *FEBS Lett.* **28**, 56-60.
62. Strasser, R.J. and Sironval, C. (1973). *FEBS Lett.* **29**, 286-288.
63. Vandermeulen, D.L. and Govindjee (1974). *Biochim. Biophys. Acta* **368**, 61-70.

ONTOGENETIC EFFECTS IN OXYGEN EVOLUTION

Z. ŠESTÁK

Institute of Experimental Botany, Czechoslovak Academy of Sciences, Praha (CSSR)

Since the discovery of the Hill reaction [9] this activity of isolated chloroplasts has been measured very often as a characteristic of photochemical capacity of the respective plant material. The importance of this characteristic increased after the formulation of the scheme of electron transport chain with two photoreactions, in which Hill activity is connected with the oxygen evolving site of the photosynthetic mechanism. As a simply determinable relation unit the chlorophyll amount in chloroplasts has mostly been used.

The first scientists who observed, that the Hill activity per chlorophyll unit is not the same in leaves of various ages or sizes, were Clendenning and Gorham [3]. They observed in New Zealand spinach and wheat leaves an increase in Hill activity till leaf maturity, followed by a decline. Since 1950 more than fifty papers appeared, in which the differences in Hill activity with leaf age were shown (for details see the review [15]). In addition to these papers dealing with leaves developing in normal daylight, further studies measured the development of Hill activity during the greening of etiolated leaves.

Depending on plant species, leaf age as well as growing and irradiation conditions, detectable Hill activity per unit chlorophyll was found after 2 to 6 h of irradiation of etiolated leaves. The first peak of activity was followed mostly by a decline, relieved afterwards by a slow increase [8].

The studies with leaves developing on a plant growing under normal daylight mostly compared only chloroplasts from leaves in two to three phases of their ontogenesis (young - mature - old) or leaves from a few insertion levels (upper - middle - lower). Such comparisons imply the general course of increase followed by decline. Nevertheless, more detailed studies with chloroplasts from French bean primary leaves [16-19] and pumpkin cotyledons [7] showed a first maximum of Hill activity after leaf unfolding, a second peak after attaining the maximum leaf area and chlorophyll content, and sometimes a third peak preceding the end of metabolic activity of the leaf (Fig. 1). The observations of Vecher *et al*. with pea and French bean chloroplasts [20,21] and Andersen *et al* with bundle sheath chloroplasts of maize [1] are in agreement with this ontogenetic course.

There are various possible reasons for the ontogenetic changes in Hill reaction rate. The most probable, but not yet exactly experimentally proved explanation is the variability of size of the photosynthetic unit connected with changes in chloroplast and thylakoid ultrastructure.

The increase in Hill activity coincides usually with the formation of more grana per plastid and more lamellae per granum and *vice versa* [2,5,10]. Important factors are certainly the consistency of the outer chloroplast membrane, determining *in vivo* the transport of substances to and from the photosynthetic centres, and in experiments with isolated chloroplasts the fragility of the outer membrane determining the penetration of electron acceptor to the photosynthetic reaction centres. De Jong and Woodlief [4] found, that the decline in Hill activity during the tobacco leaf ageing was accompanied by a decline in chloroplast fragility.

The size of the photosynthetic unit increases with leaf

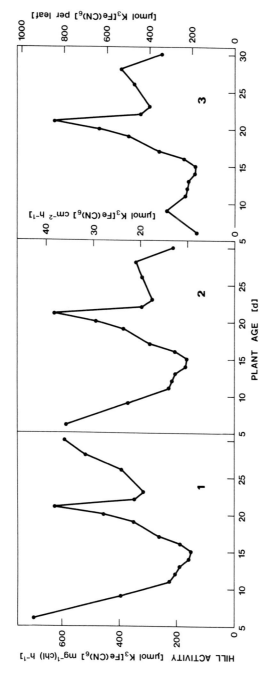

Fig. 1. Changes in Hill activity during the ontogenesis of primary leaf of *Phaseolus vulgaris* L. cv. Jantar showing a similar course with three peaks when calculated per chlorophyll amount in chloroplast suspension (1), per leaf area unit (2) or per one leaf (3). Calculation 2 lowered the third maximum of activity, calculation 3 both the first and third maxima. The calculations 2 and 3 show only hypothetical capacity values, true if all photosynthetic units in the leaf tissue were saturated with radiant energy. (For experimental details see [17]).

age: young maize leaves have units with 290 chlorophyll molecules, old leaves with 860 molecules [11]. The French press treatment of chloroplasts and ultracentrifugation in a sucrose gradient revealed relatively more particles enriched with photosystem I in chloroplasts from young spinach and radish leaves, and relatively more particles enriched with photosystem II in old leaves [13]. This explains the decreasing ratio of activities of photosystem I/photosystem II in the course of leaf ontogenesis [12,17]. Nevertheless, the changes in composition of the photosynthetic unit are more quantitative than qualitative and thus the above-mentioned shape of ontogenetic course was typical not only for the Hill activity, but also for the activities of photosystem I and non-cyclic photophosphorylation [16,17]. These findings are in agreement with ontogenetic changes in the ratio of chlorophyll forms *in vivo* (for review see [14]).

On the other hand, the discussed ontogenetic course of Hill reaction was not induced by the methods of determination (spectrophotometry or potentiometry, ferricyanide or dichlorophenol indophenol as electron acceptor [19], irradiance during measurement [17] or composition or pH of the reaction medium [18]). The shape of irradiance curves of the Hill reaction showed that with leaf ageing the chloroplasts acquired a shade character [17], but the zero intensity quantum requirements of Hill reaction did not change [6]. Unfortunately, up till now nobody has ascertained whether the oxygen evolving mechanism itself changes in the course of leaf and plant ontogenesis.

The ontogenetic changes in the rate of Hill reaction exist and are so expressed, that the age of the source leaf used for chloroplast isolation must be taken into account in all serious quantitative studies.

References

1. Andersen, K.S., Bain, J.M., Bishop, D.G. and Smillie, R.M. (1972).

Plant Physiol. **49**, 461-466.
2. Boardman, N.K., Anderson, J.M., Hiller, R.G., Kahn, A., Roughan, P.G., Treffry, T.E. and Thorne, S.W. (1972). *In* "Proceedings of the IInd International Congress on Photosynthesis Research" (G. Forti, M. Avron and A. Melandri, eds). Vol. III, pp. 2265-2287, Dr W. Junk N.V. Publishers, The Hague.
3. Clendenning, K.A. and Gorham, P.R. (1950). *Can. J. Res.* **28 C**, 114-139.
4. De Jong, D.W. and Woodlief, W.G. (1974). *Tobacco Sci.* **18**, 105-107.
5. Downton, W.J.S. and Pyliotis, N.A. (1971). *Can. J. Bot.* **49**, 179-180.
6. Drury, S. and Park, R.B. (1968). *Plant Physiol.* **43**, S-29.
7. Harnischfeger, G. (1974). *Z. Pflanzenphysiol.* **71**, 301-312.
8. Henningsen, K.W. and Boardman, N.K. (1973). *Plant Physiol.* **51**, 1117-1126.
9. Hill, R. (1937). *Nature* **139**, 881-882.
10. Horak, A. and Zalik, S. (1975). *Can. J. Bot.* **53**, 2399-2404.
11. Keresztes, Á. and Faludi-Dániel, Á. (1973). *Acta biol. Acad. Sci. Hung.* **24**, 175-189.
12. Pospíšilová, J., Zima, J. and Šesták, Z. (1976). *Biol. Plant.* **18**, 473-479.
13. Šesták, Z. (1969). *Photosynthetica* **3**, 285-287.
14. Šesták, Z. (1977). *Photosynthetica* **11**, in press.
15. Šesták, Z. (1977). *Photosynthetica* **11**, in press.
16. Šesták, Z., Čatský, J., Solárová, J., Strnadová, H. and Tichá, I. (1975). *In* "Genetic Aspects of Photosynthesis" (Yu. S. Nasyrov and Z. Šesták, eds), pp. 159-166, Dr W. Junk N.V. Publishers, The Hague.
17. Šesták, Z., Zina, J. and Strndová, H. (1977). *Photosynthetica* **11**, in press.
18. Šesták, Z. Zima, J. and Wilhelmová, N. (1978). *Photosynthetica* **12**, in press.
19. Strnadová, H. and Šesták, Z. (1974). *Photosynthetica* **8**, 130-133.
20. Vecher, A.S. and Lebedeva, T.D. (1967). *Dokl. Acad. Nauk beloruss. SSR* **11**, 727-730.
21. Vecher, A.S., Lebedeva, T.I., Raitsina, G.I. (1967). *Dokl. Acad. Nauk beloruss. SSR* **11**, 451-454.

EARLY PHOTOACTIVITY IN ILLUMINATED, ETIOLATED BEAN LEAVES
Dedicated to Edgard Lederer at the occasion of
his 70th birthday

C. SIRONVAL, M. JOUY and J.M. MICHEL

*Laboratoire de Photobiologie, Université de Liège,
Sart Tilman - Liège (Belgique).*

The Pathway to Chlorophyll-proteins

The red absorption band of many green plants (essentially algae and higher plants) has been resolved into several components. This has been generally made on the basis on an analysis of the derivative spectrum and of a reconstitution of the shape of the red absorption by addition of chosen Gaussian curves [2,6]. By changing the plant under study, by growing it in different conditions, by comparing mutant to wild type, intact chloroplasts to chloroplast fractions, etc...., the conclusion was drawn that a limited number of chlorophyll forms in suitable proportions (2 chlorophyll b forms and some 6 chlorophyll a forms) could account for the spectra.

The following principal components of the red band have often been quoted as pertaining to chlorophyll a forms:

Component:	Location (nm) of the red absorption between	
	x and	y
1	667	670
2	672	675
3	683	685
4	around 695	

Some less abundant components absorbing at $\lambda>700$ nm are to be added to the list. Component (2) is generally thought to be the most abundant; sometimes two components are introduced instead of component 3 : 3', absorbing at 680-681, and 3", absorbing around 686 nm [6].

When they start greening etiolated plants exhibit a red absorption which appears simpler than that of fully green plants. In this case, one or two chlorophyll a components account for the shape of the band. Moreover the spectrum changes in time, reflecting the successive occurrence of distinct pigment-protein complexes. Because in this case the "components" appear isolated one from the other and successively, the study of greening did help a great deal for interpreting the spectrum of fully green plants.

30 s after an etiolated bean leaf has received a short flash of white light, which reduces the bulk of the original protochlorophyllide into chlorophyll a, the red absorption is made of the superposition of three components: a band at 628 nm, which belongs to the non-photoreducible protochlorophyll(ide)-protein complex $Pr_{630-628}$; a band appearing as a shoulder at 668 nm and a main band at 683 nm, which belong respectively to the chlorophyll a-protein complexes $C_{675-668}$ (minor complex) and $C_{696-683}$ (major complex) (the complexes are usually designated by letters, Pr for protochlorophyll(ide), C for chlorophyll(ide) and P for pigment, with subscripts recalling first the wavelength of the emission band and then the wavelength of the main red absorption band, in nm).

At room temperature the position of the absorption band found at 683 nm, 30 s after a short flash, shifts slowly to 672 nm. This shift - which was first seen by Shibata [9] - lasts more than 20 min. Its speed depends on leaf age and on temperature. It is due to a change of $C_{696-683}$ into $C_{685-672}$. Thus, in a relatively short time after its illumination, the component bands 1, 2 and 3 of the fully green leaf are encountered in the spectrum of the etiolated leaf.

$C_{675-668}$ and $C_{685-672}$ are stable end products in the sense that, once they have been found, they are steadily found in the greening leaf and finally in the green leaf. On the contrary $C_{696-683}$ is a transient product, which disappears after a light flash as a consequence of the Shibata shift. It reappears, however, as soon as light is given again, as a product of the reduction of protochlorophyll(ide) molecules newly synthesized in the leaf. Thus, in the light, there is a flow from protochlorophyll(ide)-proteins to the chlorophyll a containing $C_{685-672}$ through $C_{696-683}$.

This pathway includes at least two intermediates between protochlorophyll(ide)-proteins and $C_{696-683}$:

One of these is the immediate precursor of $C_{696-683}$ which may be trapped in liquid nitrogen in times of the order of 30 ms after a 1 ms, intense flash. At that time and after such a flash a red absorption is found which culminates near 676-678 nm with an emission at 688-690 nm. This absorption shifts rapidly at room temperature and gives rise to the characteristic absorption of $C_{696-683}$ within less than 30 s [3,11]. The transient precursor of $C_{696-683}$ was called $P_{688-676}$ by Sironval and Michel [11].

In fact two pigment proteins, a major and a minor one, with two distinct fates are trapped 30 ms after a flash, if sufficiently strong. Following Litvin et al.[7], the absorption of a first photon by a protochlorophyll(ide) molecule produces a pigment-protein complex with an absorption at 676 and a fluorescence emission at 684 ($C_{684-676}$), and the absorption of another, second photon by $C_{684-676}$ transforms this species into the true, immediate precursor of $C_{696-683}$ ($C_{688-680}$). We do not intend to enter here into the discussion of the data which support this two-photon mechanism. It is, however, established

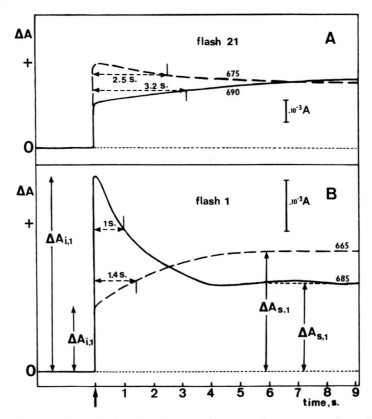

Fig. 1. Kinetics of the absorbance changes after successive, low intensity flashes. The flash is given at (o) time. Absorbances after flash 1 (B) and 21 (A) are observed respectively at 665 and 685 nm, and at 675 and 690 nm. The shift is towards 670 nm after the first flash; it is towards 690 nm after flash 21 (reproduced from Michel and Sironval, [8]; see this paper for details).

that, as shown by Litvin and Belyaeva [5] when an etiolated leaf has received a short, low intensity flash - thus at the very start of the photoreduction of protochlorophyll(ide) - the absorption band with a low absorbance near 676-678 nm, trapped immediately after the flash, does not shift to 683 nm - *i.e.* to a longer wavelength - but to a shorter wavelength - down to 667-670 nm.

Using repetitive (about 0.1 ms) low intensity flashes, we were not able to distinguish spectroscopically, 30 ms

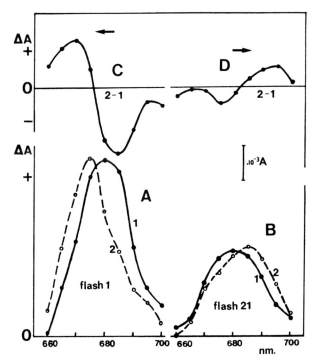

Fig. 2. Difference spectra: [absorbance 30 ms after the flash minus absorbance before the flash] (curves 1) and [absorbance 10 s after the flash minus absorbance before the flash] (curves 2) for flash 1 (A) and 21 (B) - Curves C and D are the differences: [absorbance 10 s after the flash minus absorbance 30 ms after the flash] for flash 1 and 21 respectively. In these experiments the change of the shift direction occurred around flash 4 (reproduced from Michel and Sironval [8]; see this paper for details).

after a flash, between the absorption of the product of the first flash and the absorption of the product of later flashes. But the half life time of the 30 ms product of the first flash was about 1 s, and the shift was towards 670 nm, yielding an absorption at 668 and an emission at 675 nm, the spectral characteristics of $C_{668-675}$. At later flashes, another 30 ms product appeared with a half lifetime of about 3 s, whose absorption shifted towards 685 nm, yielding $C_{696-683}$ (Figs 1 and 2; see [8]). It has been computed that the proportion of the protochlorophyll(ide)-proteins of the etiolated leaf which transforms into

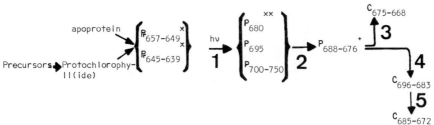

Fig. 3. The pathway to chlorophyll-proteins. Pr=protochlorophyll(ide); P=pigment; C=chlorophyll(ide). The first subscript refers to the wavelength of the main emission band of the pigment-protein complex; the second subscript refers to the wavelength of its main red absorption band; when a single subscript is given, it stays for the wavelength of the main red absorption band. (XX) P680 probably represents $P_{688-676}$ molecules with an efficient energy transfer to neighbour, long-wavelength absorbing molecules (Dujardin and Sironval [1]). (+) Litvin et al. [6] distinguish here between $C_{688-676}$ and $C_{688-680}$; they include a photoreaction leading from $C_{684-676}$ to $C_{688-680}$. The symbol $P_{688-676}$ covers both $C_{684-676}$ and $C_{688-680}$; it stays for the "energy transfer unit" (see Sironval and Kuiper [10]). The first $P_{688-676}$ molecules which appear are transformed into $C_{675-668}$ through pathway 3; some percent only of the total pigment content follows this pathway; the bulk of $P_{688-676}$ is transformed into $P_{685-672}$ through the pathway 4, 5.

$C_{668-675}$ through its 30 ms precursor at the very start of the illumination amounts to some percent only; the bulk of $P_{688-676}$ is made of the immediate precursor of $C_{696-683}$.

Intermediates which precede $P_{688-676}$ have been recently trapped in liquid nitrogen by illuminating etiolated bean leaves at temperatures from -95° to -120°C [1]. They include precursors whose life-time at room temperature is unknown, but lies below some ms. The absorption of these precursors shows a main band around 695 nm with a tail extending up to 750 nm - i.e. their absorption appears similar to that of component 4 and to that of the long wave-length components of the green spectrum. The paper of Dujardin at this symposium deals with these particular precursors. By warming up in darkness to temperatures higher than 90°C the long-wavelength pigment-proteins disappear while $P_{688-676}$ appears. It is remarkable that the transient, long-wavelength absorbing chlorophylls do not seem to emit any fluorescence at -196°C.

The scheme of Fig. 3 depicts the pathway to stable chlorophyll-proteins (in the sense already mentioned) as we know it today. Since each chlorophyll a-protein found along this pathway has spectral characteristics of a component of the "fully green" spectrum and since this pathway is seen as soon as an etiolated leaf is illuminated, we conjectured that this leaf might exhibit early photobiochemical activities similar to some extent to those of the green leaf. This has already been shown to be the case when etiolated leaves are submitted to series of 1 ms flashes which trigger the mentioned pathway; although the total illumination does not exceed 1 s, photobiochemical activities are recorded in these leaves [13].

Fluorescence Kinetics in Etiolated Leaves

A very early photobiochemical activity, distinguishable from protochlorophyll(ide) reduction, was suggested by the kinetics of the fluorescence emitted by an etiolated leaf when illuminated for the first time. The experiment was essentially as follows [4]:

A continuous light from a He-Ne laser tube (632.8 nm) was sent on an etiolated leaf. The transmission of that light and the intensity of the excited fluorescence were simultaneously monitored at room temperature using a device similar to that previously used by Strasser [12]. The measured fluorescence emission included either all wavelengths above 665 nm, or (by means of a proper interference filter) one of the following selected wavelengths: 680, 690, 702 and 740 nm.

It was found that, after an initial, rapid rise which took about 2 s, the fluorescence emission of the etiolated leaf decreased to some 50% of the maximum in a time of the order of 60 s, and finally reached a steady state somewhat below the 50% level. The same kinetics was found when receiving the red light above 665 nm in the photomultiplier,

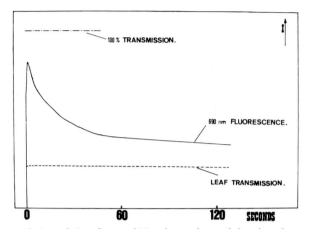

Fig. 4. An etiolated leaf was illuminated at (o) time by a red light from a He-Ne laser tube. The kinetics of the fluorescence emitted at 690 nm and the transmission of the 632.8 nm light through the leaf were recorded simultaneously. After an initial rise, the fluorescence emission decreases markedly while the leaf transmission stays constant. Illumination of the leaf at room temperature.

or any one of the selected wavelengths, showing that the decay could not be due to any spectral shift as such. On the other hand, the decay occurred while the leaf transmission remained constant at 632.8 nm. This forced us to admit that it was in essence a yield decay (Fig. 4).

The decay begins while light is still used for the reduction of protochlorophyll(ide) into chlorophyll(ide). This reduction is ended before the steady state level of the fluorescence emission is reached. Thus, the decay does not parallel protochlorophyll(ide) photoreduction.

On the other hand, it was shown by pouring liquid nitrogen around the leaf during the decay - without changing the physical conditions for excitation and reception of the fluorescence - that the -196°C emission at 690 nm increased along with the progression of protochlorophyll (ide) reduction, while the room temperature emission decreased (Fig. 5).

All these facts point to the establishment of some hitherto unnoticed photoactivity of the chlorophyll *a*

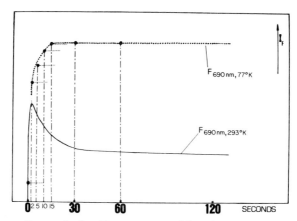

Fig. 5. Comparison of the fluorescence kinetics seen when etiolated leaves are illuminated for the first time at room temperature (full line), with the fluorescence emitted at 77 K when these leaves are frozen at increasing times during their room temperature illumination. For obtaining the points of the 77 K fluorescence (along the dotted line) liquid nitrogen was poured around the leaves during the recording of the room temperature kinetics; this occurred for successive leaves: 0, 2, 5, 10, 15, 30 and 60 s after the start of the first illumination. The room temperature and 77 K fluorescences have been measured under exactly the same physical conditions, the normalization of the leaf response being made using the room temperature spectrum. Illumination with light at 632.8 nm from a He-Ne laser. Recording of the intensity of the fluorescence emitted at 690 nm.

containing complexes which appear in the etiolated leaves when illuminated for the first time (especially of $C_{696-684}$).

The establishment of this photoactivity requires the presence in the leaf of intact protochlorophyll(ide)-protein complexes. Heating the leaf to 50°C stops the fluorescence decay; the same is true for hydroxylammonium chloride infiltration. Infiltration of ethanol-water (5% in ethanol) also suppresses the decay, but in contrast to heating or infiltration of hydroxylammonium chloride, it does not increase the fluorescence yield when the steady state level of fluorescence emission has been reached. Addition of DCMU to ethanol-water does not change this result.

Early Cytochrome Photoreduction

In order to make more explicit the suspected photoactivity of the first chlorophylls, which appear in the etiolated leaf, we investigated absorption changes in the blue region of the spectrum under a red polychromatic illumination light from 600 to 750 nm (Schott RG 630 filter). Two kinds of changes have been recorded in this region: some pertain to protochlorophyll(ide) photoreduction, others pertain to the photoactivity of chlorophyll-protein complexes resulting from this reduction. We restrict our description to the photoreduction of a cytochrome in red light.

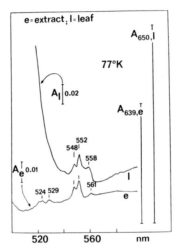

Fig. 6. Absorption spectra of an etiolated leaf and of a leaf extract showing the absorption bands of the cytochromes f and b_6. On the right, $A_{639,e}$ is the absorbance of the extract at its absorption maximum at 639 and $A_{650,l}$ is the absorbance of the leaf at its absorption maximum at 650 nm. Other explanations in the text.

Fig. 6 shows the low temperature absorption of an etiolated bean leaf and of a rather crude extract of the same leaf in the region of the α and β cytochrome absorption bands. The $α_1$ and $α_2$ bands of cytochrome f are respectively seen at 552 and 548 nm in the leaf and in the extract. The $β_1$ and $β_2$ bands of cytochrome f, respectively at 529 and 524 nm, are seen in the extract; they are hidden by the carotenoid absorption in the leaf spectrum. The α band of cytochrome b_6 is seen at 557 nm in the leaf spectrum, but it seems to be absent in the extract.

When a red actinic light is sent on an etiolated leaf, the main absorption changes between 400 and 500 nm are due to the photoreduction of protochlorophyll(ide). However at 410-420 nm other changes are superimposed to those due to photoreduction. On the contrary at 430-440 nm, the kinetics are typical of the reduction. In this spectral region the absorbance decreases steadily when the actinic light is turned on - 80 to 90% of the maximum decrease being reached about 2 s after the light is set on in our illumination conditions. A characteristic feature of the spectral changes due (in the blue region) to protochlorophyll (ide) photoreduction is their irreversibility. An actinic flash of 0.1 s given to the etiolated leaf produces a difference: (absorption before - absorption after the flash) which essentially matches the blue absorption of the protochlorophyll(ide)-protein complexes around 440 nm; isosbestic points are found at 415 and 500 nm,

Between 525 and 565 nm the kinetics are more complex. At 550 nm, the wavelength of the α absorption band of cytochrome f, the leaf absorbance decreases during about 2 s (in our conditions!) after the onset of the actinic light. This 2 s decrease does not reverse when the actinic light is turned off. If the absorbance changes are measured from 500 to 570 nm for a 2 s actinic illumination, the difference spectrum: (absorption before - absorption after the illumination) fits the difference: (absorption of the protochlorophyll(ide)-proteins - absorption of the chlorophyll(ide)-proteins). It may be controlled that the 2 s changes pertain to the exclusive reduction of protochlorophyll(ide), in subtracting from the difference spectrum measured using etiolated leaves, the spectrum measured (with the same 2 s illumination) using leaves in which protochlorophyll(ide) has been already reduced by a strong flash of light; this substraction does not alter the difference.

When the actinic illumination of the etiolated leaf is continued for times longer than 2 s, the 550 nm absorbance is seen to increase (Fig. 7). The increase starts between second 2 and second 3, apparently when the bulk of the protochlorophyll(ide) has been reduced. It is at once

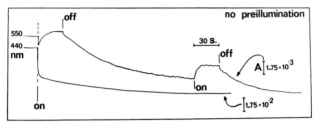

Fig. 7. Absorbance changes seen at 550 and 440 nm when an etiolated leaf is illuminated with red light for the first time. The same red actinic light was used for both traces. The changes at 440 nm are concerned with protochlorophyll(ide) photoreduction only. At 550 nm, this process dominates the kinetics during the first two seconds; after that time the changes are essentially due to the reduction of cytochrome f; the cytochrome is reduced in the light and reoxidised in darkness. At a 1 s illumination of the leaf (on the right) the trace records cytochrome changes only.

rather rapid. Within 2 to 5 s the absorbance rises over its value before leaf illumination. Later on it goes up more slowly, reaching a steady value 10 to 15 s after the start of the illumination. On turning the light off, a rapid decrease is observed. This phase lasts some seconds; it is followed by a slower dark absorbance decrease. In general a final minimum level is reached in darkness after some minutes. This minimum lies below the level reached 2 s after the start of the illumination, showing that the absorbance increase seen in the light after protochlorophyll(ide) photoreduction is fully reversible in darkness.

The 550 nm reversible changes can be repeated several times. After a first 30 s illumination of an etiolated leaf followed by a 60 s dark period, a new illumination provokes a new absorbance rise with a kinetics similar to that of the rise at the first illumination (Fig. 7). In this case the initial decrease linked to protochlorophyll(ide) reduction does not occur, however. On turning the light off, the absorbance decreases immediately as it does after the first illumination. In our working conditions, a 2 s red illumination, or else a strong 1 ms white flash suffices to suppress the initial decrease of the absorbance at 550 nm. Then light causes the absorbance rise immediately.

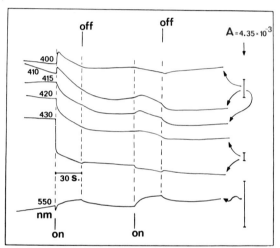

Fig. 8. Absorbance changes seen between 400 and 430 nm when an etiolated leaf is illuminated for the first time with red light. After completion of photochlorophyll(ide) reduction during the first 30 s illumination, the traces record cytochrome changes at 410-420 and in the region of 530-550 nm during a second illumination. The same actinic light was used in all experiments.

Events similar to those recorded at 550 nm are seen between 420 and 410 nm (Fig. 8). In this spectral region the absorbance changes due to protochlorophyll(ide) reduction are large at a first actinic illumination of the leaf. Opposite changes are, however, superposed to them after some seconds of illumination. After completion of protochlorophyll(ide) reduction, in particular at a 1 s illumination of the etiolated leaf, the kinetics are the same at 420-415 nm and at 550 nm. The similarity is restricted to these wavelengths, a circumstance which identifies a cytochrome. A preliminary examination of the changes seen at several wavelengths between 520 and 560 nm indicates that cytochrome f absorption changes are involved.

Because the observed changes are recorded under a red illumination of the leaf, and because they follow the formation in the leaf of the first chlorophyll(ide)-protein complexes, there is little doubt left that these complexes, or some of them (possibly $C_{696-684}$), exhibit photochemical activity as soon as they appear. The data presently available let us propose that the light they absorb serves very early for reducing cytochrome f.

Acknowledgements

The authors thank the "Fonds National de la Recherche Scientifique" for financial support. They also thank S. Sougne and R. Gysemberg for technical assistance.

Addendum

Since the presentation of this paper, additional measurements of the authors have shown that the absorption changes denoting an early photochemical activity of chlorophyllide in the illuminated, etiolated leaves, are not, or are not essentially, due to cytochrome f reduction. The data pertaining to the possible chemical nature of the products involved will be presented elsewhere (added 19.11.1977).

References

1. Dujardin, E. and Sironval, C. (1977). *Plant Sci. Lett.*, in press.
2. French, C.S. (1966). *In* "Biochemistry of Chloroplasts" (T.W. Goodwin, ed.), Vol. 1, pp. 377-386, Academic Press, London and New York.
3. Gassman, M., Granick, S. and Mauzerall, D. (1968). *Biochem. Biophys. Res. Comm.* **32**, 295-300.
4. Jouy, M., and Sironval, C. (1977). *Plant Sci. Lett.*, in press.
5. Litvin, F.F. and Belyaeva, O.B. (1971). *Biochimija* **36**, 615-622.
6. Litvin, F.F., Belyaeva, O.B., Gulaev, A.B., Karneeva, I.V., Sinetchekov, B.A., Stadnitchik, I.N. and Chubine, V.V. (1974). *In* "Chlorophyll" (A.A. Shlyk, ed.), pp. 215-230, Nauka i Teknika, Minsk.
7. Litvin, F.F., Efimtsev, E.I., Ignatov, N.V. and Belyaeva, O.B.

(1976). *Physiol. rastenij* 23, 17-21.
8. Michel, J.M. and Sironval, C. (1977). *Plant Cell Physiol.*, in press.
9. Shibata, K. (1957). *J. Biochem.* (Tokyo) 44, 147-173.
10. Sironval, C. and Kuiper, Y. (1972). *Photosynthetica* 6, 254-275.
11. Sironval, C. and Michel, J.M. (1967). Abstracts of the European Photobiology Symposium, p. 105 Hvar.
12. Strasser, R. (1974). *Experientia* 30, 320.
13. Strasser, R and Sironval, C. (1972). *FEBS Lett.* 28, 56-60.

EMERSON-LIKE EFFECT IN PROTOCHLOROPHYLL(IDE) PHOTOREDUCTION IN BEAN LEAVES
A PRELIMINARY REPORT

E. DUJARDIN

Laboratoire de Photobiologie, Université de Liège, Sart Tilman - Liège (Belgium).

Introduction

When a higher plant (angiosperm) is cultivated in darkness, biosynthesis of chlorophyll does not take place. A precursor, protochlorophyll, accumulates in the leaves in the prolamellar bodies of the etioplasts [10,11]. This precursor is in fact a mixture of protochlorophyllide (about 90%) and of the esterification products of this pigment by phytol or in some cases by geranylgeraniol [9]. The mixture is generally referred to as "protochlorophyll (ide)". Light is needed for the reaction, which reduces protochlorophyll(ide) into chlorophyll(ide). It is a stereospecific photoreduction which may be expressed as follows:

protochlorophyll(ide) + 2 RH $\longrightarrow$ chlorophyll(ide) + 2 R.

RH is an unknown reducer. The photochemical nature of the reduction is clearly evident from the dependence of the reaction rate constant on the intensity of light [17]. Protochlorophyll(ide) is the photosensitizing pigment [17].

The reaction takes place only if protochlorophyll(ide) is bound to a protein. Any treatment which denatures the protein makes the reaction impossible [1,4]. This is one of the circumstances obliging us to admit its essentially photobiochemical character. In the leaf, the reduced product - chlorophyll(ide) - leaves one site of the protein and is replaced at that site by new protochlorophyll(ide) which thus becomes reducible in light [2,18]. The protein therefore behaves similarly to an enzyme (photoenzyme) whose specific substrate is protochlorophyll(ide).

The kinetics of the reaction is not first order, but apparently second order when the formation of the product or the disappearance of the substrate is observed [17]. This may be interpreted by admitting that a certain amount of the light energy absorbed by the sensitizer (substrate of the reaction, i.e. protochlorophyll(ide)) may be transferred to the product (chlorophyll(ide)) during the reaction [8,13].

When a fresh etiolated bean leaf was illuminated at room temperature with an intense 1 ms flash, and frozen to 77 K, Sironval and Michel [15] trapped a transitory pigment-protein complex absorbing in the red at 676-678 nm and emitting fluorescence at 688 nm. This complex is called $P_{688-676}$; the first subscript referring to the main 77 K fluorescence emission and the second subscript to the main 77 K red absorption band.

$P_{688-676}$ is formed from the two photoactive protochlorophyll(ide) proteins found in the etiolated leaf.

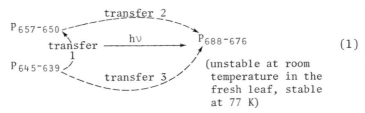
(1)

It is not stable in the fresh leaf at room temperature and transforms in darkness into a complex called $C_{696-684}$ which finally transforms into $C_{685-672}$ (see the paper of C. Sironval in this Symposium).

Sironval and Kuiper [16] described $P_{688-676}$ as an entity containing chlorophyll(ide) and protochlorophyll(ide). During the photoreduction of protochlorophyll(ide) to chlorophyll(ide), some of the non-reduced protochlorophyll(ide) molecules transfer the absorbed energy to neighbour chlorophyll(ide) molecules. The transfer is clearly demonstrated by a comparison of the characteristic absorption, emission and excitation spectra [2,8]. Its efficiency varies during the reaction according to a precise law, which supports the notion of "energy transfer units" [8,16]. This term introduced by Kahn *et al.* [8] refers to domains formed from the arrangement of several protein-pigment complexes within which the absorbed energy is wholly or partly transmitted to acceptor centres containing chlorophyll(ide). These units are enclosed in the membrane system of the prolamellar body of the etioplast.

The fact that the reaction requires light, that it requires the pigment to be bound to a protein, and that, in the leaves, the complexes form areas within which transfers between pigments are directed towards energy acceptor centres is reminiscent of certain properties of photosynthetically active green membranes. In fact, as we shall see, the similarity goes even further.

Phototransformation of the Protochlorophyll(ide) - Protein Complexes into Non-fluorescent, Long-wavelength Absorbing, Transitory Complexes.

Rubin *et al.* [14] illuminated etiolated leaves at a low temperature (~-120°C). This did not produce any change in the 77 K fluorescence emission, but upon warming the leaves above -90°C in darkness a 690 nm fluorescence band appeared.

In a similar experiment Goedheer and Verhulsdonk [7] observed, that after a short illumination of etiolated leaves at temperatures between -80°C to -75°C, during which the intensity of the 657 nm fluorescence emission due to protochlorophyll(ide) decreased, there was a dark increase of the 690 nm fluorescence emission; this dark increase was observed after illumination only, and when the leaf temperature was maintained between -120°C to -40°C.

Sironval and Kuiper [16] have also illuminated etiolated bean leaves at low temperatures (-125°C and -95°C). They found that during the low temperature illumination, part of the protochlorophyll(ide) was reduced to chlorophyll(ide), and that this was followed in some cases by a complementary dark reduction upon warming the leaves. They postulated a non-fluorescent, unknown intermediate X involved in the reduction process, with a very short lifetime at room temperature.

It is interesting to note that all of the above authors have looked at the low-temperature fluorescence emission, but have apparently neglected the examination of the absorbance of the leaves.

We found recently [5], that, when a fresh etiolated bean leaf is illuminated at -95°C by means of an intense white source during 1 min before being plunged into liquid nitrogen, the 77 K fluorescence emission at 657 nm disappears nearly completely, as shown by Fig. 1a, curve 1, although there is scarcely any emission band appearing around 688 nm. The 77 K absorption spectrum of the same leaf (Fig. 1b, curve 1) shows that a pigment absorption around 681 nm and pigments absorbing at longer wavelengths (from 690 nm $\rightarrow$ 745 nm) have been formed at the expense of the protochlorophyll(ide) complexes of the etiolated leaf.

At intense 688 nm irradiation 77 K fluoresence emission appears later upon heating the leaf in the dark (Fig. 1a, curve 2). At the same time, the long-wavelength absorption disappears, a narrow absorption band emerges at 678 nm, and the absorption of the protochlorophyll(ide)-protein complexes at 637 nm and 648 nm decreases. This shows that the long wavelength-absorbing pigment-protein complexes formed during illumination of the leaf at a low temperature behave as intermediates in the formation of $P_{688-678}$.

Rather high quantities of the non-fluorescent long wavelength absorbing intermediates are accumulated by illuminating lyophilized leaves instead of fresh leaves.

PROTOCHLOROPHYLL(IDE) PHOTOREDUCTION 515

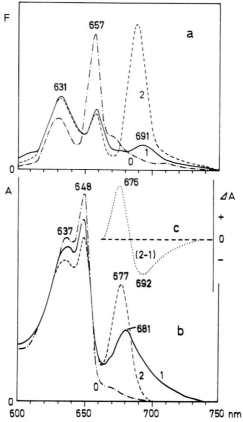

Fig. 1. 77 K fluorescence emission (a) and absorption (b) spectra of an etiolated, fresh bean leaf (curves 0). The same leaf has been heated to 178 K, then illuminated during 60 s with the polychromatic light of a 500 W projector source and finally plunged into liquid nitrogen for recording the fluorescence and absorption spectra (curves 1, a and b). Thereafter, the sample has been heated in the measuring Dewar inside the Cary spectrophotometer to a temperature of about 250 K. It remained at that temperature for 1 min. Then, without moving the sample, liquid nitrogen was poured into the Dewar, for recording the absorption spectrum (curve 2, b). The fluorescence emission spectrum (curve 2, a) was recorded just after recording of curve 2b. The fluorescence emission spectra were adjusted to identical values at 630 nm. [Taken from (5)].

The main long-wavelength absorbing precursor of $P_{688-678}$, which appears in the lyophilized leaves illuminated at -120°C to -95°C, is a pigment-protein complex with an absorption band at 695 nm; other minor long-wavelength intermediates are responsible for an absorption tail extending up to 750 nm. On heating the leaf in darkness $P_{688-678}$ appears, while the long-wavelength absorption dis disappears as in the fresh leaf. However, the dark absorption decrease at 648 nm and 635 nm - the protochlorophyll (ide) absorption bands - is much less evident in lyophilized leaves than in fresh leaves (see [5] for details).

The long-wavelength absorbing intermediates have the following remarkable properties:
1. Their lifetime is very short at room temperature (t<1 ms), but they are stable at 77 K.
2. They do not emit any measurable fluorescence at 77 K.
3. They function at a low temperature as a trap, which collects the energy absorbed by other pigments including protochlorophyll(ide). Their formation explains the data of Rubin *et al.* [14], Goedheer and Verhülsdonk [7], Sironval and Kuiper [16].

Activity of the transitory, long wavelength absorbing complexes

Since long-wavelength absorbing, transient protein complexes like those described above are likely to be present in the etiolated leaf when it is illuminated at room temperature, we tested the action of pure 694.5 nm light on these leaves. For that purpose we filtered the light from an intense white source through a Bausch and Lomb monochromator set at 694.5 nm (15 nm band-width at half-height; M_{695} in the figures) plus an interference filter transmitting at 694.5 nm (5 nm band-width at half-height; F_{695} in the figures).

Three experiments have been made:

First Experiment:

Figure 2 shows the absorption spectrum of an etiolated fresh leaf which has been frozen in liquid nitrogen (dotted line). The red absorption bands are due to protochlorophyll(ide)-protein complexes [4]. When such a leaf is illuminated by the 694.5 nm light at room temperature for

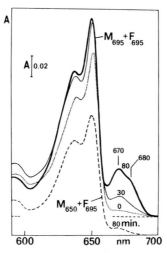

Fig. 2. 77 K red absorption spectrum of a fresh etiolated bean leaf (dotted line); of an etiolated leaf which has been illuminated during 30 min (thin full line) or during 80 min (thick full line) by pure 694.5 nm light before falling into liquid nitrogen; control: an etiolated leaf has been illuminated using light with the monochromator set at 650 nm, while the 694.5 nm interference filter remained against the leaf (dashed line). (Details see the text.)

30 minutes, a 676 nm absorption band characteristic of an chlorophyll(ide)-protein complex appears in the 77 K spectrum (thin full line). After 80 minutes, bigger amounts of chlorophyll(ide)-protein complexes have been formed (Fig. 2, thick full line).

The dashed line spectrum in Fig. 2 is a control which proves that the reduction is not due to the absorption of stray light by the photoreducible protochlorophyll(ide)-proteins: the monochromator has been set in this case at 650 nm - the position of the main absorption band of protochlorophyll(ide) in the leaf -, while the 694.5 nm interference filter remained positioned against the leaf. In this case, the quantity of chlorophyll(ide) formed after an 80 min illumination is nearly zero.

The same results have been obtained using lyophilized, etiolated leaves instead of fresh leaves. In the lyophilized leaves there is no protochlorophyll(ide) resynthesis even after hours or days; this precludes any action of the light on processes other than the photoreduction of the protochlorophyll(ide), which is present at the start of the illumination of the etiolated material.

Second Experiment:

In a second experiment the illumination of the etiolated leaf by the pure 694.5 nm light has been combined with its illumination by a polychromatic 1 ms flash which reduced ~20% of the photoactive protochlorophyll(ide) when given alone. The flash was fired either at the beginning of the 694.5 nm illumination or at its end.

Typical results are presented in Fig. 3. When the flash was fired at the beginning of a 90 min 694.5 nm illumination (dashed lines) the amount of chlorophyll(ide) formed was higher than when it was fired at the end of this illumination (Fig. 3, full line). It was also higher than the sum of the amounts of chlorophyll(ide) synthesized by illuminating for 90 minutes with the 694.5 light alone on one hand, and by firing the flash alone on the other hand.

We conclude that, when the flash occurs at the beginning of the 694.5 nm illumination, it enhances specifically the action of this illumination on the reduction of protochlorophyll(ide) into chlorophyll(ide). This effect is understood in admitting that the chlorophyll(ide)-protein complexes produced by the flash are able to use the 694.5 nm light for promoting protochlorophyll(ide) reduction.

Third Experiment:

We have studied the effect of a simultaneous illumination by the 694.5 nm light and by a 652 nm light on the quantity of chlorophyll(ide) produced. The 652 nm light came from a white source; it was filtered through a red filter plus an interference filter; the band-width at half-

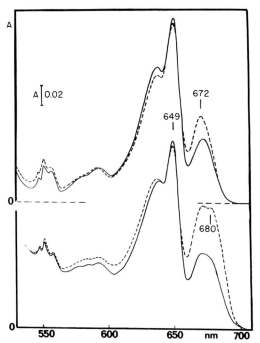

Fig. 3. 77 K absorption spectrum of two etiolated primary bean leaves of the same pair, which have received a weak 1 ms white flash at the beginning (dashed line) or at the end (full line) of a 45 min (upper spectra) or a 90 minutes (lower spectra) 694.5 nm illumination, before falling into liquid nitrogen.

height of the transmitted band was 15 nm. The 694.5 nm light was as described above.

As an example the following result was obtained using fresh etiolated leaves: 1% of the photochlorophyll(ide) present in these leaves was reduced into chlorophyll(ide) by the 694.5 nm light alone, and 40% was reduced by the 642 nm light alone. Putting together these two lights produced a 50% reduction. The yield increase was about 20% with respect to the yield obtained with the 652 nm light alone; it was 10 times with respect to the yield obtained with the 694.5 nm light alone.

The action of the combined 650 nm and 694.5 nm lights cannot be understood in terms of a phytochrome-mediated reaction, because it specifically deals with protochlorophyll(ide) photoreduction - as proved in particular when using lyophilized leaves - a reaction in which protochlorophyll(ide) is certainly the main initial photoreceptor.

On the other hand, as far as we can see at the present time, the effect is an enhancement involving the cooperation of the excitations of two chlorophyll-protein complexes. The situation can thus be tentatively likened to that encountered in their oxygen measurements by Emerson *et al.*[6] and Myers *et al.*[12]. It is an Emerson-like effect.

Acknowledgements

The author thanks the "Fonds National de la Recherche Scientifique" for financial support. She also thanks S. Sougné and R. Gysemberg for their technical assistance, and Professor K. Sauer for reading the manuscript.

References

1. Boardman, N.K. (1966). *In* "The Chlorophylls" (L.P. Vernon and G.R. Seely, eds), pp. 437-479. Academic Press, New York and London.
2. Brouers, M. and Sironval, C. (1974). *Plant. Sci. Letters* **2**, 67-72.
3. De Greef, J., Butler, W.L. and Roth, T.F. (1971). *Plant Physiol.* **47**, 457-464.
4. Dujardin, E. and Sironval, C. (1970). *Photosynthetica* **4**, 129-138.
5. Dujardin, E. and Sironval, C. (1977). *Plant. Sci. Letters*, in press.
6. Emerson, R., Chalmers, R. and Cederstrand, C. (1957). *Proc. Natl. Acad. Sci. Washington* **43**, 113-143.
7. Goedheer, J.C. and Verhülsdonk, C.A.H. (1970). *Biochem. Biophys. Res. Comm.* **39**, 260-266.
8. Kahn, A., Boardman, N.K. and Thorne, S.W. (1970). *J. Mol. Biol.* **48**, 85-101.
9. Liljenberg, C. (1974). *Physiol. Plantarum* **32**, 208-213.
10. Liro, J.I. (1909). *Ann. Acad. Scient. Fenn.*, Ser. A 1, I, 1.
11. Lubimenko, M.V. (1928). *Rev. générale Bot.* **40**, 88-94.
12. Myers, J. and Graham, J.-R. (1963). *Plant Physiol.* **38**, 105-116.
13. Nielsen, O.F. and Kahn, A. (1973). *Biochim. Biophys. Acta* **292**, 117-129.
14. Rubin, A.B., Minchenko, I.E., Krasnovsky, A.A. and Tumerman, I.A. (1962). *Biofizica* **7**, 571-577.
15. Sironval, C., Brouers, M., Michel, J.-M. and Kuiper, Y. (1968). *Photosynthetica* **2**, 268-287.
16. Sironval, C. and Kuiper, Y. (1972). *Photosynthetica* **6**, 264-275.
17. Smith, J.H.C. and Benitez, A. (1954). *Plant Physiol.* **29**, 135-143
18. Sundquist, C. (1970). *Physiol. Plantarum* **23**, 412-424.

SUBJECT INDEX

acceptor, secondary, see electron
  acceptor, secondary
acetyl-bilatriene, 386
ADRY reagents, 168, 196, 203,
  220, 245, 377, 378
  influence on fluorescence
    yield, 197
adsorption energy, 61
AgCl model, 68, 406
agranal chloroplasts, 117, 458
allomerization, 384
Ant 2s, 220, 239, 367
  influence on fluorescence
    yield, 197
antenna chlorophyll, 89
antenna size, 458, 459, 463, 490
anthraquin-cyanine, 444
anthraquinone-2-sulfonate, 94, 95
antisera
  against carotenoids, 91, 100
  against cytochrome f, 93
  against lutein, 92, 100, 103
  against neoxanthin, 93, 100
  against plastocyanin, 93
  against polypeptides, 91
  against proteins, 93
ATPase, 238

bacterial thylakoids, see
  thylakoids, bacterial
bacteriochlorophyll
  dimer, 147
  protein complexes, 85
bacteriopheophytin, radical
  anion, 153
bacteriopheophytin a, 147
bacteriopheophytin b, 147
bacteriorhodopsin, 416
benzylviologen, 207
bicarbonate
  dark fixation, 283

depletion method, 287
effect, 269, 270, 275, 379
  influence of gramicidin D,
    271
  influence of methylamine,
    271
  influence of nigericin, 271
  influence of uncouplers, 271
  on Hill reaction, 68, 72,
    269, 275
  on model systems, 68
  on oxygen evolution, 245
  on photophosphorylation, 271
  on secondary acceptor (R),
    278
  on UV absorption changes,
    274
high-affinity site, 285
low-affinity site, 285
membrane-bound, 283
pools, 285
radicals, 73
bilatriene, 386
bilayer lipid membranes, 411
  chlorophylll containing, 432
  charge generation, 432
  photoreactions, 415, 416,
    418
  preparation, 413
Bloembergen-Morgan equation, 326
blue-green algae
  thermophilic, 105
  thylakoids, 100, 105
bound manganese, see manganese,
  bound
bound water, see water, bound
2-bromo-3-isopropyl-1,4-naphtho-
  quinone (BIN), 182, 183
2-bromo-3-methyl-1,4-naphtho-
  quinone, 182
bundle sheath chloroplasts, 120

fluorescence spectrum, 123
iron content, 121
manganese content, 121
stacked regions, 120

C model, 259
C-550, 128, 137
   photoreduction, 158
$C_{668-675}$, 499, 500
$C_{675-668}$, 496, 497
$C_{684-676}$, 497
$C_{685-672}$, 496, 497, 512
$C_{696-683}$, 496, 497, 499, 500
$C_{696-684}$, 503, 508
^{14}C incorporation
   into carotenes, 179
   into membrane constituents, 179, 180
   into phylloquinone, 179
   into plastoquinone-9, 180
   into prenyllipids, 179, 180
calcium, in thylakoids, 122
carbonylcyanide-phenyl-hydrazones, 196
carboxidismutase, 81
α-carotene
   ^{14}C incorporation, 179
β-carotene
   as antioxidans, 387
   ^{14}C incorporation, 179
   in Mn-protein complex, 373
carotenoids
   antisera, 91, 100
   as antioxidans, 387
   ^{14}C incorporation, 179
   epoxides, 73
   in Mn complexes, 373
   photooxidation, 196
   triplet, 264
Carr-Purcell-Maiboom-Gill pulse sequence, 324
cartilage
   sorption isotherms, 25
   structure temperature, 25
CCCP
   influence on fluorescence yield, 197
   influence on oscillations, 338
   influence on oxygen uptake, 201
   influence on water-splitting enzyme, 168

cells, electrochemical, 47, 49, 409, 411
α-centers (reaction centers PS II), 466, 468, 469, 470
β-Centers (reaction centers PS II), 466, 468
chaotropic reagents, 167
charge accumulation, 158, 168, 195, 229, 250, 321, 346
   disturbance by hydroxylamine, 348
   in manganese complexes, 393
charge, deactivation, 345
charge delocalization, 231
charge separation, intra-molecular, 49
charge transfer
   absorption, 49, 383
   energy requirement, 67
   quenching, 395
chelates
   N_4-, 440, 444
   N_2O_2-, 446
   N_2S_2-, 447
   O_4-, 447
   S_4-, 447
chemical bond, zero point energy, 68
chloride ions
   replacement by fluoride ions, 339
   role in oxygen evolution, 91, 321
2-chloro-3-methyl-1,4-naphthoquinone, 182
chlorophyll
   absorption changes (after flash treatment), 498
   dimer, 388
   fluorescence quenching, 186, 187, 188
      by manganese (II) complexes, 394
   fluorescence yield, 150
   midpoint potential, 383
   of antenna, 89
   protein complexes, 504
   redox potential, 65
chlorophyll a
   dimer, 65, 135
   derivative spectrum, 495
   forms, 495
   photooxidation, 233

protein complexes, 496, 501
radical cation, 126
redox potential, 127
chlorophyll b
   biosynthesis, 457, 470, 484
   ^{14}C-incorporation, 180
chlorophyll c, 376
chlorophyllide, 511, 512
   protein complex, 517, 518
chlorophyllin, 417
chloroplast
   agranal, 117, 458
   bundle sheath, 120
   development, 453, 484
   glutaraldehyde fixation, 336
   granal, 117
   membrane, 490
   mesophyll, 120
   TRIS treatment, 96, 97
chloroproteins, 85, 87, 88
chondroitin sulfate
   sorption isotherms, 25
   structure temperature, 25
coacervation, 22
cobalt-phthalocyanine, 442
cobalt(III)-porphyrins, 388
cobalt-tetraazaannulene, 450
collagen
   sorption isotherms, 25
   structure temperature, 25
conformational changes, 167, 168
connector molecule (D), 242
conservative misses, 243, 244, 252, 256
Coombs test, 92
cooperative mechanism, 9
copper-phthalocyanine, 442
coupling factor, 81
   protein, 272
critical temperature, 1, 6
cryptohydroperoxide, 232, 234
cryptohydroxyl, 226, 232, 233, 245
cryptoperoxide, 226, 232, 245
cryptosuperoxide, 232, 233, 234
cyano compounds, photodissociation, 34
cyclic electron flow in PS II, 168, 177, 209, 318, 319
cytochromes, 440
   photoreduction, 504
   redox potential, 168

cytochrome b_{559}, 234
   photooxidation, 128
cytochrome f,
   absorption changes, 508
   antiserum, 93
   photoreduction, 508

d-δ* transition, 50
d-π* transition, 49, 50, 54
D_1, see electron donor, primary
D_2, see electron donor, secondary
D'_1, see electron donor, slow
D_2O
   photosynthesis rate in ..., 71, 299
   structure, 14
damping parameter, 257
DBMIB, 110, 181, 183, 218, 275, 276
   influence on fluorescence induction, 186
DCMU
   influence on fluorescence induction, 186
   influence on fluorescence yield, 206
   influence on oxygen consumption, 200
   influence on Q^- reoxidation, 358
DCPIP photoreduction
   inhibition by NphSH, 158
decomposition potential, 61
delayed light emission, 67, 348
   oscillations, 166
   quantum yield, 349
desalination membranes, 20
desorption energy, 63
detergent treatment, 88
deuterated water (HOD)
   IR spectra, 4
deuterium isotope effect, 300, 302
diaminodurene, 275, 276
diborane pyrolysis, 33
2,2-diethylchromone, 34
digitonin fragments, 148, 149
dihydro-dibenzo-tetraazaannulene, 444, 446
diketogulonate oxidation, 208
2,3-dimethyl-5-dihydroxy-6-phytol-benzoquinone, 181

diphenylcarbazide, 93, 94, 321
discontinuity point, 111
dispersion forces, 7
dithionite, 138, 152
DNA helix, 20
donor, see electron donor
double hits, 222, 243, 252, 303, 336
duroquinone, 69

electrochemical cells, 409
electrodes, semiconductor-, 68
electrolysis, 60
electron
  hydrated, 63, 64
  transport routes, 318
electron acceptor
  primary, 97
  secondary (R), 135, 270, 273, 322
    influence of bicarbonate, 278
electron affinity, 60
electron back flow, see cyclic electron flow
electron density distribution, 84
electron donor M, 232
electron donor
  first secondary (=fast, D_1), 127, 129, 131
  primary, 126
    redox potential, 66
  secondary (D_2), 132, 143
  slow (D'_1), 131
Emerson-like effect, 511
energy transfer
  between PS II units, 463
  ontogenetic development, 464
  unit, 500, 513
enhancement effect, 520
enzyme protein I, 89
epoxidation, 73
equilibrium isotope effect, see isotope effect
E-P model, 259, 261, 262, 263
EPR signal
  II_f, 132, 141, 144
  II_{vf}, 132, 136, 214, 225, 226
  $2T_{II}$, 112
etioplast, 453, 511, 513

differentiation, 484
phototransformation, 471
excitation energy
  distribution among photosystems, 471, 473
  transfer chlorophyll→carotenoids, 180
excited states, chemistry, 31
exciton, 434
  annihilation, 264
  trapping, 138, 139, 349
  yield, 349, 359
extended Hückel calculation, 38

fast donor, see electron donor
FCCP, influence on fluorescence induction, 198
Fermi energy level, 431
ferricyanide
  influence on fluorescence induction, 198
  photoreduction, temperature dependence, 109
first secondary donor, see electron donor
flash clusters, 164
fluorescence, low temperature, 161
fluorescence depolarization, 37
fluorescence induction, 139, 186, 348
  in etiolated leaves, 501
  in greening leaves, 459
  in NH_2OH-treated cells, 366
  influence of antisera, 98
  influence of DBMIB, 186
  influence of DCMU, 186
  influence of FCCP, 198
  influence of ferricyanide, 198
  influence of halogenated naphthoquinones, 186
  influence of hydroxylamine, 198
  influence of NpHSH, 160, 161, 163
  influence of PCB^-, 198
  kinetics, 149
fluorescence spectra
  of bundle sheath chloroplasts, 123
  of mesophyll chloroplasts, 123

Subject Index

fluorescence yield, 349
   influence of ADRY reagents, 197
   influence of Ant 2s, 197
   influence of CCCP, 197
   influence of DCMU, 206
   influence of n-propylgallate, 198
   influence of PCB^-, 197
   of methincyanines, 36
   oscillations, 165, 166
fluoride ions, 339
formaldehyde photolysis, 32
formylbiliverdinate, 386
formylethylidene-hydroxy-chlorin, 387
free molar enthalpy, 44
fucoxanthin, 376
fuel cells, 60, 61, 439, 440

gas solubility, 10
gels
   formation, 11
   melting points, 24
Goldman equation, 353
gramicidin D
   influence on bicarbonate effect, 271
gramicidin J, 113
grana stacking, 471
grana thylakoids, see thylakoids
granal chloroplasts, see chloroplasts
greening process, 496
   changes in fluorescence induction, 459
   phylloquinone synthesis during .., 175
   PS II formation during .., 455
guanidine, 167

H_2O^+, 63, 64, 230
heat of melting, 1, 5
heat of sublimation, 1, 5
heat of vapourization, 6, 7
hemoglobin, 440
hexachlorobenzene, 33
Hill activity
   age dependence, 489
   bicarbonate effect, 68, 72, 269
   during greening, 489
   inhibition by PCB^-, 207
   ontogenetic changes, 489, 491
   quantum requirement, 492
Hofmeister series, 16, 18
hydrated electron, 63, 64
hydration
   energy, 67
   numbers, 16
   water, 71
      isotope ratio, 71
hydrogen bonds, 3, 11
hydrogen evolution, 409
   in model systems, 405, 408
hydrogen fuel cells, 60, 61, 439, 440
hydrogen isotope effect, 71, 300, 302
hydrogen peroxide, 208
hydrogenase, 409
hydroperoxides, organic, 208
hydroxyl radicals, 402
hydroxylamine
   extraction of cells, 354
   influence on charge accumulation, 348
   influence on fluorescence induction, 198
   influence on luminescence decay curves, 360
   influence on oscillations, 338

Ice
   heat of melting, 1, 5
   heat of sublimation, 1, 5
   structure, 3
icosaborane-16, 33
inner sphere complexes, 231
intermediate X, 514
intermolecular pair potential, 1
intersystem crossing, 35
intervalence absorption, 52
intramolecular charge separation, 49
intramolecular motions, 36
inversion recovery method, 324
ion permeability, 353
ionization energy, 60
iron
   in bundle sheath chloroplasts, 121
   in thylakoids, 122

iron-phthalocyanine, 441, 442
iron(III)-porphyrins, 388
isotope effects, 68, 69, 70
   equilibrium, 69
   hydrogen, 71, 300, 302
   in chemical reactions, 68
   in oxygen evolution, 290, 291, 292, 406
   in photosynthesis, 68
   in respiration, 70

Janusgreen, 377

Kautsky effect, see fluorescence induction
Kok model, 297

leucothionine, 396
linolenic acid, 106, 112, 113
lipid bilayer, 88
liposomes, 421
   chlorophyll-containing, 423, 425
   formation, 413
   pH gradient, 425
   oxygen evolution, 422
liquid crystalline state, 111
liquids, hole-model, 1
low temperature fluorescence, 161
luminescence decay curves, 360
   influence of hydroxylamine, 360
lutein
   antiserum, 92, 100, 103
   ^{14}C incorporation, 180
   chlorophyll b complex, 103
   in manganese-protein complex, 375
   role in PS II, 103
lyotropic ion series, 16

M, see water-splitting enzyme
magnesium, in thylakoids, 122, 479
magnesium-octaethylchlorin, 383, 386
   magnesium-octaethyl-porphyrins, 386
   magnesium-porphyrinate, 386, 387
manganese
   bound, 322, 326, 345, 346, 371
   removal by heat treatment, 326
   EPR signal, 358
   function in PS II, 91, 214, 321, 346
   in bundle sheath chloroplasts, 121
   in thylakoids, 122
   membrane bound, 345, 346
      release by hydroxylamine, 356, 358
   photoredox reactions, 393
   removal by TRIS washing, 326
   removal by TRIS-acetone washing, 326
manganese complexes
   models, 336
   redox potentials, 72
   valence states, 244
manganese (II) complexes
   as fluorescence quenchers, 394
   photooxidation, 394
manganese (IV) complexes, 393, 401
   binuclear, 394, 403
manganese-containing factor, 91
manganese-gluconate complexes, 396
manganese-phthalocyanines, 397, 398, 399
manganese-porphyrin complexes, 397
manganese-protein complexes, 109, 231, 371
   isolation, 372
   green, 372, 374
      absorption spectrum, 375
      light-induced oxygen uptake, 378
   yellow, 372
Markoff process, 253
matrix recurrence relation, 253
Mehler reaction, 94, 196, 377, 407
Meiboom-Gill phase modification, 325
membrane-bound bicarbonate, see bicarbonate
membrane-bound manganese, see manganese
membrane lipids, 106
   fatty acid composition, 107

Subject Index 527

menadione (vitamin K_3), 177, 182, 183
meso-oxyporphyrin, 386
mesophyll chloroplasts, see chloroplasts
metallochlorins, 383
metalloporphyrins, 381
  midpoint potentials, 382
  radical dimers, 384
metals, in thylakoids, 121
methincyanine, 36
methylamine, 113
  as uncoupler, 120
  influence on bicarbonate effect, 271
methylfluoride, 34
2-methyl-3-hydroxy-1,4-naphthoquinone (phthiocol), 182
2-methyl-3-phytyl-1,4-naphthoquinone, 182
methylviologen reduction, 407
  influence of naphthoquinones, 182
  influence of NphSH, 158
misses, 222, 230, 239, 242, 252, 303, 336
  conservative, 243, 244, 252, 256
  photochemical, 252, 263
model membranes, 409, 412
model reactions, 68, 390
  bicarbonate effect, 68
  silver chloride, 68, 406
  titanium dioxide, 68, 405
  tungsten trioxide, 68, 406
  zinc oxide, 68, 405
molar heat, 6
molar surface tension, 10
molecular electrolysis, 405

n-π* transition, 50
N_4-chelates, 440, 444
N_2O_2-chelates, 446
N_2S_2-chelates, 447
naphthoquinones
  function in photosynthesis, 171
  halogenated, 171, 181, 189
1,4-naphthoquinone, 182, 183
naphthoquinone "K", 174
naphthoquinone K_1, 177
negative staining, 80
Nemethy-Scheraga model, 13

neoxanthin
  antiserum, 93, 100
  in manganese-protein complex, 373
  role in PS II, 103
neutral red, 215, 216, 239
NH_2OH, see hydroxylamine
nigericin
  influence on bicarbonate effect, 271
p-nitrothiophenol, see NphSH
NphSH, 167, 245
  binding, 157
  combined action with DCMU, 158
  combined action with DPC, 158
  incorporation to chloroplast proteins, 162
  influence on C-550 photoreduction, 158
  influence on DCPIP photoreduction, 158
  influence on fluorescence induction, 160, 163
  influence on low temperature fluorescence, 161
  influence on methylviologen reduction, 158
  influence on oxygen evolution, 158
nuclear magnetic dipols, 323

O_4-chelates, 447
O_2 ions, 450
O-D bond, rupture, 301
OH^- anions, discharge, 62
O-H bond
  rupture, 61, 71, 301, 306
    activation energy, 301
  types, 12
  zero point energy, 301
OH radical, 230, 232
  formation, 407
olefine peroxidation, 388
oleic acid, 113
orientational effects, 3
oscillations, 249, 265, 333
  damping, 130, 322
  disturbance by CCCP, 338
  disturbance by hydroxylamine, 338
  disturbance by TPB, 338
  hydrogen isotope effect, 303

of delayed light emission, 166
of fluorescence yield, 165, 166
of oxygen evolution in H_2O/D_2O, 303
of thermoluminescence, 166
of transverse spin-spin-relaxation rates, 322
overvoltage, 61, 434
oxoniaporphyrin, 386
oxygen
  adsorbed, 62
  as electron acceptor, 199
  cathodic reduction, 439
  diffusion coefficient, 297
oxygen
  lone pair electrons, 2
  reduction on chelates, 439, 450
  singlet, 207, 208
oxygen electrode, 63, 65, 295, 296, 309
oxygen evolution, 229
  bicarbonate effect, 245
  blockage
    by TRIS treatment, 129
    by light flashes, 221
  activation energy, 299, 307
  in model systems, 405, 408
  ontogenetic changes, 489
  rate constants, 307
  rate-limiting reaction, 307, 317
  manganese requirement, 91
  solid state mechanism, 430
  chloride requirement, 91
  as energetic problem, 59
  inhibition by hydroxylamine, 129, 130
  inhibition by NphSH, 158
  temperature dependence, 108
  isotope effects, 290, 291, 292, 406
oxygen-evolving enzyme, see water-splitting enzyme
oxygen precursor, 71, 72, 289, 291, 299
oxygen release, see oxygen evolution
oxygen uptake, CCCP-induced, 201

$P_{645-639}$, 512
$P_{657-650}$, 512
$P_{688-676}$, 497, 500, 512
$P_{688-678}$, 514, 516
$PR_{630-628}$, 496
pair potential, intermolecular, 1
palmitic acid, 113
palmitoleic acid, 113
Pariser-Parr-Popel calculation, 38
PCB^-
  influence on fluorescence induction, 198
  influence on fluorescence yield, 197
  influence on Hill reaction, 207
2,2-pentacyanoferrate complex, 51
peroxicarbonic acid, 379
peroxidation
  in chloroplasts, 202
  of thylakoid lipids, 201
peroxidicarbonic acid, 73
peroxy radicals, 207, 208
Pfeiffer complexes, 446
pH gradient
  in liposomes, 425
  light-induced, 185
phase transition temperature, 112
1,10-phenanthroline, 167
pheophytin
  photoreduction, 147, 152
  radical anions, 152
phlorin, 384
photocatalysts, 405
photochemical misses, 252, 263
photochemical reactions
  redox potentials, 59
photodissociation, 34
photogalvanic cells, 47, 49
photoinactivation, 289
photolysis of formaldehyde, 32
photolyte, 72
photon, energy content, 67
photooxidation, 195
photooxygenation, 387
photophosphorylation
  bicarbonate effect, 271
photoredox reactions, 31, 39, 43, 44
  efficiency, 43
  iodine type, 44

of manganese, 393
  thionine type, 39
photosensitization, 405
photosynthesis models, see
  model reactions
photosynthetic oxygen, see
  oxygen evolution
photosynthetic unit, 490
  composition, 492
photosystem I particles, 178
photosystem II
  bicarbonate effect, see
    bicarbonate effect
  formation, 455, 456
    during greening, 455
  inactive form, 458
  particles, 137, 178
  quantum yield, 252, 255
  reaction center
    assemblage, 458
    two types, 143, 466
  units, 139
    assemblage, 453
photovolta cells 411
phthalocyanines, 440, 446
  catalase activity, 451
  monomeric, 442
  polymeric, 442
phylloquinone (vitamin K_1),
  171, 174, 182
  biosynthesis during
    greening, 175
  ^{14}C incorporation, 179
photochrom-mediated reaction,
  519
pigment-protein complexes, see
  chlorophyll-protein
  complexes
plastocyanin, antiserum, 93
plastoglobuli, 176, 178
plastohydroquinone, 176
plastoquinone-9, 176
  ^{14}C incorporation, 180
plastoquinone pool, 256, 270
plastosemiquinone, 137
  anion, 135
PMS, 196, 210
polarization constant, 353
polypeptides, 480, 483
  antisera, 91, 94
porphyrin
  di-cation, 384
  dimers, 383

prenyllipids,
  ^{14}C incorporation, 179, 180
primary electron acceptor, see
  electron acceptor
primary electron donor, see
  electron donor
primary thylakoid, see thylakoid
prolamellar body, 453, 471, 511,
  513
n-propylgallate
  as antioxidans, 202
  influence on fluorescence
    induction, 198
protein antisera, 93
protein-manganese complexes, see
  manganese-protein complexes
protochlorophyll(ide), 511
  photoreduction, 484, 496,
    497, 502, 504, 507, 514,
    518
  Emerson-like effect, 511
  protein complexes, 496, 497,
    503, 514
protochloroplast, 453, 457, 463,
  471
proton release, 213, 214, 225,
  236
  during flash pattern, 221
    extrinsic, 236, 238
    intrinsic, 236, 238
  oscillation, 339
  relaxation rates, 326, 329
PS I, see photosystem I
PS II, see photosystem II
pseudoisocyanine, 11
pyrazine-pentacyanoferrate
  complex, 49
pyrolysis, diborane, 33

Q
  as exciton trap, 138
  flash-induced redox changes,
    140
  relative values, 138
Q^-
  decay, 263
  formation, 347, 363
  reoxidation, 141, 270, 272,
    347, 348, 366
    influence of DCMU, 359
Q function, 83

quantum yield
    of delayed light emission, 349
    of fluorescence, 36, 349
    of Hill reaction, 492
    of PS II, 252, 255
quencher, see Q

R, see electron acceptor, secondary
radicals, adsorption energies, 61
radiolysis, 64, 65
Raoult's Law, 1
reaction center I, see photosystem I, reaction center
reaction center II, see photosystem II, reaction center
reciprocity law, 264
redox potential, 59, 66
    of chlorophyll a, 127
    of thionine, 39
reduced temperature, 7
reduction degree, 31, 44
respiration, isotope effect, 70
ribonuclease, transition temperature, 18
rotational diffusion, 36

S states, 131, 214, 230, 234, 250, 251, 321, 335, 336, 346
    -life times, 251, 365, 367
S_4-chelates, 447
salt-in effect, 15
Scatchard plot, 283, 284
Second Principle, 67
secondary electron acceptor, see electron acceptor
secondary electron donor, see electron donor
semiconductors, 411
    electrodes, 68, 434
semithionine, 396
sensitization, 405, 435
SH group requirement, 166
Shibata shift, 496
silicagel, 11
silicomolybdate, 275
silver chloride model, see model reactions
single oxygen, 207, 208

slow donor, see electron donor
Solomon-Bloembergen equation, 325
sorption isotherms, 25
specific heats, 6
spillover, 97, 465, 475, 483
stacking, 475
Stern-Volmer equation, 138
stroma thylakoids, see thylakoids
structure breakers, 15, 16
structure makers, 15, 16
structure temperature, 14, 25
superoxide dismutase, 208, 374, 379
superoxide radicals, 207, 230
surface tension, 9

temperature dependence
    of ferricyanide photoreduction, 109, 113
    of oxygen evolution, 108
tetrachloroethylene, 33
tetra-(p-methoxy-phenyl)-porphyrine, 444
tetramethylbenzidine, 93, 94, 95
tetramethyl-1,2-dioxetane, 33
tetraphenylboron, 136, 141
    oxidation, 137
tetraphenylporphyrin, 444
thermoluminescence, 166
thiathiophthene, 38
thionine, 396
    energy level scheme, 39
    Fe (II) system, 41, 42
    redox potential, 39, 40
thylakoids
    bacterial
        electron density distribution, 82
        freeze fracture picture, 80
        model, 87
        negative staining picture, 80
        surface view, 79
    calcium content, 122
    electron density distribution, 84
    grana, 453, 471
    in blue-green algae, 100
    iron content, 122

Subject Index

lipids, 201
magnesium content, 122, 479
manganese content, 122
membrane
  discontinuity point, 111
  electron flow, 218
  fatty acid composition, 107
  inner surface, 78
  ion permeability, 353
  liquid crystalline state, 111
  model, 77, 86
  outer surface, 78
  pH gradient, 299
  phase transition, 107
  phase transition temperature, 112
  potential difference, 299, 353
  proton conductivity, 215
  proton flow, 218
  replica pictures, 79
  structural changes, 347
    after manganese extraction, 368
  thermostability, 106
metal content, 121
models, 412
primary, 453, 471
stroma, 453, 471
structure, 88
unstacked, 100, 453
volume, 65, 353
X-ray diffraction pattern, 82

titanium dioxide, 405
  electrode, 68, 435
  "membrane", 409
  model system, 68
α-tocopherol, 176
  ^{14}C incorporation, 180
α-tocoquinone, 176

TPB, influence on oscillations, 338
transition coefficient, 253
transition matrices, 258
tridymite structure, 3
triplet quenching, 395
TRIS block, 130
TRIS-washed chloroplasts
  absorbance changes, 135
  fluorescence changes, 135
  UV absorption changes, 141
tungsten trioxide, see model systems
tunnelling, 433
turbidity temperature, 17

ubiquinone
  iron complex, 147
  photoreduction, 264
unstacked thylakoids, see thylakoids
urea, 167
UV absorption changes, 130, 136, 273, 276, 277, 322
  bicarbonate effect, 274
  in TRIS-washed chloroplasts, 141

violaxanthin
  in manganese-protein complex, 375
vitamin K_1, see phylloquinone
vitamin K_3, see menadione

water, 1
  anomalous properties, 3
  bound, 22, 27, 28, 71
    to chlorophyll, 385
    to manganese, 346
  clusters, 9, 10, 11
  decomposition, see water splitting
  deuterated (HOD), 4
  electrolysis, 60
    energy requirement, 60
  excited states, 64
  hydration, 71
  intracellular, 28
  molar heat, 6
  molar surface tension, 10
  photooxidation (UV), 390
  proton relaxation rates, 245, 322, 323, 346
  radiolysis, 64, 65
  splitting, 213, 229
    four-step process, 235
    in liposomes, 422
    intermediates, 231
    minimum energy, 433
    proton release, 236
    rate-limiting reaction, 295

splitting enzyme, 135, 168, 207, 214, 223, 226, 229, 230, 231, 233, 241, 458, 484
  influence of CCCP, 168
structure, 1, 3, 13
  influence of electrolytes, 14
surface tension, 9
UV absorption bands, 64

X_{320}, 97
  photoreduction, 273
xanthin oxidase reaction, 73
xanthophyll epoxide cycle, 375

Z, see electron donor
zero point energy, 68, 301
zinc-formylbilatriene, 386
zinc-octaethylchlorin, 386
zinc-octaethyl-5,15-dioxoporphodimethene, 388
zinc-octaethyl-formylbiliverdinate, 386
zinc-octaethylporphyrin radical, 383
zinc oxide, see model reactions
zinc-porphyrinate, 386

π radicals, 383
π-π*transition, 50
σ analysis, 252, 254, 264

092434